Guojian Wang, Junjie Yuan
Polymer Synthesis

Also of Interest

Multicomponent Polymers.
Guojian Wang, Junjie Yuan, 2020
ISBN 978-3-11-059632-8, e-ISBN 978-3-11-059633-5

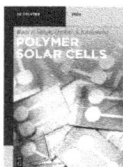

Polymer Solar Cells.
Ram P. Singh, Omkar S. Kushwaha, 2020
ISBN 978-3-11-065268-0, e-ISBN 978-3-11-065270-3

Processing of Polymers.
Chris Defonseka, 2020
ISBN 978-3-11-065611-4, e-ISBN 978-3-11-065615-2

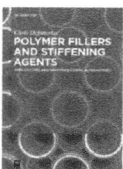

Polymer Fillers and Stiffening Agents.
Applications and Non-traditional Alternatives
Chris Defonseka, 2020
ISBN 978-3-11-066989-3 e-ISBN 978-3-11-066999-2

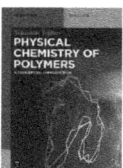

Physical Chemistry of Polymers.
A Conceptual Introduction
Sebastian Seiffert, 2020
ISBN 978-3-11-067280-0, e-ISBN 978-3-11-067281-7

Guojian Wang, Junjie Yuan

Polymer Synthesis

—

Modern Methods and Technologies

DE GRUYTER

同济大学 出版社
TONGJI UNIVERSITY PRESS

Authors
Prof. Guojian Wang
Department of Polymer Materials
School of Materials Science and Engineering
Tongji University
Shanghai
China

Junjie Yuan
Department of Polymer Materials
School of Materials Science and Engineering
Tongji University
Shanghai
China

ISBN 978-3-11-059634-2
e-ISBN (PDF) 978-3-11-059709-7
e-ISBN (EPUB) 978-3-11-059714-1

Library of Congress Control Number: 2020947687

Bibliographic information published by the Deutsche Nationalbibliothek
The Deutsche Nationalbibliothek lists this publication in the Deutsche Nationalbibliografie;
detailed bibliographic data are available on the Internet at http://dnb.dnb.de.

Preface

Since Staudinger found polymer theory in 1920, polymer chemistry has been the most active part in the field of polymer science. Early polymer chemistry was a realm of necessity. Although many useful polymer compounds have been prepared through polymer chemical reactions, only little is known about their intrinsic nature and law. For many years, polymer chemists have been working hard to find the ways to prepare polymer compounds with specific structure and molecular size. The emergence of coordination polymerization technology in the early 1950s opened a new field of stereoscopic polymerization, leading to the birth of several high-quality new materials such as linear polyethylene and stereopolypropylene, which led to the development of the entire petrochemical industry. In the mid-1950s, the development of living anionic polymerization technology provided the possibility for molecular design and synthesis of polymer materials with controllable structure and composition. Especially in the second half of the twentieth century, various new methods of polymer synthesis have been emerging to adapt to the rapid development of science and technology. Polymer chemistry is moving from the realm of necessity to the realm of freedom.

This textbook was originally a handout of the course "polymer chemistry progress" for the master's degree students in Tongji University of China. After years of supplement, modification, and improvement, combined with the authors' own practical experience in scientific research, the authors' understanding for the importance of new polymer synthesis technology is also deepened. They think that it is necessary to summarize and introduce these achievements of predecessors. Therefore, in 2004, the postgraduate teaching materials of *New Technology of Polymer Synthesis* was compiled in China to provide reference for teachers, postgraduates, and senior students in universities, researchers in scientific research institutions, and technicians in factories when they are engaged in teaching, scientific research, and technical work. After the publication of *New Technology of Polymer Synthesis*, it has been favored by teachers, students, and scientists. So, it was revised and republished in Tongji University Press in 2013, and renamed as *The Modern Methods and Technology of Polymer Synthesis*. Today, we hope this textbook can go out of China and provide help for the study of students and reference of scientists and technicians all over the world. Therefore, with the help of Tongji University Press in China and De Gruyter Press in Germany, we compile this English version of textbook.

Part of the materials used in the compilation of this textbook come from the authors' experience of teaching and scientific research and the scientific research achievements for many years, as well as many domestic and foreign documents are consulted, and relevant works are referred. We would like to express our gratitude to the authors of the works involved.

As the new methods and technologies of modern polymer chemistry involve many disciplines such as chemistry, physics, materials, electricity, biology, and medicine,

https://doi.org/10.1515/9783110597097-202

and due to the authors' lack of talent and learning, although the authors tried to be correct and accurate in the process of compilation, there must be many omissions and fallacies in the book. We hope that readers will be free to correct them.

Authors at Tongji University in October 2019

Acknowledgments

The publication of this textbook was supported by Tongji University Press in China and De Gruyter Press in Germany. The master's graduate students of the School of Materials Science and Engineering of Tongji University, Siyu Yan, Jiasheng Wu, and Tengri Wang, have made great contributions to the translation of Chinese characters. To them we owe many thanks.

https://doi.org/10.1515/9783110597097-203

Contents

List of Figures

https://doi.org/10.1515/9783110597097-205

List of Tables

https://doi.org/10.1515/9783110597097-206

Chapter 1
Ionic living polymerization

1.1 Introduction

The addition polymerization of vinyl monomer under the action of foreign force is usually chain polymerization. The basic characteristics of chain polymerization are that the reaction needs active center, for example, free radical, negative ion, and positive ion; and the whole process could be divided into three elementary reactions. The activation energy of each step has a large difference. Time has little effects on molecular weight, mainly affecting the conversion rate. So far, the most widely used method in chain polymerization is free radical polymerization, which is mainly due to the wide range of monomers for free radical polymerization, various synthesis methods, simple preparation process, and low cost of industrialization. At present, about 70% of the polymer materials are products of radical polymerization.

Free radical polymerization process can be divided into chain initiation, chain growth, chain termination, chain transfer, and other elementary reactions. The activation energy is 105–150 kJ/mol, propagation activation energy is 16–33 kJ/mol, and termination activation energy is 8–21 kJ/mol. Therefore, slow initiation, fast growth, and rapid termination are basic characteristics of free radical polymerization dynamics.

In addition, radical polymerization has the following characteristics: The initiator concentration is very low, usually 10^{-7}–10^{-9} mol/L. The initiator concentration varies greatly with the conversion rate, and it is a first-order kinetic process. The time has little effect on the molecular weight. Prolonging the polymerization time is mainly to increase the conversion rate. There are a large number of chain transfer reactions and chain termination reactions during the polymerization process, including termination of coupling and termination of disproportionation. Therefore, the molecular weight and molecular weight distribution cannot be controlled.

Like radical polymerization, ionic polymerization can be divided into three steps: chain initiation, chain growth, and chain termination. The fundamental difference between ionic polymerization and free radical polymerization is that the reactive species of the polymers are different. The reactive species of ionic polymerization are charged ions, usually carbon cations or carbon anions. Therefore, ionic polymerization can be divided into two major categories of cationic polymerization and anionic polymerization. The stability and reactivity of ions depends not only on the structure and properties of the ions themselves, but also on their environment, such as temperature, pressure, and solvent properties. The same ions can show completely different properties in different environments. In addition, the requirements of experimental conditions for ionic polymerization are very harsh, and traces of water, air, or impurities will have a strong influence on the polymerization process. Therefore, the

https://doi.org/10.1515/9783110597097-001

experimental repeatability is poor. These factors make ionic polymerization much more complex than radical polymerization.

Cationic polymerization is characterized by rapid initiation, rapid growth, easy transfer, and difficult termination. Its industrial implementation is relatively difficult, so there are few practical applications. Anionic polymerization has developed rapidly in recent years. The basic characteristics of its dynamics are rapid initiation, slow growth, and no termination. The so-called slow growth is relative to initiation. In fact, its growth rate is much faster than radical polymerization. The most striking feature of anionic polymerization is the absence of termination, which leads to the concept of living polymerization. Living polymerization is one of the greatest discoveries in the history of polymer development. The concept is proposed to create a new era in polymer design, making it develop from the realm of freedom to the realm of necessity.

It has been nearly 50 years since the discovery of living polymerization. It has already become one of the most academic and industrially valuable research directions in the field of polymer chemistry. According to Szwarc's initial definition of living polymerization, the so-called living polymerization refers to those polymerization reactions that do not cause any stoppage of the chain extension reaction or irreversible side reactions. However, few reaction systems fully satisfy such conditions. Since the group transfer reaction was discovered in the early 1980s, it was found that, although such polymerizations have chain transfer and chain termination reactions, they are negligible relative to the chain growth reaction, and thus various expected structures can still be obtained. The polymer and the relative molecular mass can be designed within a certain range, the relative molecular mass distribution index is less than 1.1 and has the characteristics of obvious living polymerization. This greatly expands the concept of living polymerization. In order to distinguish them from real living polymerizations, these macroscopic effects are similar to those of living polymerizations, but there are actually chain termination or chain transfer polymerizations, which are generally referred to as living/controllable polymerizations. This chapter focuses on ionic living polymerizations that have a real meaning of living polymerization. After that, some following chapters will then introduce other types of living/controllable polymerization.

1.2 Anionic living polymerization

1.2.1 Characteristics of anionic living polymerization

Anionic living polymerization was first discovered by people and is currently the only active polymerization method that has been industrially applied. In 1956, Szwarc et al. studied the epoch-making polymerization reaction of naphthalene sodium in the polymerization of styrene initiated in tetrahydrofuran. Szwarc et al. found that under anhydrous, oxygen-free, impurity-free, low-temperature conditions, using tetrahydrofuran as a solvent, naphthalene sodium initiator initiated anionic polymerization of styrene did not havè any chain termination reaction and chain transfer reaction. And at low temperatures and high vacuum conditions for several months, the active concentration of the resulting polymer solution remains unchanged. If styrene is added again, the polymerization reaction can continue to obtain higher molecular weight polystyrene. If a second monomer, such as (butadiene), is added, a pure styrene–butadiene block copolymer can be obtained. Based on this finding, Szwarc et al. first proposed the concept of living polymerization.

In his experiments, Szwarc found that in tetrahydrofuran solvent, the sodium of the naphthalene sodium initiator first transfers the outer electrons to naphthalene, forming a green naphthalene sodium complex. The unshared electron pair on the oxygen atom in tetrahydrofuran forms a relatively stable complexing cation with the sodium ion, which is more favorable for the styrene polymerization initiated by naphthalene radical anion. After the polymerization started, the green solution immediately changed to a bright red one, which is a characteristic of the styrene anion, and did not fade until all monomers were consumed. The initiation and polymerization processes are shown in Fig. 1.1.

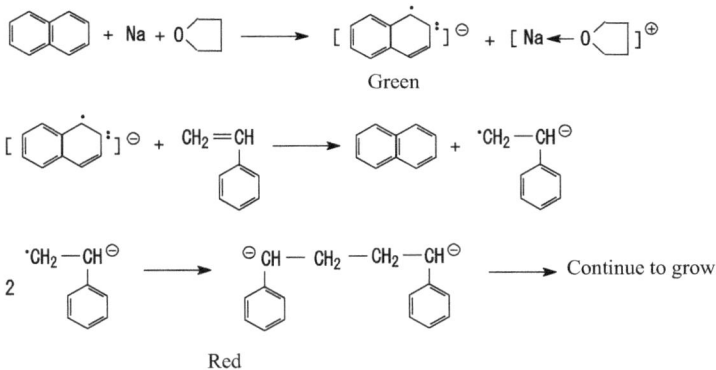

Fig. 1.1: Living polymerization of styrene initiated by naphthalene sodium.

Compared with radical polymerization, anionic polymerization has the following obvious features.

(1) Polymerization is extremely fast.
In the initiating step, the initiator can be rapidly converted to active species at concentrations up to 10^{-2} to 10^{-3} mol/L. Therefore, anionic polymerization is extremely fast and can usually be completed within minutes.

(2) Monomer has strong selectivity to initiators.

In free radical polymerization, the initiator is essentially universal for various monomers. For anionic polymerization, the monomer has a strong selectivity to the initiator. For both A and B monomers capable of anionic polymerization, some initiators can initiate polymerization of monomer A, but they do not necessarily initiate polymerization of monomer B. For example, alkyllithium can initiate the polymerization of styrene and nitroethylene, while pyridine can only initiate the polymerization of nitroethylene.

The anionic polymerization initiator can be regarded as a Lewis base, and the stronger the alkalinity is, the more active it is. The monomer can be considered as Lewis acid, and its acid value is expressed as pK_a, where pK_a is the negative logarithm of the ionization equilibrium constant. The smaller the value, the stronger the acidity and the more stable the corresponding anion. Therefore, the smaller the pK_a of the monomer is, the more active it is.

Strongly basic initiators can cause weakly acidic and strongly acidic monomers, and weakly basic initiators can only lead to strongly acidic monomers. Examples are shown in Fig. 1.2.

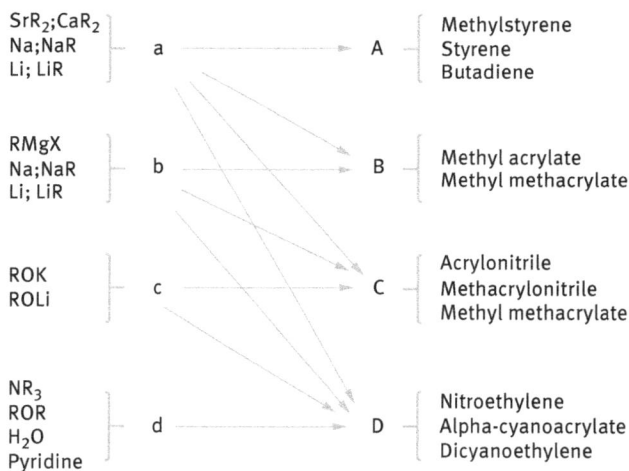

$SrR_2;CaR_2$		Methylstyrene
Na;NaR	a → A	Styrene
Li; LiR		Butadiene
RMgX		Methyl acrylate
Na;NaR	b → B	Methyl methacrylate
Li; LiR		
ROK		Acrylonitrile
ROLi	c → C	Methacrylonitrile
		Methyl methacrylate
NR_3		Nitroethylene
ROR	d → D	Alpha-cyanoacrylate
H_2O		Dicyanoethylene
Pyridine		

Fig. 1.2: Initiator and monomer matching in anionic polymerization.

(3) There is no chain termination reaction.

The most typical feature of anionic polymerization is that there is no chain termination reaction under certain conditions. In radical polymerization, the chain termination reaction rate constant is approximately 10^4 times greater than the chain growth reaction rate constant, the activation energy of coupling and disproportionation reactions is close to zero, and the average life span of radicals is only 1 to several seconds, so the chain termination reaction is unavoidable. While the anionic polymerization active segment carries the same charge, it is impossible for the coupling or disproportionation termination reaction to occur. Moreover, it is very difficult to remove H^- ion from the active chain. When the alkyllithium is used to initiate styrene and butadiene polymerization in aliphatic hydrocarbons, benzene, or ether solvents, basically there is no chain transfer reaction, so active species do not disappear automatically.

Anion active chains are easily terminated by compounds with active hydrogen such as water, acids, and alcohols. Therefore, if the system is impure, no active polymer can be obtained.

(4) Coexistence of multiple active species

Another feature of anionic polymerization is the simultaneous presence of two or more active species in the same reaction system.

For example, in polar solvents, there may be covalent bonds, tight ion pairs, loose ion pairs, and free anions. The order of polymerization activity is: free ion > loose ion pair > tight ion pair > covalent bond. As the dielectric constant and the electron donation index of the solvent increase (polarity increases), the stroke of the loose and free ions is favored.

In nonpolar solvents, initiators can often exist in both mono- and complexes. Monomer polymerization activity is greater than the complex. As the polarity of the solvent increases, the equilibrium shifts to the left, favoring the form of a single monomer.

Changes in the nature of the solvent affect the type and relative amount of active species and therefore have a great influence on the rate of polymerization. For example, n-butyllithium initiates the polymerization of styrene, and polymerization in tetrahydrofuran is about 1,000 times greater than the apparent rate of polymerization in cyclohexane. At the same time, the change of the solvent also affects the microstructure of the polymer product:

$$nA^-X^+ \rightleftharpoons (A^-, X^+)_n$$

Chemical reaction equation 1

(5) Relatively narrow molecular mass distribution

In free radical polymerization, the relative molecular mass distribution of the polymerization product is very wide since the instantaneous radical concentration changes as the conversion rate increases, and the polymerization active chain has various forms of termination and transfer.

Since there are many active species in anionic polymerization, the growth rate constants of each active species are different. If the rate of conversion between active species is slower than the rate of increase, multiple peaks will appear in the final relative molecular mass distribution. However, under normal circumstances, the conversion rate between active species is much greater than the chain growth rate, so the presence of multiple active species does not affect the relative molecular mass distribution of the product.

The relative molecular mass distribution of many anionic polymerizations is close to the Poisson distribution. The preconditions are:
(1) Relative to the chain growth reaction, the rate of initiation of the reaction is very fast.
(2) No chain termination reaction or chain transfer reaction.
(3) The chain depolymerization rate is very slow compared with the chain growth reaction.
(4) The reagents in the system can be effectively mixed.

Poisson's distribution is very narrow, with an average molecular weight of 500,000 polymers, and 95% of the polymer has a molecular weight that is within 10% of the average relative molecular mass. According to this calculation, the relative molecular mass distribution index is equal to 1.002. In fact, the relative molecular mass distribution index is ≤1.1.

1.2.2 Elementary reactions of anionic living polymerization

1.2.2.1 Initiator and chain initiation reaction
The initiation process of anionic polymerization is actually the process of charge transfer from the initiator to the monomer and can generally be divided into several types.

(1) Metal alkylate
Alkyl metal complexes are an important class of anionic polymerization initiators, and their initiating activity is related to the metal's electronegativity. The stronger the polarity of the M–C bond formed by the metal and the carbon atom is, the more likely it is to form an ionic bond, and the greater the initiator activity is, the easier the initiation of anionic polymerization is.

The electronegativity of sodium and potassium are 0.8 and 0.9, respectively. They belong to strong alkali metals. The Na–C and K–C bonds are to some degree ionic, so alkyl sodium and alkyl potassium are very active anionic polymerization initiators. Because of its high activity, it is easy to cause side reactions, so it is rarely used. The electronegativity of lithium is 1.0, and the Li–C bond formed with the carbon atom is a polar covalent bond, so the initiation activity of the alkyllithium is relatively moderate and is currently the most important anionic living polymerization initiator. The electronegativity of magnesium is relatively large, reaching 1.2 to 1.3, so the polarity of Mg–C is weaker and generally cannot be directly used to initiate anionic polymerization. The polarity of the Mg–C bond in Grignard reagent RMgX is greater, and the activity is increased, but usually only very active monomer is initiated.

Thus, alkyllithium is currently the most important and most commonly used anionic living polymerization initiator. Alkyllithium is a polar covalent compound, and this compound undergoes bond polarization in the solvent. As the polarity of the solvent increases, the covalent bond transforms to an ionic bond.

Therefore, such organic metal compounds can be dissolved in polar solvents such as ethers and can also be dissolved in nonpolar solvents such as hydrocarbons. This property of alkyllithium makes it widely used in the anionic polymerization of various solvent systems.

Alkyllithium reacts quickly with water, carbon dioxide, and ethanol and loses its activity, and the reaction rate is far greater than the rate at which the monomers are initiated. Therefore, the requirement for moisture and impurities in the living polymerization system is high and the operation should normally be performed under high vacuum or inert gas protection, otherwise the polymer cannot be obtained. The reaction of butyllithium-initiated styrene can be illustrated as follows:

$$BuLi + CH_2 = CH \longrightarrow BuCH_2 - CH^- Li^+$$

Chemical reaction equation 2

The reason why alkyllithium initiators are widely used is because they can be dissolved in hydrocarbon solvents, while other alkali metal alkyl and aryl compounds do not dissolve well in hydrocarbon solvents. However, in nonpolar solvents, alkyllithium is usually present in associated form, affecting the rate of initiation and growth reactions.

Linear alkyllithiums are generally present in six-associates. Branched alkyllithiums such as s-BuLi and t-BuLi all form tetra-associates in hydrocarbon solvents. Therefore, in the initiation process, the association body with the alkyllithium needs to be first deassociated and form a single-membered body and react with the monomer. As in the benzene solvent, the rate of styrene initiated by the linear BuLi (n-BuLi) is proportional to one-sixth of the concentration of butyllithium, and the initiation reaction can be considered approximately as a single unit contribution.

In certain nonpolar solvent systems where the induction period is very long, even monomers cannot be initiated, appropriate amounts of polar additives such as dimethoxyethyl ether and tetramethylethyldiamine can be added to promote lithium alkylation. The disassociation of association leads to a dramatic acceleration of the initiation reaction. If about 2% of tetrahydrofuran is added to the n-BuLi hydrocarbon solvent, the initiation efficiency can be greatly increased.

(2) Alkali metal

Alkali metals used as an anionic polymerization initiator mainly include potassium, sodium, lithium, cesium, and rubidium. Alkali metals are generally insoluble in common solvents for anionic polymerization and are, therefore, heterogeneous initiators.

The initiation rate of the heterogeneous initiator is proportional to the surface area of the initiator. Therefore, in order to increase the surface area of the initiator and increase the initiator efficiency, alkali metals are generally made into "mirrors." There are two methods of mirror making in the laboratory: (1) Direct sublimation method, that is, direct heating and sublimation of the alkali metal to cool it into a mirror on the reactor wall. (2) The solution method, that is, dissolving the alkali metal in a suitable solvent, and then evaporating the solvent so that the alkali metal is deposited on the reactor wall as a mirror. Alkali metals such as potassium, sodium, rubidium, and cesium are usually prepared using the first method of making mirrors. The vapor of metallic lithium is usually used to make mirrors because it easily reacts with glass. For example, lithium is dissolved in liquid ammonia and the ammonia is evaporated into a mirror. Industrial large-scale mirror making generally adopts the melting method, that is, the alkali metal is heated to boiling in an inert hydrocarbon, and then quenched under strong stirring to obtain an alkali metal particle mirror surface.

In an alkali metal-initiated styrene polymerization system, an alkali metal is in direct contact with styrene, and styrene, which has an easily polarized π electron, captures electrons from an alkali metal atom to form a radical anion, and the radical anion undergoes a coupling reaction. A double anion is formed and the initiation reaction starts from both ends of the double anion, forming a double active species molecular chain. The polymerization process is shown in Fig. 1.3.

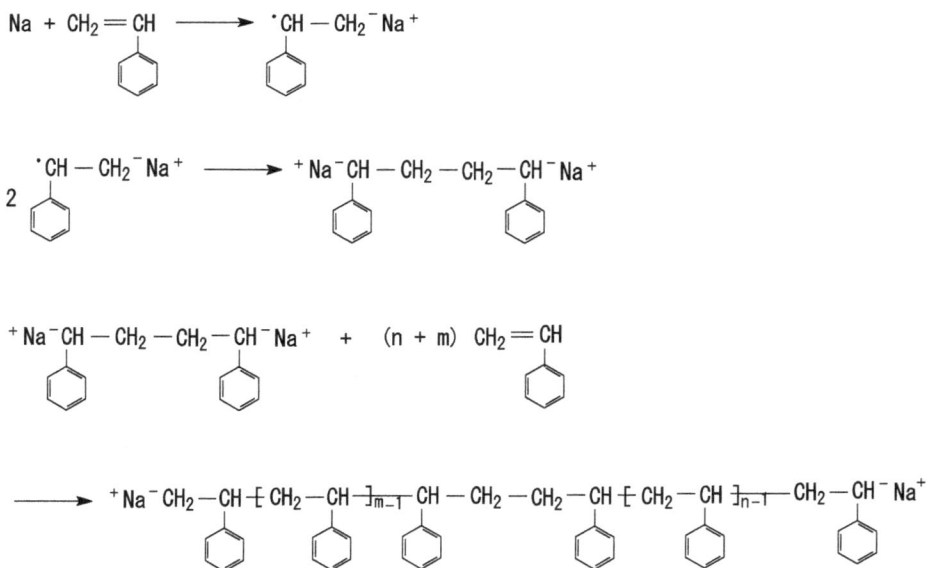

Fig. 1.3: Schematic diagram of styrene polymerization process initiated by alkali metal.

(3) Alkali metal complexes

In polar solvents such as ethers, alkali metals can form complexes with aromatics such as naphthalene, anthracene, or ketones (e.g., benzophenone). The electrons on the alkali metal are transferred to the lowest orbits of aromatics such as naphthalene and anthracene to form anion radicals. When a monomer, such as styrene, is present in the system, the anion transfers the free radicals to the styrene monomer, forming the styrene anion radical, and the naphthalene is reduced. The anion radical of styrene further couples the double anion. For example, the reaction of sodium and naphthalene-initiated styrene polymerization is shown in Fig. 1.4.

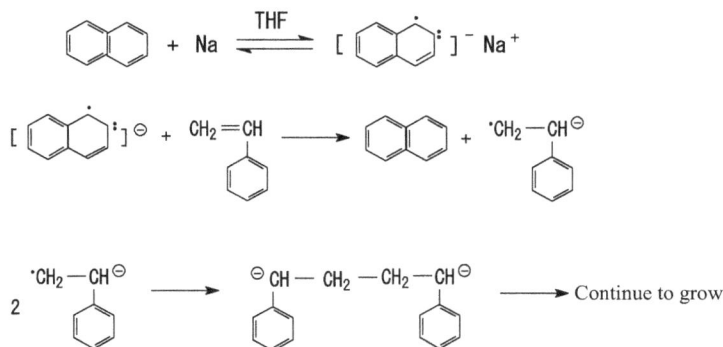

Fig. 1.4: Schematic diagram of polymerization process of styrene initiated by sodium naphthalene.

It can be seen that the difference between the alkali metal complex initiator and the metal alkyl compound initiator is that the former only undergoes electron transfer reaction when the monomer is initiated, and the latter anion portion forms the polymer front end after initiation. The concentration of formed anion radicals depends on the activity of the alkali metal, the solvating power of the solvent, and the reducing ability of the aromatic compound. For example, naphthalene sodium can obtain 95% free radical anion in normal temperature tetrahydrofuran, and only less than 1% free radical anion can be obtained in ethyl ether at room temperature.

Rhodium can capture two electrons from the alkali metal to form a double anion. The dianion does not show the nature of free radicals, that is, it is not a paramagnetic material.

When an alkali metal complex is used as an initiator, the reaction can easily proceed as long as the relative electron affinity of the initiator and the monomer are matched. Typical examples are sodium naphthalene initiators initiate styrene reactions in tetrahydrofuran. Adding styrene in a green sodium naphthalene tetrahydrofuran solution instantly turns the system red, forming a styrene radical anion.

When an alkali metal is used as an anionic polymerization initiator, it has a strong selectivity to monomers. For example, potassium benzophenone ($[C_6H_5\text{-}CO\text{-}C_6H_5]^-K^+$) can cause methyl methacrylate (MMA) to initiate but not styrene. However, in tetrahydrofuran solvent, potassium benzophenone can accept the second electron to form a double anion and initiate styrene polymerization.

(4) Lewis base

For some very active monomers, Lewis base can directly initiate its anionic polymerization. For example, trialkylphosphine can initiate the polymerization of nitroethylene or acrylonitrile (as shown in the following formula), and the formation of anions can further increase the monomer:

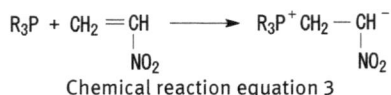

$$R_3P \ + \ CH_2 = \underset{\underset{NO_2}{|}}{CH} \ \longrightarrow \ R_3P^+ CH_2 - \underset{\underset{NO_2}{|}}{CH}^-$$

Chemical reaction equation 3

Pyridine can also initiate the anionic polymerization of the abovementioned monomers.

Lewis base-initiated anionic polymerization is only active polymerization at very low temperatures (−100 to −70 °C). At room temperature, the chain transfer reaction easily occurs during the Lewis base-initiated polymerization, so generally, only oligomers having a very low degree of polymerization are obtained, and the degree of polymerization is independent of the monomer concentration. At low temperatures, the chain transfer reaction is inhibited, and the relative molecular mass of the polymer increases with the conversion rate, showing the characteristics of active polymerization.

1.2.2.2 Chain propagation reaction

In the case of a very pure polymerization system, there is no chain termination reaction in the anionic polymerization. That is, anionic polymerization has only chain initiation reaction and chain propagation reaction. The chain initiation reaction is an addition reaction of an initiator to a monomer, and the chain propagation reaction is an addition reaction of an active chain to a monomer, and the nature of both is the same. However, because the initiator structure is different from the active chain, the rate and reaction route of the initiation reaction and the growth reaction may be quite different. In addition, external conditions such as solvent and temperature may also make the initiation reaction and growth reaction rate different.

For example, anionic polymerization is initiated with alkyllithium, and in a polar solvent such as tetrahydrofuran, the chain initiation reaction is much faster than the chain extension reaction and is almost instantaneously completed. In nonpolar solvents such as hydrocarbons, the initiation reaction is equivalent to the rate of the growth reaction. In a relatively long polymerization time, the chain initiation reaction is accompanied by the chain growth reaction. Therefore, it is often only anionic polymerization in polar solvents that is living polymerization.

1.2.2.3 Chain transfer reaction

Under certain conditions, there is the possibility of an anionic polymerization chain transfer reaction. For example, when there are substances in the system that contain hydrogen atoms that are easily robbed, the chain transfer reaction may occur, as shown in the following formula:

$$P^-X^+ + R^-H \longrightarrow P-H + R^-X^+$$

Chemical reaction equation 4

As a result of the chain transfer, the original active chain is terminated, the degree of polymerization is reduced, and the relative molecular mass distribution is broadened. If the activity of the newly formed active strand is the same as that of the original active strand, the polymerization rate will not change. If the active agent of the newly formed active chain is weakened, the polymerization rate becomes slower and acts as a slow polymerization. In the extreme case that the newly formed anion does not own initiate activity, the abovementioned chain transfer reaction actually forms a chain termination reaction. Regardless of the abovementioned results of chain transfer, these are deviating from the basic characteristics of active polymerization and, therefore, must be avoided in living polymerization.

The chain transfer reaction in the anionic polymerization process mainly includes the transfer to the solvent, the transfer to the monomer, and the transfer to the macromolecule.

(1) Transfer to the solvent

The anionic polymerization solvents are generally benzene and aliphatic hydrocarbons, and the hydrogen atoms in these solvent molecules are relatively stable and are generally not easily captured by the active chain, so the possibility of chain transfer is small. However, when substituted aromatics such as toluene, xylene, and ethylbenzene are used, the active center will take hydrogen atoms from these molecules and transfer them. For example:

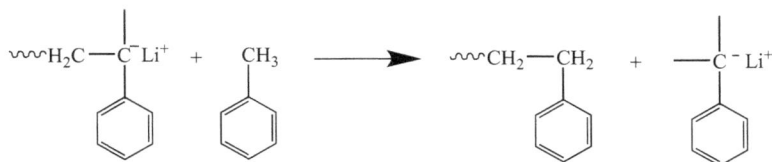

Chemical reaction equation 5

The solvent transfer reaction, which is undesirable during anionic living polymerization, can be used to prepare oligomers. Anionic living polymerization can also produce oligomers in the presence of a large amount of initiators, but the use of a small amount of initiators to achieve the same effect using chain transfer reactions is clearly more economical and convenient.

The compound as a chain transfer agent has as large a chain transfer activity as possible, and the newly formed living anion should have the same activity as the original active chain. In the same time, the toxicity and cost should be considered. Aromatic compounds are commonly used chain transfer agents such as diphenyl, xylene, and cumene. Their chain transfer activity order is: toluene > xylene > cumene. Toluene anion has an equivalent initiating activity to polystyrene anion, so it is an ideal chain transfer agent.

The chain transfer reaction is not only related to the activity of the transfer agent, but also related to the nature of the active species. For the same chain transfer agent, the larger the counterionic radius of the active species, the easier the chain transfer occurs. Because the larger the radius of the counterion, the easier it is to be solvated, which results in the larger ratio of the loose pair of ions owing a large reaction activity and freer anion ratio, the chain transfer reaction occurs more easily. For example, common counterions in anionic polymers include K^+, Na^+, and Li^+. Their order of influence on chain transfer reactions is $K^+ > Na^+ > Li^+$. Table 1.1 shows the chain transfer constants for various solvents and counterions. From the table it can be seen that toluene has the greatest chain transfer activity.

Tab. 1.1: Chain transfer constants of solvent and counterion.

Counterion	Temperature (°C)	Chain transfer agent	Chain transfer constant ($\times 10^4$)/(L/mol s)
Na^+	20	Toluene	5.9
		Xylene	4.6
		1-Hexene	2.5
		Ethylbenzene	1.8
		Cumene	1.2
		Benzene	0.2
K^+	60	Pentene	0.25
		1-Decene	1.55

Anionic active species usually associate with each other, such as polystyrene active chain. The addition of polar substances to the polymerization system and complexing with counter ions can promote the disassociation of the association and the transformation of the association to the monomer, which is advantageous to the occurrence of chain transfer reaction. The chain transfer constants after adding different polar substances are shown in table 1.2.

(2) Transfer to monomer

Hydrogen atoms on monomers such as styrene and diolefins are relatively stable and are not easily captured by living chains. Therefore, it is generally difficult for

Tab. 1.2: Effect of polar material on chain transfer reaction.

Counterion	Polar additives	Temperature (°C)	Chain transfer constant ($\times 10^4$)/ (L/mol s)
Li$^+$	None	60	0.01
	Dimethoxyethyl ether		0.08
	Tetramethylethyldiamine		2.00
	Hexamethylphosphoramide		3.90

anionic polymerization to occur in reactions that transfer to monomers. However, for some of the polar monomers, such as acrylonitrile, the hydrogen atoms on the monomer molecules are more active, and the chain transfer reaction may still occur at higher temperatures. Therefore, in order to ensure the normal progress of the living polymerization and avoid the occurrence of the transfer reaction to the monomer, the polymerization reaction should be carried out at the lowest possible temperature.

(3) Transfer to large molecules

The alpha-H on the polystyrene chain is more active than the alpha-H on the styrene, so the possibility of transferring to the polystyrene macromolecular chain is greater than the transfer to the monomer. However, under normal circumstances, in the anionic polymerization of styrene using alkyllithium as an initiator, it is still very difficult to transfer macromolecules.

The greater the activity of the initiator is, the greater the possibility of transfer to macromolecules is. For example, when sec-butyllithium is used as an initiator, the rate of transfer to macromolecules is much greater than when n-butyllithium is used as an initiator.

When an alkali metal as an initiator directly initiates anionic polymerization, the proportion of transfer to macromolecules is high, which is why alkali metals are less directly used for living polymerization.

1.2.2.4 Chain termination reaction

Anionic polymerization is characterized by the absence of chain termination reactions. However, under certain conditions, the anionic active chain may still be inactivated and terminated. The most important reason is the reaction with the terminator and the isomerization of the active segment group.

(1) Termination reaction with impurities

Trace impurities in the polymerization system, such as water, carbon dioxide, alcohols, acids, and amines, can easily stop the anion. Therefore, the anionic polymerization must be carried out under high vacuum or inert atmosphere and the reagents and glassware are very clean. Traces of water can terminate the growing chain by proton transfer and the resulting OH⁻ does not initiate activity. Oxygen and CO_2 in the air can be added to the growing active chain anion to form a peroxycarbonyl anion and no activity is also induced.

In anionic living polymerization, after the monomers have all been reacted, water, alcohol, acid amine, and other chain terminators are usually added to terminate the active chain.

During the living polymerization process, carbon dioxide, ethylene oxide, diiso-cyanate, polyhydric alcohols, and polybasic acids were purposely added to terminate the living chain of the polymer, resulting in the formation of carboxyl, hydroxyl, iso-cyanato, and other specific end groups at the chain ends that can be further used to prepare special structural polymers such as block and rocker.

(2) Active chain isomerization

Under the condition of completely eliminating the influence of impurities, the active polymer chain may still gradually lose its activity after being left for a long time, which is often caused by the isomerization of the living polymer chain. For example, in the polymerization of α-methylstyrene initiated by alkyllithium, poly-α-methylstyryllithium gradually loses lithium hydride and loses its activity in benzene solution.

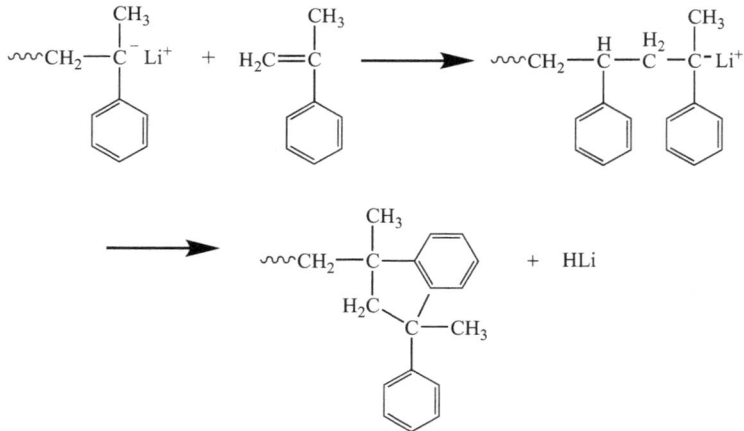

Chemical reaction equation 6

For polar monomers, it is also possible that active end groups migrate inside the molecule and lose activity:

$$\text{\char`\~\char`\~\char`\~} \{ CH_2 - \underset{\underset{COOCH_3}{|}}{\overset{\overset{CH_3}{|}}{C}} \}_{\overline{n-1}} CH_2 - \underset{\underset{COOCH_3}{|}}{\overset{\overset{CH_3}{|}}{C^-}} Li^+ \rightleftharpoons \text{\char`\~\char`\~\char`\~} \{ CH_2 - \underset{\underset{COOCH_3}{|}}{\overset{\overset{CH_3}{|}}{C}} \}_{\overline{n-1}} CH_2 - \underset{\underset{\underset{O}{\parallel}}{\overset{|}{C}-O^-Li^+}}{\overset{\overset{CH_3}{|}}{C}} - CH_3$$

Chemical reaction equation 7

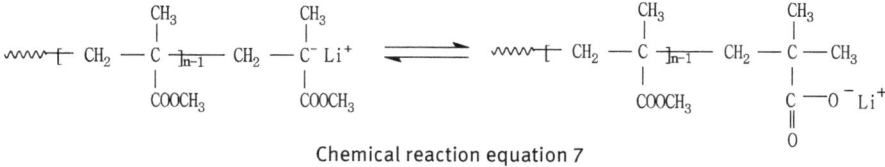

As a result of the transfer of the active site, the more active alkyllithium is converted into a less active carboxyl anion so that chain extension reaction with the monomer is no longer possible.

Under stringent polymerization conditions, the isomerization chain termination rate is much lower than the chain growth rate. And therefore, it is generally not considered in the preparation of living polymers.

1.2.3 Anionic living polymerization kinetics

1.2.3.1 Anionic living polymerization rate
In the anionic living polymerization of the elementary reaction, the chain initiation rate is far greater than the growth rate. Therefore, the rate of anionic living polymerization can be simply expressed by the chain growth rate, the relationship between the polymerization rate and the monomer concentration, and the kinetic relationship, as follows:

$$[N] = [M^*] = [C] \tag{1.1}$$

In the abovementioned formula, $[M^-]$ is the anionic active center concentration. Since the chain initiation rate of the anionic living polymerization is greater than the growth rate, that is, before the initiation of polymerization, the initiator has been quantitatively converted into the active center, $[M^-]$ is equal to the initiator concentration, and therefore, eq. (1.2) can be expressed as

$$R_p = k_p[C][M] \tag{1.2}$$

where $[C]$ is the initiator concentration and k_p is the polymerization rate constant.

k_p of anionic polymerization is basically equivalent to that of radical polymerization, but the concentration of active centers in anionic polymerization is much greater than that of free radical polymerization. Generally, the concentration of active centers of free radical polymerization is 10^{-9}–10^{-7} mol/L, and anionic polymerization is performed. The concentration of active centers can be as high as 10^{-3} to 10^{-2} mol/L, so the rate of anionic polymerization is usually 10^5 to 10^8 times larger than radical polymerization.

1.2.3.2 The degree of polymerization of anionic living polymer
If anionic living polymerization meets the following conditions:
(1) Initiators all quickly become active centers.

(2) Good mixing, uniform monomer distribution, and all growth chains begin to grow at the same time.
(3) Pure system without impurities.
(4) No chain transfer reaction.
(5) No depolymerization reaction.

Then when the monomer conversion is 100%, the degree of polymerization of the living polymerization should be equal to the number of monomers added per active site, that is, the ratio of the monomer concentration to the concentration of the active center.

$$\bar{X}_n = \frac{[M]}{\frac{[M^-]}{n}} = \frac{n[M]}{[C]} \tag{1.3}$$

where $[C]$ is the initiator concentration, n is the number of active centers per macromolecule, n is the active chain of the dianion, and n is 1.

The molecular weight distribution of the anionic living polymerization product follows the Poisson's distribution. Theoretical studies have shown that the ratio of the weight average polymerization degree to the number average polymerization degree has the following relationship:

$$\frac{\bar{X}_w}{\bar{X}_n} = 1 + \frac{\bar{X}_n}{(\bar{X}_n + 1)^2} \approx 1 + \frac{1}{\bar{X}_n} \tag{1.4}$$

1.2.3.3 Effect of solvents on anionic living polymerization
Solvents have a very important influence on anionic living polymerization. In different solvents, the growth-active species can be in different states such as covalent bonds, ion pairs (loose or tight), and free ions, which have a great influence on the polymerization rate.

The effect of the solvent on the anionic living polymerization can be quantified by the dielectric constant of the solvent and the electron donation index. The dielectric constant reflects the polarity of the solvent, the electron donation index indicates the electron donating ability of the solvent, and the electron donation index has a large solvent, which easily makes the counterion solvate. The dielectric constant and the electron donation index reflect two different properties of the solvent. There is a certain relationship between the two, but they are not necessarily the same. Table 1.3 shows the dielectric constant and electron donation index of some solvents. Tab. 1.4 shows the effect of the polarity of the solvent on the rate constant of styrene anion polymerization.

As shown in Tab. 1.4, the greater the polarity of the solvent is, the larger the k_p is. However, the dielectric constant of 1,2-dimethoxyethane is smaller than that of tetrahydrofuran, but k_p is much larger than that of tetrahydrofuran. This is due to the different solvating power of the solvent.

Tab. 1.3: Partial solvent dielectric constant and electron donation index.

Solvent	Dielectric constant	Electron donation index	Solvent	Dielectric constant	Electron donation index
Nitromethane	35.9	2.7	Tetrahydrofuran	7.6	20.0
Nitrobenzene	34.5	4.4	Dimethyl sulfoxide	45.0	29.8
$(CH_3CO)_2O$	20.7	10.5	Dimethylformamide	35.0	30.9
Acetone	20.7	17.0	Pyridine	12.3	33.1
Ethyl ether	4.3	19.2	Hexamethylphosphoramide	30.0	38.9

Tab. 1.4: Effect of solvent polarity on rate constant of styrene anion polymerization.

Solvent	Dielectric constant	k_p/(L/mol s)	Solvent	Dielectric constant	k_p/(L/mol s)
Benzene	2.2	2	Tetrahydrofuran	7.6	550
Dioxane	2.2	5	1,2-Dimethoxyethane	5.5	3,800

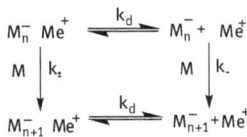

Fig. 1.5: Ion pair and free ion dynamic equilibrium.

Because anionic polymerization can coexist in many different forms of active centers, and the polymerization rates of various active centers vary greatly, the polymerization rate R_p is the sum of the growth rates of various active centers.

For example, the polymerization of styrene in the polar solvent tetrahydrofuran is a chain growth reaction in which both ion pairs and free ions participate in the process. In the polymerization process, ion pairs and free ions are in a dynamic equilibrium, as shown in Fig. 1.5.

In the formula, k_+ and k_- represent the growth rate constants of ion pairs and free ions.

Assuming that the difference between loose pairs and tight pairs in an ion pair is not considered, the aggregate apparent rate can be expressed as

$$K_p = k_-\gamma + k_\pm (1-\gamma) \tag{1.5}$$

where γ represents the fraction of free ions in the active center, and $(1-\gamma)$ represents the fraction of ion pairs.

The ionization equilibrium constant is represented by K_d, and $[C]$ indicates the concentration of the active center:

$$K_d = \frac{\gamma[C] \cdot \gamma[C]}{(1-\gamma) \cdot [C]} = \frac{\gamma^2[C]}{1-\gamma} \tag{1.6}$$

If γ is far less than 1, there is

$$K_d = \gamma^2[C], \quad \gamma = \left(\frac{K_d}{[C]}\right)^{1/2} \tag{1.7}$$

Bring it in eq. (1.5), then

$$K_p = k_\pm + (k_- + k_\pm)\left(\frac{K_d}{[C]}\right)^{1/2} \tag{1.8}$$

Assuming $k_- \gg k_\pm$, eq. (1.8) can be simplified to

$$K_p = k_\pm + k_-\left(\frac{K_d}{[C]}\right)^{1/2} \tag{1.9}$$

According to eq. (1.9) K_p is plotted against $[C]^{-1/2}$ to obtain a straight line with a slope of $k_- K_d^{1/2}$ and an intercept of $k_\pm$. Since K_d can be obtained by the dynamic method or the conductance method, k_- can also be obtained. When styrene is used as a monomer and sodium naphthalene is used as initiator in tetrahydrofuran polymerization, $k_\pm \approx 135$ L/(mol · s), $k_- \approx 65{,}000$ L/(mol · s), it can be seen that the growth anion of free anion is more than 500 times that of ion pair. However, when styrene is used as a monomer and naphthalene sodium is used as an initiator in tetrahydrofuran polymerization, there are no free ions and only ion pairs in the system.

1.2.4 Anionic polymerization method

1.2.4.1 Refining of reagents

The active center of living anionic polymerization is an anion with very strong nucleophilicity, and it is very easy to inactivate hydrogen atoms from active hydrogen substances (such as water, alcohols, acids, and amines). Oxygen molecules in the air, due to the presence of unshared electron pairs, are also susceptible to termination by the active center coupling. In addition, aldehydes, ketones, and other substances can also be deactivated by nucleophilic side reactions with the anion ends. Therefore, in the living anionic polymerization, the abovementioned substances should be removed as much as possible, otherwise the living polymerization cannot proceed smoothly.

For example, to prepare 5 g of polystyrene with Mn = 5.0×10^4, the monomer concentration is 5% and the total volume is 100 mL. The amount of initiator can be calculated as 1.0×10^{-4} mol. If the system contains 1.0×10^{-4} mol water, that is, 1.8 mg water, the initiator will be completely destroyed. If the water content is between 0 and 1.8 mg, although the system can be polymerized, the actual molecular weight will be higher than the designed molecular weight.

Monomer content in anionic polymerization is generally about 10%, and another 90% is solvent, so the refining of the solvent is particularly important. The commonly used anionic polymerization solvents are cyclohexane, benzene, and tetrahydrofuran. Nonpolar cyclohexane is most easily refined, and it can be refluxed in the presence of CaH_2 for several hours at atmospheric distillation.

Benzene is first refined with concentrated sulfuric acid, then neutralized, washed with water to neutrality, dried over anhydrous $MgSO_4$, then dried over CaH_2, and distilled under reflux in the presence of metallic sodium. This is the initial purification of benzene. Distillation in the presence of deep red 1,1-diphenylhexyllithium is still required before use. The purification of tetrahydrofuran is more complicated. First, CaH_2 needs to be left to dry for a whole day and night, and then pressed into the sodium wire to continue drying for more than 24 h. Then the sodium salt (which can also be used as sodium naphthalene) is refluxed for 1 h and then distilled out. This is the initial purification of tetrahydrofuran.

The final refinement is generally carried out before use by distillation in the presence of a red-colored α-methylstyrene tetramer disodium salt. Since α-methylstyrene tetramer disodium salt is a typical active anion, it is often used as an anionic polymerization initiator. As long as its color does not disappear during purification, it means that all the impurities in tetrahydrofuran have been removed and the polymerization will not occur. It also inactivates initiator or anionic active centers. In the purification of the monomer, the purification methods of the nonpolar monomer and the polar monomer are different from each other. The nonpolar monomer styrene was refined by first washing with aqueous NaOH solution (10%) to remove the polymerization inhibitor hydroquinone, washing with water to neutrality, drying with $CaCl_2$ and vacuum distillation. For general use, the styrene is distilled before it is polymerized in the presence of sodium salt of octyl benzophenone (blue-violet, which does not initiate polymerization in the absence of a solvent). If high-molecular-weight polymers are to be synthesized, triphenylmethyllithium (red) is further used as a refined preparation, and it is distilled under reduced pressure at a relatively low temperature.

Purification of isoprene is generally carried out by distillation purification using *n*-BuLi as a detection agent in the absence of a solvent. The polar monomer, such as MMA, is first purified with aqueous NaOH (10%), dried over anhydrous $MgSO_4$ for several hours, then dried overnight with CaH_2, and then in the presence of CaH_2. Vacuum distillation is distillation performed under reduced pressure in the presence of trialkylaluminum (usually $AlEt_3$) prior to polymerization. Trialkyl aluminum can form a light-yellow complex with the monomer, as long as there is a yellow color, indicating that the impurities have been removed.

1.2.4.2 Experimental operation

There are mainly two methods for implementing the living anionic polymerization, namely the inert gas protection method and the high vacuum method. The inert gas protection method is easier to operate. The polymer bottle is preloaded with a magnetic stirrer rotor, the antiplug is sealed, and the manifold on the polymer bottle is connected to the vacuum system with a hose. Boil under a high temperature flame, then fill in high purity N_2, and cool to room temperature. This is repeated three times. The feed was injected using a dry syringe. The high vacuum (1.333×10^{-4} Pa) anionic polymerization method is very difficult to operate. First of all, we should master the fine glass technology (about six months of practice). All experimental devices are designed and blown by the experimenter. The experimental intensity is strong, and the cycle is long. In the experiment, the use of a high-temperature flame of high-pressure oxygen-supported combustion to cut off glass devices containing organic solvents or monomers is a great danger. However, the credibility of the data obtained by this method is much higher than the inert gas protection method.

1.2.5 Anionic living polymerization of polar monomers

1.2.5.1 Problems and their causes

Earlier anionic active polymerization studies focused on the polymerization of styrene and conjugated diolefins. On the other hand, anionic polymerization of polar monomers (MMA, acrylates, etc.) has long been considered incapable of active polymerization due to the following reasons:

(1) Monomer refining is difficult. Due to the presence of oxygen atoms, the hygroscopicity of these monomers is much stronger than that of nonpolar monomers, and it is more difficult to refine them to high purity suitable for anionic polymerization. In addition, since the polymerization activity of the polar monomer is much higher than that of the nonpolar monomer, it is difficult to avoid polymerization during the purification process.

(2) Due to the presence of a more reactive carbonyl group in the acrylic ester monomer, nucleophilic side reactions with the anionic initiator can easily occur, so that the initiation efficiency is greatly reduced, and it is not even possible to initiate polymerization.

(3) Due to the low conjugation of α,β-unsaturated esters, the stability of the formed carbon anions is much lower than that of styrene and conjugated dienes. The polymerization reaction is strongly influenced by temperature and often requires cryogenic temperatures under the polymerization. It was found that when *n*-butyllithium was used to initiate MMA polymerization with typical anionic polymerization initiators, many side reactions occurred due to small steric hindrance and high reactivity.

In the first few seconds of the initiation phase, n-butyllithium can attack monomer's α-position double bond and cause monomers to initiate polymerization. It can also attack the carbonyl on the monomer to generate α,β-unsaturated ketones. Figure 1.6(a) shows that with the occurrence of this side reaction, a large amount of byproduct CH_3OLi is produced. The resulting α,β-unsaturated ketone can also be copolymerized with MMA to enter the polymer chain, so the resulting product is a copolymer containing two structures. When MMA is polymerized in toluene, cyclization of the end groups as shown in Fig. 1.6(b) can also occur, leading to the termination of polymerization and the formation of oligomers.

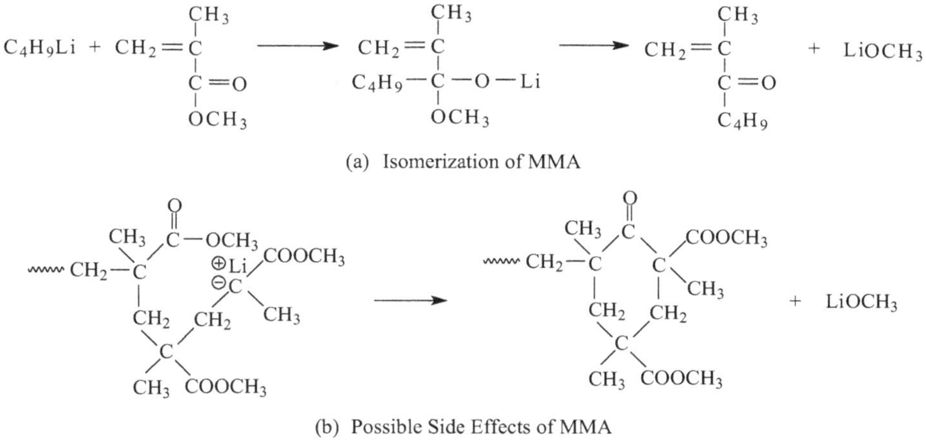

(a) Isomerization of MMA

(b) Possible Side Effects of MMA

Fig. 1.6: The cyclization reaction end groups of MMA, (a) isomerization of MMA, (b) possible side effects of MMA.

According to the data in Tab. 1.5, it is obvious that in MMA polymerization initiated with butyllithium, ca. 50% monomers could only form oligomer; ca. 40% monomers react with butyllithium to produce α,β-unsaturated ketones; other ca. 10% monomers' H^+ are captured by butyllithium to form butane or further become butane by elimination polymerization.

1.2.5.2 Solutions

Based on former discussion, the main side effect during polymerization of polar acrylic monomers initiated with alkyllithium is the nucleophilic addition of carbonyl. Study shows that this side effect can be restrained by certain method, thereby the living polymerization can be realized.

Tab. 1.5: The product of MMA polymerization initiated with butyllithium.

The consumption of butyllithium	Ratio (%)
Butene, butane	9.8
CH_3 $\sim\sim\sim \text{CH}_2-\overset{\mid}{\underset{\mid}{\text{C}}}{}^{\ominus}\ \text{Li}^{\oplus}$ $\overset{\mid}{\text{C}}=\text{O}$ $\overset{\mid}{\text{O}}\text{CH}_3$	Oligomer 48.4
	Polymer 2.0
CH_3 $\sim\sim\sim \text{CH}_2-\overset{\mid}{\underset{\mid}{\text{C}}}{}^{\ominus}\ \text{Li}^{\oplus}$ $\overset{\mid}{\text{C}}=\text{O}$ $\overset{\mid}{\text{C}}_4\text{H}_9$	Oligomer 36.4
	Polymer 2.1

(1) Use steric hindered, less active initiators. When 1,1-diphenyl hexyllithium (or triphenyl methyllithium) initiates the polymerization of MMA in tetrahydrofuran, neither α,β-unsaturated ketones nor LiOCH_3 can be produced, which shows that this initiator selectively reacts with alpha double bond of MMA to initiate the polymerization of MMA.

(2) Use proper additives, stabilizing carbon anion of polar monomers. For example, inorganic salt LiCl can greatly improve the stability of anionic active center of polar monomers. According to the study, in tetrahydrofuran at −78 °C, using active anionic of α-methyl styrene to initiate the polymerization of MMA, rate of polymerization after adding LiCl is slightly slower than before, but the molecular weight distribution is greatly narrowed (<1.1) and the efficiency is greatly improved (Tab. 1.6). It illustrates that the use of LiCl inhibits side effect during initiation phase as well as stabilizes active center.

Tab. 1.6: Effect of lithium salt additives on MMA anionic polymerization.

Additive	Polymerization time (s)	Conversion rate	Molecular weight ($\times 10^{-3}$)	Distribution index	Initiation efficiency (%)
	25	32	14.0	1.30	0.64
No additives	50	52	22.0	1.22	0.66
	70	81	33.0	1.20	0.66
	100	99	41.0	1.20	0.66

Tab. 1.6 (continued)

Additive	Polymerization time (s)	Conversion rate	Molecular weight ($\times 10^{-3}$)	Distribution index	Initiation efficiency (%)
	25	25.6	8.0	1.18	0.86
	41	56.5	18.0	1.09	0.86
LiCl	80	70.7	22.0	1.09	0.86
	100	99.5	32.5	1.09	0.85

Another excellent additives of polar monomer anionic living polymerization is crown ether, in which dibenzo-18-crown-6 has the best effect especially, as shown in Fig. 1.7.

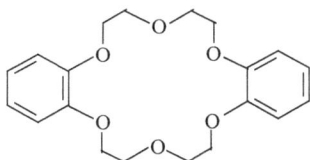

Fig. 1.7: A schematic diagram of two benzo-18-crown ethers-6.

For example, under the existence of dibenzo-18-crown-6, initiating polymerization of MMA with ph_2CHN_2, living polymerization can be achieved and polymerization temperature can be increased to above 0 °C, no matter in polar solvents (tetrahydrofuran) or nonpolar solvents(toluene) (Tab. 1.7).

Tab. 1.7: Anionic polymerization of MMA in the presence of crown ethers.

$[CE]_0/[I]_0$	Solvent	Temperature (°C)	Conversion rate (%)	Molecular weight ($\times 10^{-3}$)	Distribution index	Initiation efficiency (%)
1.0	THF	0	100	8.5	1.15	0.91
2.0	THF	0	100	8.1	1.05	0.96
2.0	THF	25	100	10.5	1.40	0.74
4.0	THF	25	100	10.2	1.10	0.75
1.0	Toluene	0	98.0	8.5	1.20	0.91
2.0	Toluene	0	100	8.5	1.05	0.91
1.0	Toluene	25	99.0	10.5	1.40	0.74
2.0	Toluene	25	100	10.5	1.30	0.74
4.0	Toluene	25	100	10.0	1.15	0.78

Use low-temperature reactions. Polymerizing under a little low temperature (e.g., −78 °C), the cyclization of end group forming oligomers can be completely avoided. Under the abovementioned conditions, the polymerization owns the whole characteristics of living polymerization.

1.2.5.3 Effect of polar monomer structure on living polymerization

Besides abovementioned factors including initiators, additives, solvents, and so on having great influence on polar monomer anionic polymerization, polymerization results are also different due to different molecular structures of polar monomers. Especially synthesizing high-molecular weight polymer, this difference is more pronounced.

For example, in tetrahydrofuran at −78 °C, when 1,1-diphenyl hexyllithium acts as initiator, the anionic polymerization results of MMA and tert-butyl methacrylate (TBMA) are listed in Tab. 1.8.

Tab. 1.8: Anionic polymerization of different methacrylate.

Monomer	Polymerization time (min)	Conversion rate (%)	Theoretical molecular weight ($\times 10^{-4}$)	Theoretical molecular weight ($\times 10^{-4}$)	Distribution index	Initiation efficiency (%)
MMA	13	100	9.9	12	1.09	0.83
MMA	13	96	9.5	11	1.08	0.83
MMA	13	100	14.7	23	1.28	0.64
TBMA	150	100	6.2	6.8	1.08	0.91
TBMA	330	100	13.9	17	1.08	0.82
TBMA	40	98	13.9	18	1.07	0.77

According to the data in chart, when $M_n < 10 \times 10^4$, the anionic polymerization of MMA can be well controlled, meanwhile the obtained polymers have narrow molecular weight distribution (<1.10) and initiation efficiency is higher (>80%); however, when $M_n > 15 \times 10^4$ around, molecular weight distribution becomes wider and initiation efficiency decreases.

If methyl is replaced by tert-butyl, carbonyl side effects can still be prevented although polymerization rate of TBMA is significantly slower than MMA due to steric inhibition of tert-butyl. The obtained polymers have narrower molecular weight distribution all the time. Even if $M_n > 18 \times 10^4$ around, it keeps monodisperse.

1.2.5.4 Refinement of polar monomer

The traditional way to refine polar monomer is CaH_2 dehydration drying method, which is difficult to refine completely due to following two reasons.

(1) Because CaH_2 is solid, heterogeneous solid–liquid reaction is hard to totally remove active hydrogen species from polar monomer. Salt, alkali (e.g., $Ca(OH)_2$),

and so on formed in the early stage of refining usually deposit on the surface of CaH_2 which prevents the further refinement.

(2) Monomer synthesis is often accomplished by exchange reaction of esters and alcohols; therefore, a small amount of alcohol is easily left in monomer. The reaction of CaH_2 and H_2O is severe, yet the reaction of CaH_2 and alcohol is relatively mild. When carbon chain of alcohol becomes a little longer, CaH_2 can even be used as alcohol's desiccant which leads to the incomplete removal of alcohol from monomer by it.

The improved technology can choose trialkyl aluminum (e.g., $AlEt_3$) to refine polar monomer. Trialkyl aluminum can violently react with all active hydrogen species at room temperature, except the reaction with methacrylate monomers or initiating polymerization. Meanwhile, the liquid state of trialkyl aluminum such as $AlEt_3$ let them have homogeneous reactions with impurities in monomers. Trialkyl aluminum will not react with carbonyl of monomers and form yellow complexes to instruct the coming to terminate until the impurities do not exist anymore. It could show that monomers have been as pure as the demand of anionic polymerization once the yellow color appears.

1.2.6 Application of anionic living polymerization

From the time Szwarc put forward the concept of living polymerization in 1956 to now, polymer chemistry researchers and relative industries have paid much attention and have been interested in anionic polymerization. People gradually realized that living polymerization has unique merits compared with common polymerization methods, which provides strong strategies for molecular design of polymers. During the past 40 years, anionic living polymerization has not only had great improvement in theoretical study but also in commercial production application with series of products, expanding application ranges and generally significant benefits to economy and society.

1.2.6.1 Synthesis of narrow relative molecular mass polymers

In recent years, anionic living polymerization is still the first choice to produce narrow relative molecular mass polymers, although a lot of living/controllable polymerization methods have come up. Up to now, the narrowest relative molecular mass distribution of polymers produced by anionic living polymerization is 1.04, which is close to homogeneous relative molecular mass distribution. This provides helpful material basement for studying the relationship between polymer relative molecular mass distribution and performance and reliable reference for measuring other polymers' relative molecular mass by gel permeation chromatography method.

1.2.6.2 Synthesis of end group functionalized polymers

End-group functionalized polymers are very important for the preparation of functional polymers. End-group functionalized polymers are the polymers with functional groups which can be either one or both ends of molecular chains. Common end functional groups in polymers include halogen atom, hydroxyl(-OH), carboxyl(-COOH), carbonyl(-COR), and acyl halide(-COX) Reaction of anionic living chains can introduce functional groups into polymer chains' end easily.

1. Hydroxylation and carboxylation

Reaction of anionic living chains and ethylene oxide under acidic conditions can make polymer end hydroxylation.

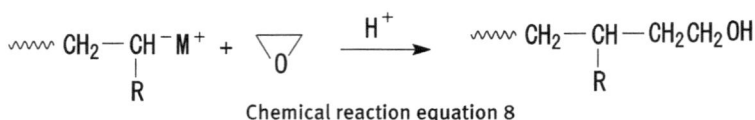

$$\text{\sffamily www } CH_2-\underset{R}{\overset{}{CH^-}}M^+ \ + \ \overset{\triangle}{O} \quad \xrightarrow{H^+} \quad \text{\sffamily www } CH_2-\underset{R}{\overset{}{CH}}-CH_2CH_2OH$$

Chemical reaction equation 8

Different counterions lead to different reactions of ethylene oxide. When counterion is Li^+, it can give rise to terminal hydroxylation. When counterion is Na^+, reaction is more active to make polymerization of ethylene oxide.

Side effects of anionic living chains hydroxylation are relatively less by using ethylene oxide. If ethylene oxide reacts with double anionic living chains, they can produce dihydroxyl telechelic polymers with a functionality close to 2.

Certain number of CO_2 can react with end of anionic living chains. In order to guarantee higher degree of carboxylation, CO_2 should be excessive, and the reaction should be carried out in a polar solvent. The reaction formula shows as follows:

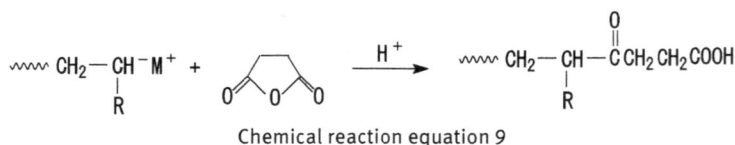

$$\text{\sffamily www } CH_2-\underset{R}{\overset{}{CH^-}}M^+ \ + \ \text{(anhydride)} \quad \xrightarrow{H^+} \quad \text{\sffamily www } CH_2-\underset{R}{\overset{}{CH}}-\overset{O}{\overset{\|}{C}}CH_2CH_2COOH$$

Chemical reaction equation 9

Study results show that polystyrene lithium in benzene solvent carboxylated by gaseous CO_2 could only gain 60% end carboxyl, in addition to 28% ketone and 12% alcohol. If polystyrene lithium is added to solid CO_2, it can obtain over 80% end carboxyl, in addition to less than 20% total amount of ketone and alcohol by-products. And the yield of polymer-terminated carboxyl groups is closely related to the purity of CO_2.

Anhydride can also react with anionic living chains which carboxylates the polymer chains end. Reaction formula shows as follows:

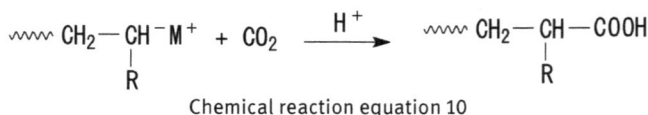

$$\text{\sffamily www } CH_2-\underset{R}{\overset{}{CH^-}}M^+ \ + \ CO_2 \quad \xrightarrow{H^+} \quad \text{\sffamily www } CH_2-\underset{R}{\overset{}{CH}}-COOH$$

Chemical reaction equation 10

2. Halogenation

There are two methods of halogenating anionic living chains: direct halogenation with halogens and reaction with dihalides.

Direct halogenation with halogens is easy to trigger coupling reaction of polymer chains. For example, adding polystyrene lithium in benzene solvent into benzene solution with excess bromine can produce 58% bromine end group polystyrene, in addition to 42% polystyrene couplings. Reaction formula is shown as follows:

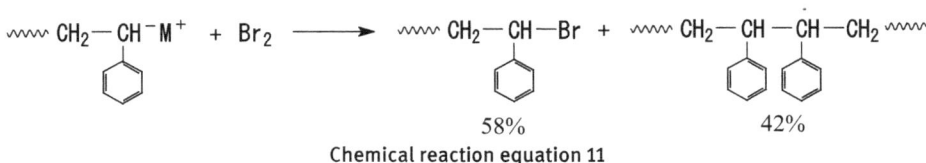

$$\text{www } CH_2-CH^-M^+ \;+\; Br_2 \;\longrightarrow\; \text{www } CH_2-CH-Br \;+\; \text{www } CH_2-CH-CH-CH_2 \text{ www}$$

58%   42%

Chemical reaction equation 11

Using dihalides, reacting with anionic living chains can end polymer chains having bromine atom, for example, reaction of dibromomethyl benzene and polystyrene lithium. Reaction formula is shown as follows:

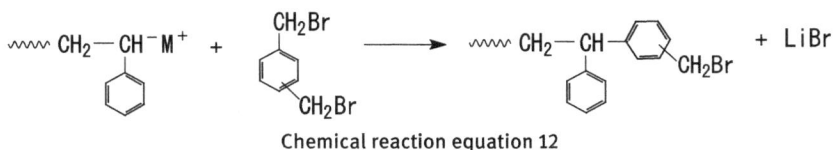

$$\text{www } CH_2-CH^-M^+ \;+\; \begin{array}{c} CH_2Br \\ \\ CH_2Br \end{array} \;\longrightarrow\; \text{www } CH_2-CH-\!\!\left\langle\right\rangle\!\!\begin{array}{c} \\ CH_2Br \end{array} \;+\; LiBr$$

Chemical reaction equation 12

This reaction also easily leads to the coupling reaction of polymer.

3. Carbonylation and Cl-acylation

Overdose esters or acyl chlorides reacting with anionic living chains can make the end of polymer carbonylated, for example, reaction of methyl benzoate or benzoyl chloride and anionic living chains, as shown in Fig. 1.8(a). In the case of a large excess of phosgene, it can cause the acyl chlorination of the end group of polymers, as displayed in Fig. 1.8(b).

a)

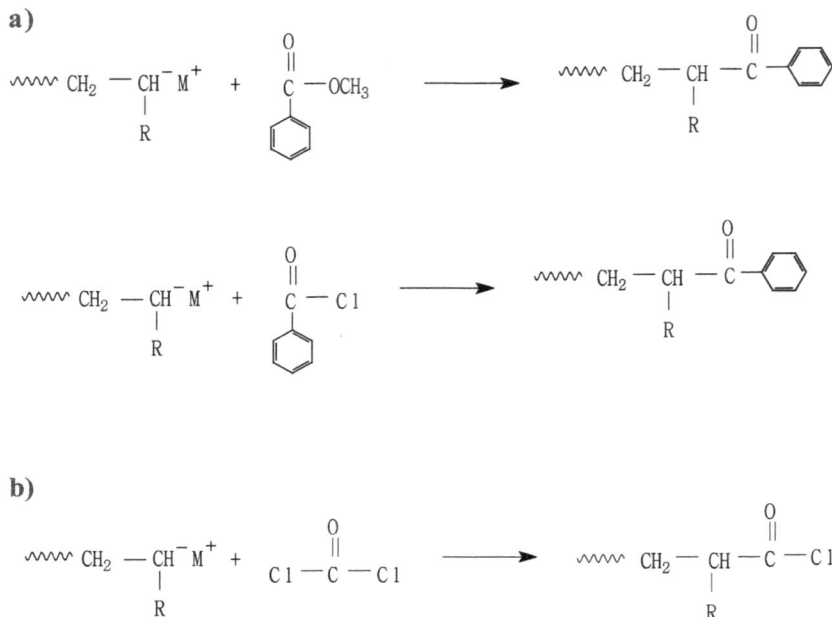

b)

Fig. 1.8: Carbonylation and acylation of anionic living chain: (a) carbonylation of anionic living chain and (b) acyl chlorination of anionic living chain.

4. Aminolation

Reaction of anionic living chains and isocyanate can aminolate the end group of polymers.

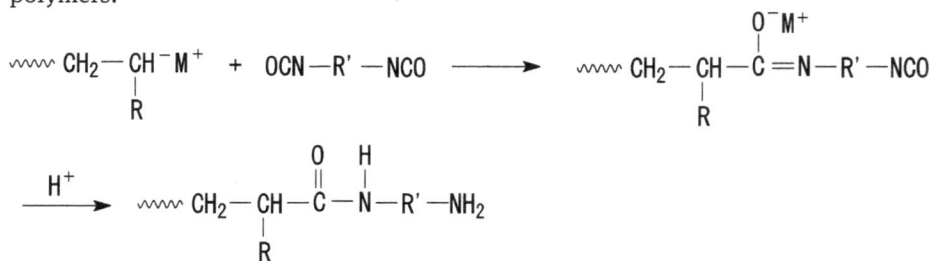

Chemical reaction equation 13

Trimethylsilyl derivatives of aniline react with anionic living chains. After acid hydrolysis, it produces aminoterminated polymers.

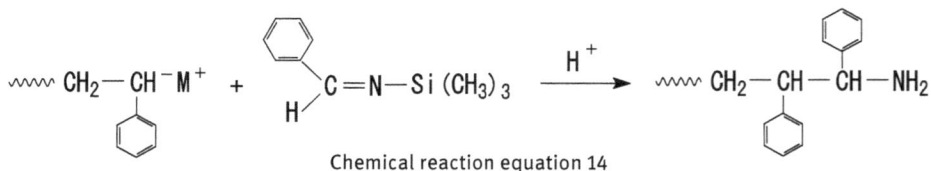

Chemical reaction equation 14

Initiators with protected amino give rise anionic living polymerization of mono-mers. Then remove to protect and gain aminoterminated polymers. For example,

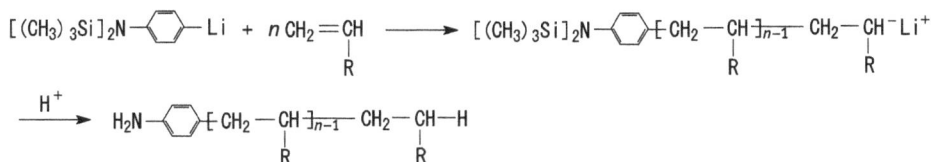

$$[(CH_3)_3Si]_2N-\langle\bigcirc\rangle-Li \ + \ n\,CH_2{=}\underset{\underset{R}{|}}{CH} \longrightarrow [(CH_3)_3Si]_2N-\langle\bigcirc\rangle{\text{-}}CH_2-\underset{\underset{R}{|}}{CH}\text{-}_{n-1} \ CH_2-\underset{\underset{R}{|}}{CH}^-Li^+$$

$$\xrightarrow{\ H^+\ } \ H_2N-\langle\bigcirc\rangle{\text{-}}CH_2-\underset{\underset{R}{|}}{CH}\text{-}_{n-1} \ CH_2-\underset{\underset{R}{|}}{CH}-H$$

Chemical reaction equation 15

If the abovementioned anionic living chains are coupled by dichlorodimethyl silane before hydrolysis, double-ended amino polymers are obtained.

Similar methods also can be used to produce hydroxyl-ended polymers.

5. Peroxidation

Reactions of anionic living chains and oxygen are used to produce macromolecular peroxide, which can give rise to radical polymerization of other monomers forming block copolymers. Reaction of oxygen and anionic living chains has a large amount of side effects, and most of which can be eliminated by choosing proper reaction conditions. For example, at the temperature as low as −78 °C, adding living polysty-rene into oxygen-saturated tetrahydrofuran solution can make products including over 90% macromolecular peroxide and less than 10% polystyrene conjugate.

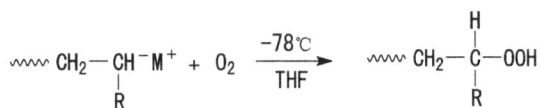

$$\text{www}\ CH_2-\underset{\underset{R}{|}}{CH}^-M^+ \ + \ O_2 \ \xrightarrow[\text{THF}]{-78\,°C} \ \text{www}\ CH_2-\overset{\overset{H}{|}}{\underset{\underset{R}{|}}{C}}-OOH$$

Chemical reaction equation 16

6. Preparation of macromolecular monomers

Double bond-ended molecular chains of polymers can be produced by anionic liv-ing chains, which can be named as "macromolecular monomer" or "large mono-mer" because of having the ability to further participate in aggregation

Ethylene oxide reacts with polystyrene lithium and then terminates with meth-acryloyl chloride, the double-bond-ended polystyrene can be obtained, as shown in Fig. 1.9.

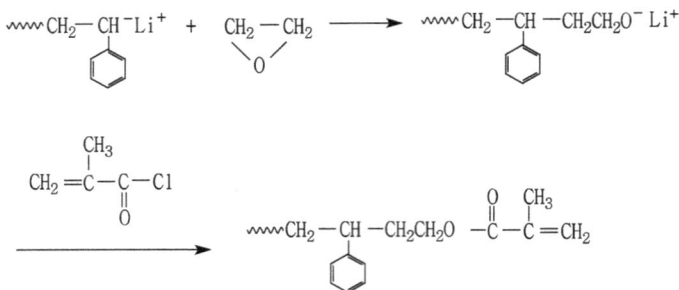

Fig. 1.9: Schematic illustration of preparation of macromolecular monomers.

Using macromolecular technology can produce a lot of special characteristic or functional comb polymers. For example, when 5–10% α-methyl polystyrene is copolymerized with MMA, the glass transition temperature of plexiglass can increase around 20 °C.

1.2.6.3 Synthesis of block, star, and comb polymers
It is the most brilliant result of anionic living polymerization study, nothing could beat the successful synthesis of block copolymers. Block polymers and star polymers have unique properties which other homopolymers and copolymers do not have, leading to be widely appreciated over the years. Anionic living polymerization can be used to produce any length or any number of block copolymers or star polymers.

1. Synthesis of block copolymers
Block copolymers produced by anionic living polymerization have hundreds of species until now, among which the most famous and meaningful product is terpolymer of SBS and SIS. Terpolymer of SBS and SIS is a kind of thermoplastic elastomer, in which S represents styrene block; B represents butadiene block; and I represents isoprene block. Because these three segments are incompatible, microphase separation occurs. Polystyrene segments play the role of physical cross-linking points in the system which results in the similar mechanical properties at room temperature between SBS/SIS and vulcanized rubber. However, when the temperature is higher than the glass transition temperature of polystyrene, polystyrene segments turn soft and physical cross-linking points are damaged, so that the system can be processed like thermoplastic elastomers.

There are three synthesis methods of SBS and SIS: monofunctional initiator initiation method, bifunctional initiator initiation method, and even legal method. The following takes the preparation of SBS as an example to introduce the synthesis of block copolymer.

(1) Monofunctional initiator initiation

Common initiator to produce SBS is butyllithium. Preparation process: first, initiate the polymerization of styrene; once the reaction of styrene finishes, add butadiene; after the reaction of butadiene finishes, add styrene; finally, terminate reaction with terminator. During synthesis, cyclohexane and benzene are generally used as solvents at 70 °C polymerization temperature. The reaction process can be represented by the following equation shown in Fig. 1.10.

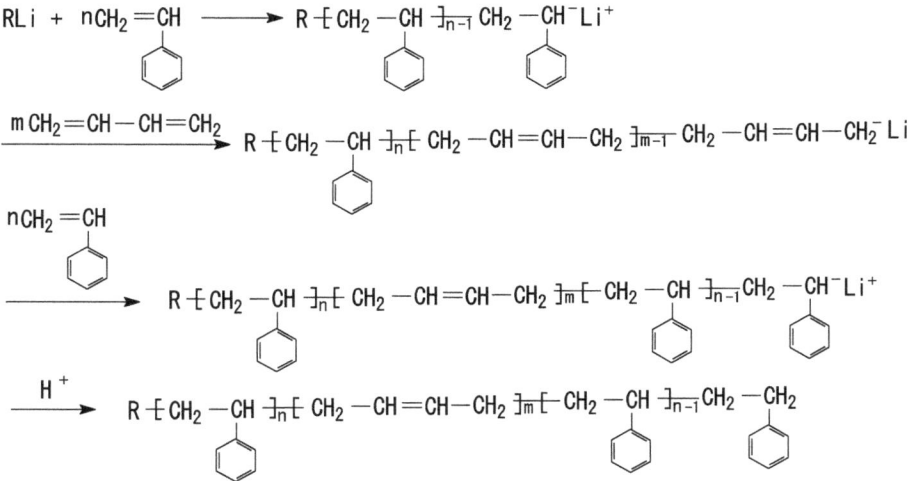

$RLi + nCH_2=CH \longrightarrow R(CH_2-CH)_{n-1}CH_2-CH^-Li^+$

$\xrightarrow{m\,CH_2=CH-CH=CH_2} R(CH_2-CH)_n(CH_2-CH=CH-CH_2)_{m-1}CH_2-CH=CH-CH_2^-Li$

$\xrightarrow{nCH_2=CH} R(CH_2-CH)_n(CH_2-CH=CH-CH_2)_m(CH_2-CH)_{n-1}CH_2-CH^-Li^+$

$\xrightarrow{H^+} R(CH_2-CH)_n(CH_2-CH=CH-CH_2)_m(CH_2-CH)_{n-1}CH_2-CH_2$

Fig. 1.10: Schematic illustration of synthesis triblock copolymer.

Among them, polybutadiene segments have *trans*-1,4- and 1,2-structures besides the abovementioned *cis*-1,4-structure.

(2) Bifunctional initiator initiation

Monofunctional initiators to produce SBS need three times feeding and every time it is unavoidable to bring in impurities which make part of the active chains stop. The product contains a certain amount of styrene homopolymer or SB diblock copolymer, which has an adverse effect on the performance of SBS. Using bifunctional initiators to initiate polymerization can reduce feeding times. For example, lithium naphthalene initiator system is a common bifunctional initiator. It formed double ions living centers; first, initiate the polymerization of butadiene, and then initiate the polymerization of styrene. Finally, SBS can be obtained. The process of initiating reaction and forming block copolymers is shown in Fig. 1.11.

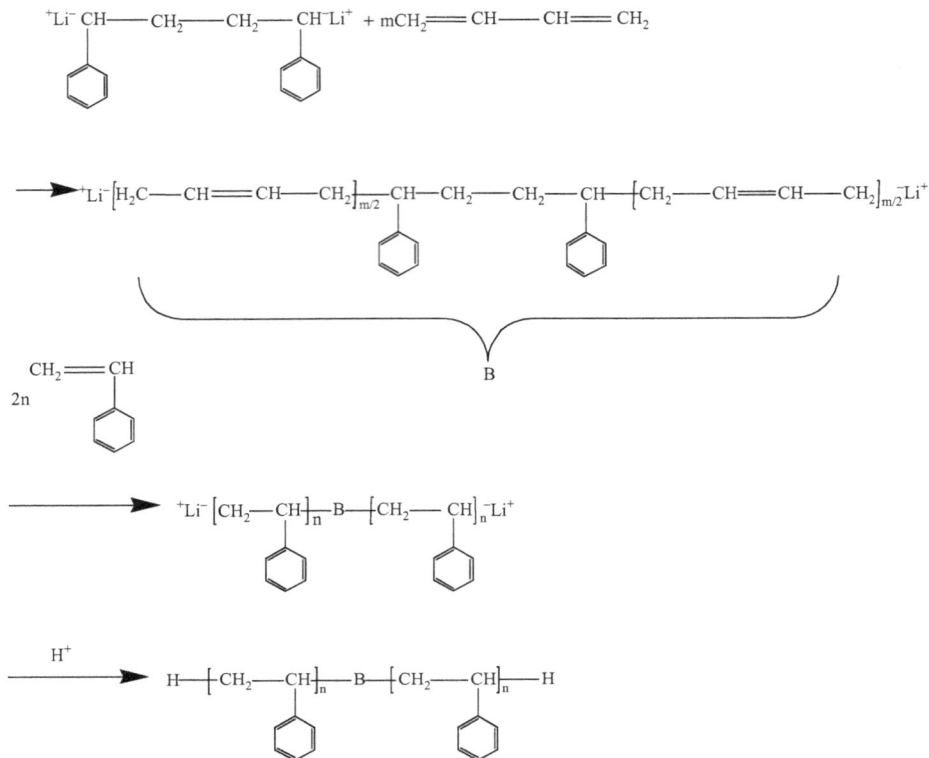

Fig. 1.11: Schematic diagram of preparation of triblock copolymer by bifunctional initiator.

(3) Coupling method

First, use monofunctional initiators to produce SB diblock copolymers, and then react with difunctional coupling agent which can also produce SBS triblock copolymers. For example, 1,6-dibromohexane is used as a coupling agent to produce SBS triblock copolymers. The reaction equation is shown in Fig. 1.12.

Anionic living chains and terminal cationic living chains can have coupling reactions that are able to effectively produce block copolymers. For example, at −58 °C temperature, the reaction of tetrahydrofuran initiated by $Ph_3 C^+SbF_6^-$ is living polymerization (living polymerization of cyclic compound will be introduced in the following part). Mixing these tetrahydrofuran cationic living chains with MMA or tetrabutyl methacrylate anionic living chains can quickly produce block copolymers. PMMA-b-PTHF and PtBMA-b-PTHF diblock copolymers can be produced by this method, and it is the same situation with PtBMA-b-PTHF-b-PtBMA and PMMA-b-PTHF-b-PMMA triblock copolymers.

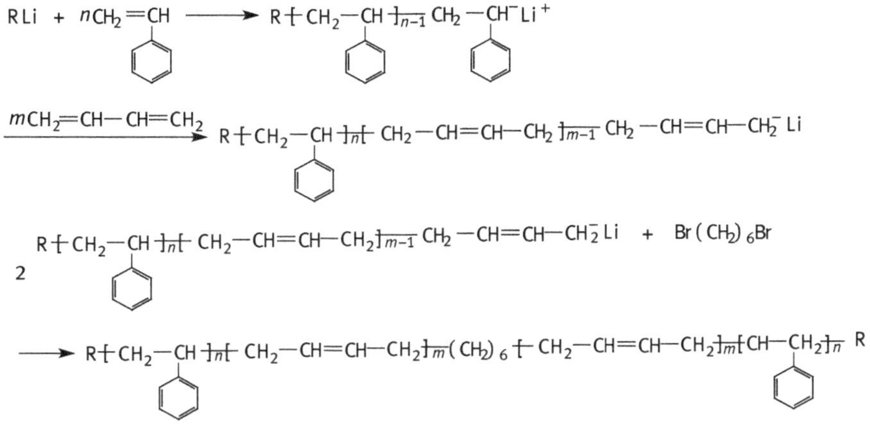

$$RLi + nCH_2=CH\underset{\bigcirc}{\overset{}{|}} \longrightarrow R\left[CH_2-CH\underset{\bigcirc}{\overset{}{|}}\right]_{n-1} CH_2-CH^-\underset{\bigcirc}{\overset{}{|}}Li^+$$

$$\xrightarrow{mCH_2=CH-CH=CH_2} R\left[CH_2-CH\underset{\bigcirc}{\overset{}{|}}\right]_n CH_2-CH=CH-CH_2]_{m-1} CH_2-CH=CH-CH_2^- Li$$

$$2\ R\left[CH_2-CH\underset{\bigcirc}{\overset{}{|}}\right]_n CH_2-CH=CH-CH_2]_{m-1} CH_2-CH=CH-CH_2^- Li\ +\ Br(CH_2)_6Br$$

$$\longrightarrow R\left[CH_2-CH\underset{\bigcirc}{\overset{}{|}}\right]_n CH_2-CH=CH-CH_2]_m(CH_2)_6\left[CH_2-CH=CH-CH_2\right]_m[CH-CH_2]_n\ R$$

Fig. 1.12: Schematic diagram of SBS triblock copolymer prepared by coupling method.

From previous discussion, we could know that compared with the second monomer, anionic living chain is a kind of macromolecular initiator. However, theoretical and experimental study both show that not all anionic living chains are able to initiate the polymerization of another monomer. According to selective principle of initiator in anionic polymerization, anionic living chains initiate the second monomer successfully or not mainly depending on relatively alkali between M_1^- and M_2, which means the electron donating ability of M_1^- and the electron affiliating ability of M_2. For example, PS$^-$ can initiate the polymerization of MMA, but MMA$^-$ is impossible to initiate the polymerization of styrene. PS$^-$ has higher alkali than MMA$^-$.

Generally, pK_d represents the relative size of monomer alkalinity. If anionic living chain and monomer are regarded as conjugate acid–base pair, K_d will be the ionization equilibrium constant of acid–base pair defined as $pK_d = -lgK_d$. According to experimental study, anion formed by monomer with high K_d can initiate monomer with low K_d. Conversely, it is unavailable. Table 1.9 shows a few kinds of monomers' pK_d.

Tab. 1.9: pK_d value of several anionic polymerized monomers.

Monomers	pK_d
Styrene, butadiene	40–42
Acrylate	25
Acrylonitrile	24
Epoxide	15
Nitroethylene	11

pK_d will be slightly different when influenced by the polar of solvent. But it has no influence on the relative size of pK_d.

Although styrene and butadiene have the same level of pK_d, styrene anion ability of initiating butadiene monomer is relatively better than butadiene anion ability of initiating styrene initiating styrene monomer which has relationship with the stability of butadiene anion.

2. Synthesis of star polymers

It is acknowledged that difunctional coupling agents reacting with SB diblock copolymers can produce SBS triblock copolymers. If the function of coupling agent is higher than 3, it can be used to produce star polymers. For example, if $SiCl_4$ or $SnCl_4$ is used as coupling agent, four-arm SB star copolymers can be obtained, as displayed in Fig. 1.13.

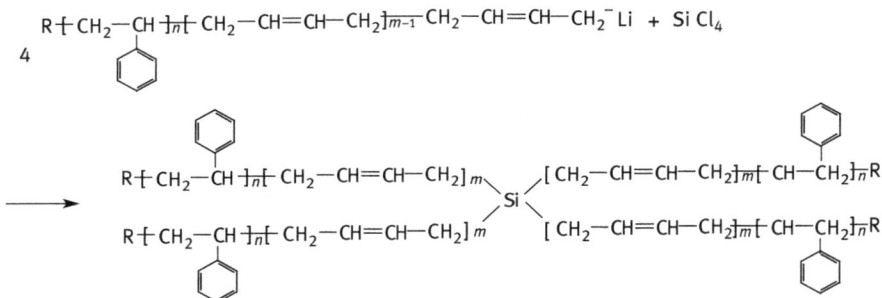

Fig. 1.13: Schematic diagram of synthesis of four-arm SB star copolymers.

Star polymers with over 30 arms have been obtained by similar methods. Special structural star copolymers have triblock copolymer as an arm or variously structural arms.

The most obvious characteristic of star polymer is that melt viscosity has no relationship with the whole relative molecular weights of polymer just depending on relative molecular weight of each arm. As a result, if the whole relative molecular weights are the same, viscosity of star polymer is obviously lower than linear polymer of the same type, leading to its good process ability. Till now, the star polymers used in industry have arms no more than six. Star polymers with over six arms are mainly used in scientific study.

3. Synthesis of comb polymers

Certain structure of comb polymers can also be synthesized by anionic living polymerization. For example, when styrene anionic living chains solution was added to solution with PMMA, living polystyrene living chains will react with the ester of PMMA and hang on the PMMA main chains to form comb polymers, as illustrated in Fig. 1.14.

$$\text{www}-\!\!\left[\!\text{CH}_2-\!\!\underset{\underset{\displaystyle \text{OCH}_3}{\overset{\displaystyle \text{C}=\text{O}}{|}}{\overset{\displaystyle \overset{\displaystyle \text{CH}_3}{|}}{\text{C}}}\!\right]_{\!m}\!\!-\text{www} \quad + \quad \text{R}-\!\!\left[\!\text{CH}_2-\underset{\displaystyle \bigcirc}{\text{CH}}\!\right]_{\!n\text{-}1}\!\!\text{CH}_2-\underset{\displaystyle \bigcirc}{\text{CH}}^{-}\text{Li}^{+}$$

$$\longrightarrow \quad \text{www}-\!\!\left[\!\text{CH}_2-\!\!\underset{\underset{\displaystyle \text{CH}-\text{CH}_2\,[\text{CH}-\text{CH}_2]_{n\text{-}1}\,\text{R}}{\overset{\displaystyle \text{C}=\text{O}}{|}}}{\overset{\displaystyle \overset{\displaystyle \text{CH}_3}{|}}{\text{C}}}\!\right]_{\!m}\!\!-\text{www}$$

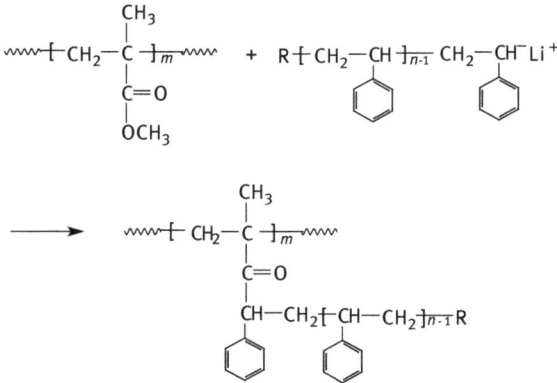

Fig. 1.14: Schematic diagram for the preparation of PMMA-g-PS comb copolymer.

1.3 Cationic living polymerization

1.3.1 Characteristics of cationic living polymerization

Cationic living polymerization is another important branch of linkage polymerization, and its invention can date back to early twentieth century. The industrial products of cationic living polymerization such as polyisobutylene and butyl rubber were found in the 1940s. Although both cationic living polymerization and anionic living polymerization belong to ionic polymerization, a large amount of studies show that cationic polymerization is not as easy to control as anionic polymerization, which leads to the fact that the stability of its living center is very poor. As a result, since anionic living polymerization was invented in 1956, the exploring research of cationic living polymerization has been struggled without notable improvement. This situation once led researchers to lose confidence until 1984. Higashimura first reported cationic living polymerization of alkyl vinyl ether, and then Kennedy developed cationic living polymerization of isobutylene. Cationic living polymerization made breakthrough across the time. In the following years, polymerization mechanism, initiation system, monomer, and synthetic applications of cationic living polymerization have made great process.

Higashimura et al. used HI/I_2 as an initiator in cationic living polymerization of alkyl vinyl ether, and they found that the process had typical characteristic of living polymerization which was similar to anionic living polymerization. That is why it is regarded as the first cationic living polymerization and its characteristics mainly include the following:

(1) Number relative molecular mass has linear relationship with monomer conversion.
(2) Adding monomer to finished polymerization system, number relative molecular mass will have proportional growth.

(3) Polymerization rate has positive ratio with initial concentration $[HI]_0$.
(4) Concentration of I_2 in initiator just has relationship with polymerization rate instead of relative molecular mass.
(5) At any conversion rate, relative molecular mass distribution of product will keep narrow, $M_w/M_n < 1.1$.

Using HI/I_2 to initiate the system at -40 °C temperature in toluene, the experiment adds more monomer during the polymerization of 2-acetoxyethyl vinyl ether (AcOVE) as shown in Fig. 1.2. Before and after adding more monomer, polymerizations are both quantitative, and relative molecular masses of polymers increase proportionally with the increase of conversion rate. At any conversion rate, relative molecular mass distribution will keep narrow ($M_w/M_n < 1.2$). After adding more monomer, smooth linear increase of M_n shows that all polymer chains formed in the first stage are active which can be called polymerization without induct period. Meanwhile amount of living centers keep constant during the polymerization.

1.3.2 Principle of cationic living polymerization

For common cationic living polymerizations, instability of carbon cation living center has inherent disadvantage. Therefore, choosing and designing nucleophilic counterion to improve the stability of carbon cation with increasing chains is the key to realize cationic living polymerization. Usually, higher the nucleophilicity of counterion, more stable the carbon cation. But if nucleophilicity is too strong, carbon cation will lose its activity, as shown in Fig. 1.15.

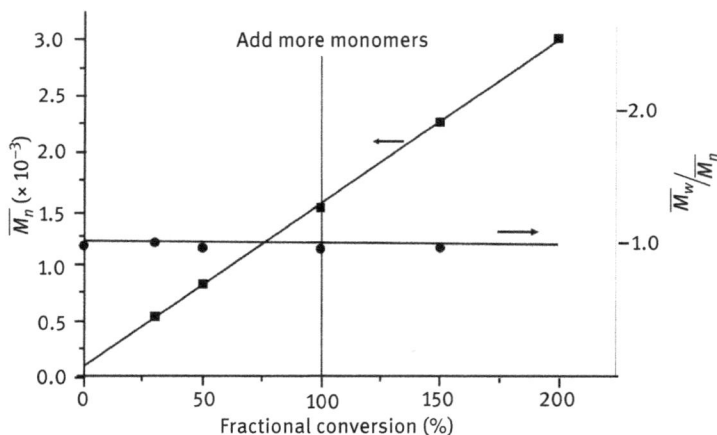

Fig. 1.15: The relationship between the conversion of monomer and M_n and M_w/M_n in AcOVE polymerization with HI/I_2 as an initiator.

In the system of ionic polymerization, existence of many kinds of living centers that are in the dynamic balance is common as shown in Fig. 1.16.

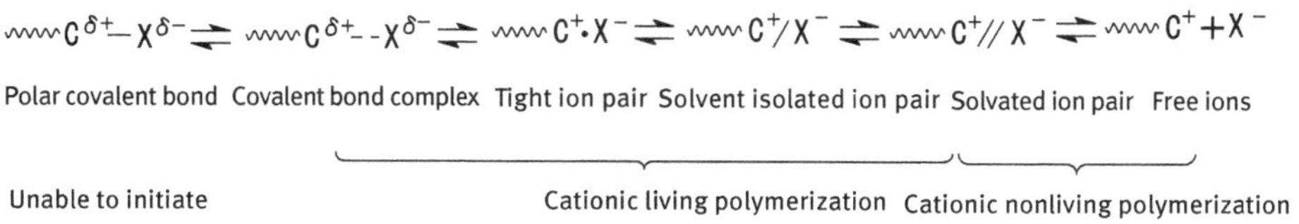

$$\sim\!\!\sim C^{\delta+}\!\!-X^{\delta-} \rightleftharpoons \sim\!\!\sim C^{\delta+}\!-\!-X^{\delta-} \rightleftharpoons \sim\!\!\sim C^{+}\!\!\cdot X^{-} \rightleftharpoons \sim\!\!\sim C^{+}\!/X^{-} \rightleftharpoons \sim\!\!\sim C^{+}\!//X^{-} \rightleftharpoons \sim\!\!\sim C^{+}\!+X^{-}$$

Polar covalent bond   Covalent bond complex   Tight ion pair   Solvent isolated ion pair   Solvated ion pair   Free ions

Unable to initiate   Cationic living polymerization   Cationic nonliving polymerization

Fig. 1.16: Dynamic equilibrium of multiple active centers in ionic polymerization system.

In the abovementioned living center spectrum, polar covalent bonds do not have initiating activity and the activities of free ions are so strong that they alone can initiate cationic polymerizations including chain transfer and chain termination.

Only the active species in middle part have proper ions to initiate cationic polymerization. Adding shared anions into the abovementioned balanced system by a proper method to decrease the concentration of polar covalent bonds and strong ionic living species, it is possible to change normal cationic polymerization into cationic living polymerization. A study by Higashimura et al. has shown that cationic living polymerization can be realized by the following two ways: (1) to design matched nucleophilic counterions; and (2) to add extra Lewis alkali. Taking the first way as an example, Higashimura et al. chose HI/I_2 to initiate cationic living polymerization of alkyl vinyl ether, whose principle is shown in Fig. 1.17.

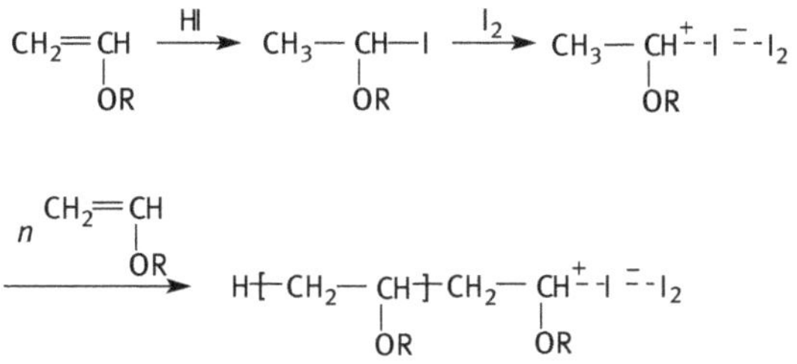

Fig. 1.17: Cationic living polymerization of alkyl vinyl ether initiated by the HI/I_2 system.

As shown in the equation, in the reaction system, HI initiates monomers to produce living end groups; meanwhile, I_2 named as an activator or coinitiator can form I^- . . . I_2 complexes with I^- by a nucleophile, which weaken the nucleophile of I^- resulting in not only the increase of living centers' activity but also the originally unstable carbon cation in the living state.

The second way was developed based on the initiator diethyl aluminum dichloride ($EtAlCl_2$), which produces anionic ions having too weak nucleophilicity to make carbon cationic increasing living chains stable. Adding moderate Lewis alkali in the abovementioned system can change it into a living polymerization system. For example, the principle of initiating polymerization of alkyl vinyl ether by EtAlCl$_2$/ethyl acetate is shown in Fig. 1.18.

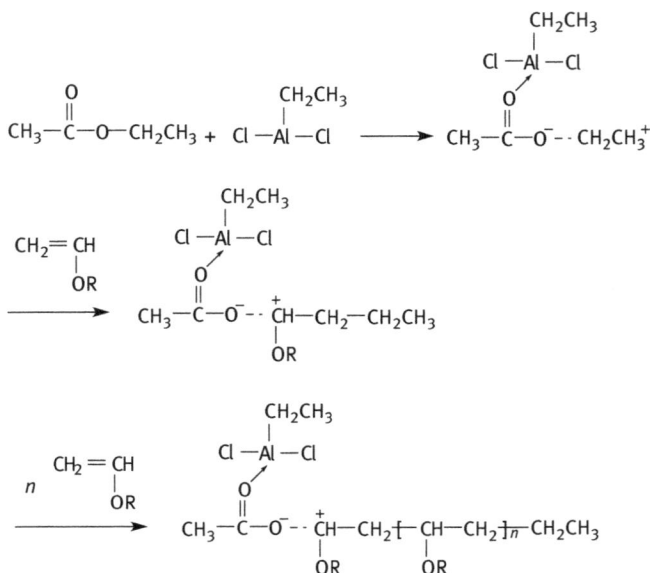

Fig. 1.18: Cationic living polymerization of alkyl vinyl ether by EtAlCl$_2$/ethyl.

Lewis alkali includes ester, ether, and heterocyclic. It is common to use dioxane, tetrahydrofuran, ethyl acetate, anhydride, sulfur sulfone, and so on.

1.3.3 Monomer and initiator of cationic living polymerization

In cationic polymerization, the activity and stability of living centers are the integrated result of interaction between carbon cation and counterions which depend not only on the nucleophilicity of carbon cationic ions and monomer molecules, but also have relationship with the electrophilicity of carbon cation.

The electrophilicity of carbon cation at the end of living chain is determined by the structure of monomer which forms carbon cation. Therefore, monomers with different chemical structures and properties need different initiating system to start cationic living polymerization.

1.3.3.1 Alkyl vinyl ether monomers

1. HI/I$_2$ initiating system

Higashimura et al. once used I$_2$ to initiate cationic living polymerization of alkyl vinyl ether. Relative molecular mass distribution of produced polymer is too wide to be named as living polymerization. Replacing I$_2$ by HI to initiate alkyl vinyl ether reaction for 48 h at −15 °C temperature still only obtain adducts of both, which means that polymerization of alkyl vinyl ether cannot be initiated only by HI, but if we add I$_2$ basing on HI, the reaction rate is almost proportional to I$_2$ concentration. In 1984, Higashimura et al. used HI/I$_2$ as initiator to obtain cationic living polymerization of isobutyl vinyl ether (IBVE), which first realized cationic living polymerization of alkyl vinyl ether monomers and put forward the reaction principle mentioned in Section 1.3.2 of this book.

2. HI/ZnX$_2$ or HI/SnX$_4$ initiating system

Both HI/ZnX$_2$ and HI/SnX$_4$ systems are effective in initiating cationic living polymerization of alkyl vinyl ether. For HI/ZnX$_2$ initiating system, when [HI]/[ZnX$_2$] = 1–50, initiating polymerization of IBVE with CH$_2$Cl$_2$ solvent in the temperature range of −40 to 0 °C or with toluene in the temperature range of −40 to 40 °C can obtain living PIBVE with so narrow relative molecular mass distribution ($\overline{M_w}/\overline{M_n} = 1.05 - 1.10$). Compared to HI/I$_2$, this kind of initiating system can give rise to cationic living polymerization of alkyl vinyl ether even at room temperature. Experiments prove that HI is an initiator and ZnX$_2$ or SnX$_4$ as coinitiator in this initiating system.

3. Monobasic phosphate ester/ZnX$_2$ initiating system

Monobasic phosphate ester/ZnX$_2$ initiating system initiates the polymerization of alkyl vinyl ether in toluene solvent at 0 °C temperature, which is also a cationic living polymerization proved by experiments with narrow relative molecular mass product ($\overline{M_w}/\overline{M_n} \approx 1.10$). The reaction principle is explained in the following part. In this initiating system, monobasic phosphate ester is an initiator and ZnX$_2$ is a coinitiator.

4. Trimethylsilicon compound/ZnX$_2$/electron donor initiating system

ZnX$_2$/trimethylsilyl iodide initiating system initiates polymerization of IBVE in toluene solvent at room temperature, which does not have living polymerization characteristics. However, after adding some electron donor carbonyl compounds (e.g., acetone and benzaldehyde), reaction displays living characteristics, and relative molecular mass distribution of products is narrow, too ($\overline{M_w}/\overline{M_n} \approx 1.10$). Reaction principle is shown in Fig. 1.19.

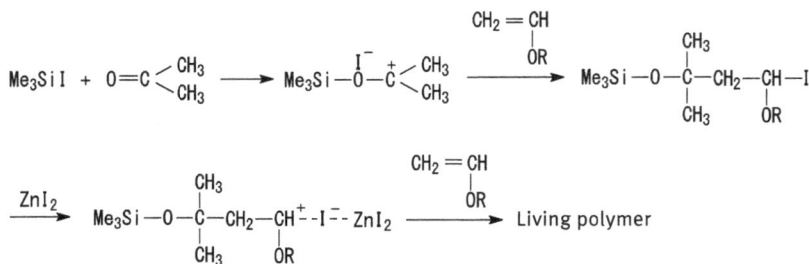

Fig. 1.19: Cationic living polymerization of alkyl vinyl ether by trimethylsilicon compound/ZnX$_2$/ electron donor.

5. Carboxylic acid/ZnCl$_2$ initiating system

Single carboxylic acid cannot initiate cationic living polymerization of alkyl vinyl ether except adding extra ZnCl$_2$. If trifluoroacetic acid/ZnCl$_2$ initiates polymerization of IBVE at 0 °C temperature, then products have extremely narrow relative molecular mass distribution, $\overline{M_w}/\overline{M_n} = 1.06$.

6. Lewis acid/Lewis alkali initiating system

Lewis acid such as EtAlCl$_2$ and TiCl$_4$ cannot directly initiate cationic living polymerization of alkyl vinyl ether except adding moderate electron donors (Lewis alkali), such as ester, ether, and heterocyclic. EtAlCl$_2$ can quickly initiate polymerization of alkyl vinyl ether and activity is strong enough to have obvious chain transfer reaction which cannot obtain living polymers. The addition of dioxane decreases the initiation rate resulting in emergence of living characteristics. If volume ratio of dioxane is higher than 10%, living polymers with relative molecular mass distribution index $\overline{M_w}/\overline{M_n} < 1.1$ are obtained. The most valuable advantage of this kind of initiator is that cationic living polymerization can be initiated at room or a little higher temperature. Initiation principle has been introduced in Section 1.3.2 of this book.

1.3.3.2 Isobutylene monomers

In 1986, Kennedy et al. first used organic tertiary alkyl esters/organic tertiary hydrocarbons to finish the cationic living polymerization of isobutylene in halogenated hydrocarbons whose polymerization temperature was in the range of −50 to −10 °C. The initiation system of isobutylene polymerization is organic tertiary hydrocarbons/Lewis acid.

In the initiation system of organic tertiary hydrocarbons/Lewis acid, Lewis acid mainly include boron trichloride (BCl$_3$) and titanium tetrachloride (TiCl$_4$). Organic tertiary hydrocarbons include organic tertiary chloride compounds, organic tertiary ester compounds, organic tertiary ether compounds, organic tertiary alcohol compounds, and organic tertiary peroxides. Kennedy et al. put forward that organic tertiary alkyl esters/BCl_3 system obeys the following mechanism if living polymerization is initiated in methyl chloride or dichloromethane, as shown in Fig. 1.20.

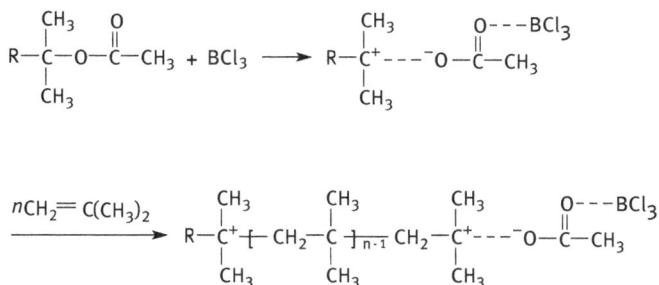

Fig. 1.20: Living polymerization mechanism of organic tertiary alkyl ester (acid)/BCl_3 initiation system in haloalkanes.

According to the mechanism, the complexes generated by organotertiary hydrocarbyl initiator and BCl_3 initiate the reaction of Isobutylene. The antiparticles in the active center are acetate ions complexed by BCl_3, whose composition is similar to the foregoing $I^- \cdots I_2$, can stabilize the growing isobutene cations. The mechanism of the tertiary ether/BCl_3 is also the same, the antiparticle is $-OCH_3 \cdots BCl_3$. When using methanol as a terminator, the end of the products, without exception, become [-$C(CH_3)_2Cl$].

The addition of electron-donating nucleophiles such as dimethyl sulfoxide, dimethylformamide, triethylamine, pyridine, and derivatives thereof in abovementioned initiation systems can further improve the living characteristics of the polymerization, improve the efficiency of initiator, reduce side effects, and reduce the molecular weight distribution.

1.3.3.3 Styrene and its derivatives

For styrene and its derivatives, living polymerization is difficult to take place because they cannot form stable carbon cation-increasing active center.

The difference in substituents and positions on the benzene ring leads to a marked difference in the reactivity of the monomers, so the initiation system used is also different. In 1987, Higashimura et al. used HI/ZnI_2 initiation system for P-methylstyrene and P-tert-butoxystyrene and found that cationic living polymer can be obtained at room temperature whose relative molecular mass distribution is narrow and relative molecular mass is quite huge ($\overline{DP_n} \geq 1,000$). Kennedy used I_2 to initiate the cationic polymerization of P-methylstyrene at –15 °C. Kennedy also achieved the cationic polymerization of styrene in methyl chloride at –30 °C by ethyl p-methylphenyl/BCl_3 initiation system, but the polymer relative molecular mass distribution is quite broad. Table 1.10 shows some cationic polymerization systems with styrene and its derivatives.

Tab. 1.10: The cationic living polymerization system of styrene and its derivative monomers.

Monomer	Initiation system
Styrene	p-Methylphenylacetate/boron trichloride, methanesulfonic acid/tin tetrachloride/tetrabutylammonium chloride, chloroethylbenzene/tin tetrachloride/tetrabutylammonium chloride, chloroethylbenzene/titanium trichloride/tetrabutylammonium chloride
p-Alkyl styrene	p-Methylphenylacetate/boron trichloride, acetyl sulfonic acid/ tetrabutylammonium chlorate, HI/ZnI_2/tetrabutylammonium chloride, HI/ZnI_2
p-Halo styrene (chloro, fluoro)	2-Methyl-4,4-dimethyl-2-chloropentane/$TiCl_4$/electron donor reagent
p-Methoxy styrene	Iodine, HI/ZnI_2, HI/ZnI_2/tetrabutylammonium chloride, HI/I_2/ tetrabutylammonium chloride, $YbOSO_2CF_3$
α-Methyl styrene	$CH_3CH(OCH_2CH_2Cl)/SnCl_4$, $(CH_3)_3CH_2C(CH_3)_2CH_2C(Ph)_2Cl/SnCl_4$, 2-methyl-4,4-dimethyl-2-chloropentane/$TiCl_4$/electron donor reagent

1.3.3.4 Diolefins and cycloolefin monomers

The diolefin monomer used in cationic polymerization generally is 1,3-pentadiene. Its cationic polymerization is usually initiated by Brønsted acid or Lewis acid. When initiated by Brønsted acid, only oligomers with small relatively molecular weight can be obtained generally. While initiated by Lewis acid like $AlCl_3$, $AlBr_3$, $BF.OEt_2$, $TiCl_4$ and $SbCl_5$, polymers with large relatively molecular weight can be obtained. If the reaction is initiated by Lewis acid alone, many side reactions exists and cross-linked polymer may also be formed. Adding Lewis base into Lewis acid initiators can inhibit the cross-linking reaction and other side reactions to improve the activity characteristics during polymerization.

Cyclic olefin monomers which are capable of carbocation polymerization mainly include indene and α-quinone. The living polymerization of indene can be initiated by the cumyl ether or cumyl chloride/$TiCl_4$ initiating system. α-quinone is a monomer which is very hard to polymerize, and we usually use system based on Lewis acid (such as $AlCl_3$ and $TiCl_4$) to initiate its polymerization. Its polymerization includes many steps such as ring opening, isomerization, and polymerization reaction. The mechanism of the cationic polymerization of these monomers has not been systematically studied.

1.3.4 Apparent living polymerization and quasiliving polymerization

With more researches on cationic living polymerization, it was found that many so-called cationic living polymerizations are not truly living polymerizations. The chain transfer reaction and chain termination reaction have not been completely eliminated, instead, they have been masked to some extent. So, these reactions can be classified into apparent living polymerization and quasiliving polymerization. The difference between the two is that the former includes a certain degree of transfer to the monomer chain, and the latter includes reversible chain transfer reaction and chain termination reaction exists in the system.

1.3.4.1 The dynamic characteristics of living polymerization

The basic characteristics of living polymerization have been discussed above, but in order to compare the difference between living polymerization and apparent living polymerization and quasiliving polymerization, the dynamic characteristics of living polymerization are summarized:

1. The chain initiation reaction rate is much higher than the chain growth reaction rate: $R_i \gg R_p$.
2. No chain termination reaction occurs: $R_t = 0$.
3. Chain transfer reaction is completely eliminated: $R_{tr} = 0$.

In the system that satisfies the abovementioned conditions, the chain initiation reaction quickly quantifies the active center, and the chain growth reaction occurs synchronously. After the reaction, the active center of the molecular chain still exists. Therefore, the polymer concentration produced in the system is equal to the active center concentration and the initiator concentration.

$$[N] = \lfloor M^* \rfloor = [C] \tag{1.10}$$

The resulting polymer has a very narrow molecular weight distribution and the degree of polymerization is directly proportional to the number of reacted monomers:

$$\overline{DP} = \frac{[M]_0 - [M]}{[N]} = Y \times \frac{[M]_0}{[C]} \tag{1.11}$$

Equation (1.11) shows a linear relationship between the degree of polymerization of the living polymerization product and the conversion: Y.

1.3.4.2 Apparent living polymerization

After deep study of cationic polymerization, Sigwalt et al. believe that cationic living polymerization is not a true living polymerization, but rather a polymerization system that does not completely eliminate the chain transfer reaction and has the characteristics of living polymerization.

According to eq. (1.11), the molecular weight of the polymer can be fully controlled by adjusting $[M]_0/[C]$, but the actual situation is not as dramatic. Sigwalt et al. found that most of the cationic active systems reported were carried out under the condition that the $[M]_0/[C]$ is relatively small. When these polymerizations are carried out under relatively large $[M]_0/[C]$ ratio conditions, the linear relationship of $\overline{DP}$~Y shows obvious deviations. In Fig. 1.21, at a lower conversion rate, curve 2 and living polymerized straight line 1 completely coincide. However, at a higher conversion rate, curve 2 deviates below the straight line.

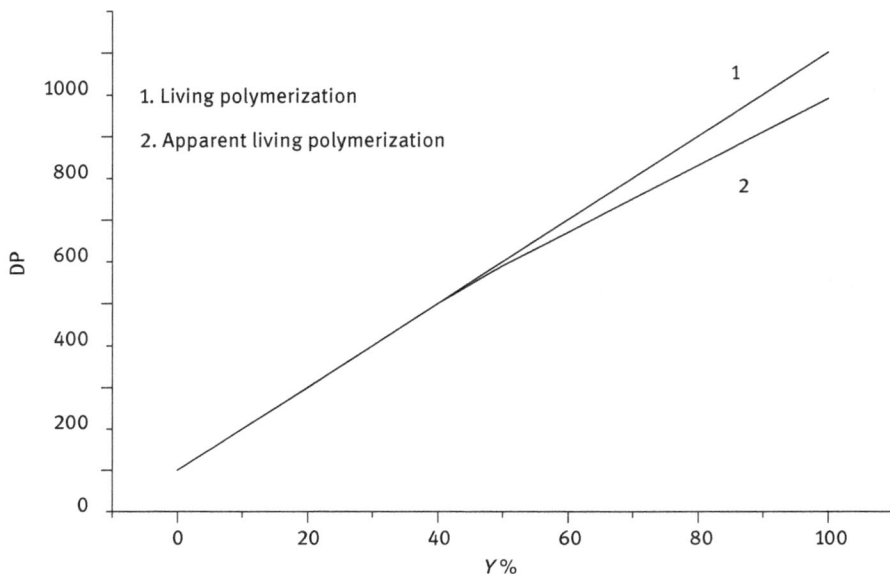

Fig. 1.21: The relationship between the degree of polymerization and conversion rate for living polymerization and apparent living polymerization.

Based on these experimental facts, Sigwalt et al. proposed the concept of apparent active polymerization to describe the type of polymerization reaction reflected by curve 2 in Fig. 1.4 and pointed out that this deviation is due to the fact that there is still some monomer chain transfer reactions. In other words, the apparent living polymerization system satisfies the first. And second condition of the three conditions of the dynamic characteristics of the living polymerization and does not completely satisfy the third condition. Almost all active cationic polymerization systems have been found belonging to apparent living polymerization. The reason why the phenomenon of chain transfer has not been observed is that $[M]_0/[C]$ is small.

In fact, chain transfer reactions to monomers in chain polymerization are extremely difficult to eliminate, and this chain transfer reaction will obviously lead to an increase in the number of macromolecules in the system, $[N] > [M^*]$. If the increase in macromolecule concentration in the apparent living polymer is marked as $[N]_{tr}$, $[N] = [N]_0 + [N]_{tr} = [M^*] + [N]_{tr}$. Depending on the reaction process, the monomer transfer chain can be divided into a zero-order reaction or a first-order reaction. According to the phenomena, both reactions generate the same product, but they are different in the reaction process and can be identified based on the relationship between [N] and conversion rate Y.

In the zero-reaction chain-transfer reaction to monomers, the chain growth active center spontaneously decomposes into a neutral macromolecule and an initiator fragment. The latter initiates a new initiation reaction with the monomer and forms a new active center. The reaction process and the dynamic equation of the zero-order reaction are as follows:

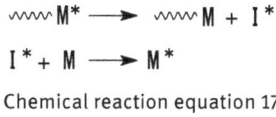

$$\text{\footnotesize www} M^* \longrightarrow \text{\footnotesize www} M + I^*$$
$$I^* + M \longrightarrow M^*$$

Chemical reaction equation 17

The dynamic equation is

$$\frac{d[N]\text{tr0}}{dt} = k_{tr0}\left[M^*\right] = k_{tr0}[N]_0 \tag{1.12}$$

The first-order chain transfer reaction to the monomer is a bimolecular reaction that occurs between the active site of growing chain and the monomer, resulting in the transfer of the active center to the monomer. Its reaction process is as follows:

$$\text{\footnotesize www} M^* + M \longrightarrow \text{\footnotesize www} M + M^*$$

Chemical reaction equation 18

Its dynamic equation is

$$\frac{d[N]_{tr1}}{dt} = k_{tr1}\left[M^*\right] = k_{tr1}[N]_0[M] \tag{1.13}$$

In addition, the chain reaction of the living polymerization reaction is a first-order reaction to both monomer and the active center. Its dynamic equation is

$$\frac{d[M]}{dt} = k_p\left[M^*\right][M] = k_p[N]_0[M] \tag{1.14}$$

After integration

$$\ln \frac{[M]_0}{[M]} = k_p[N]_0 t \tag{1.15}$$

or

$$[M] = [M]_0 e^{-k_p[N]_0 t} \tag{1.16}$$

Integrating eq. (1.12), we obtain

$$[N] = k_{tr0}[N]_0 t \tag{1.17}$$

Combining eqs. (1.15) and (1.17), we obtain

$$[N]_{tr0} = \frac{k_{tr0}}{k_p} \ln \frac{[M]_0}{[M]} = \frac{k_{tr0}}{k_p} \ln \frac{1}{1-Y} = C_{tr0} \ln \frac{1}{1-Y} \tag{1.18}$$

C_{tr0} is the transfer constant for zero order.

Combining eqs. (1.13) and (1.17), we obtain

$$[N]_{tr1} = \frac{k_{tr1}}{k_p}([M]_0 - [M]) = \frac{k_{tr1}}{k_p}[M]_0 Y = C_{tr1}[M]_0 Y \tag{1.19}$$

C_{tr1} is the transfer constant for first order.

Figure 1.22 shows the $[N]$–Y curve of apparent living polymerization systems for living polymerization, zero-order and first-order chain transfer reactions. From the picture, the difference between those two chain transfers can be easily observed. For zero-order reaction, $[N]$ gradually increases with Y. For first-order reaction, the relationship between $[N]$ and Y is a straight line.

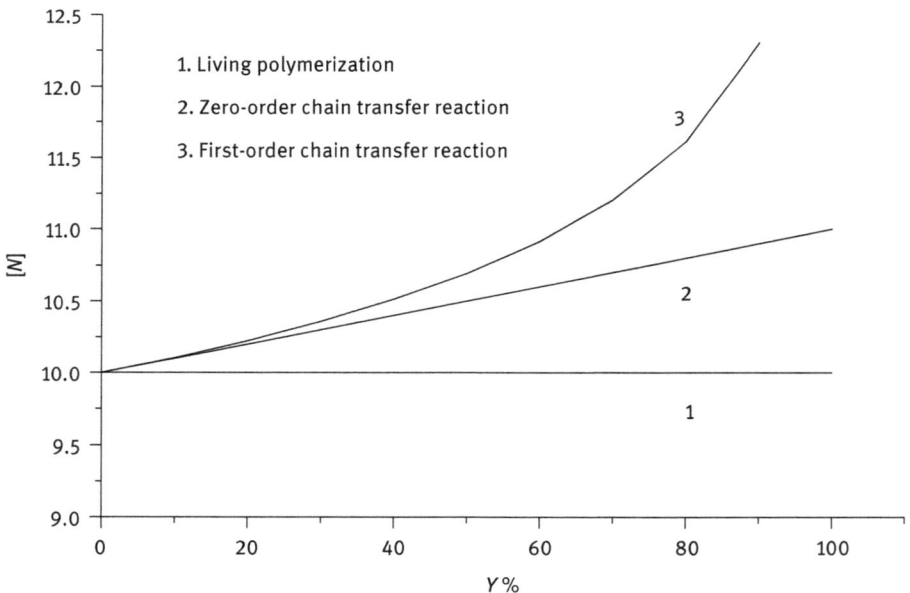

Fig. 1.22: The $[N]$–Y curve of apparent living polymerization systems for living polymerization, zero-order and first-order chain transfer reactions.

1.3.4.3 Quasiliving polymerization

Before the discovery of cationic living polymerization, Kennedy et al. had discovered an effective method to control the relative molecular mass and relative molecular mass distribution of polymers in cationic polymerization. When using PhC $(CH_3)_2Cl/BCl_3$ to initiate the polymerization of α-methylstyrene, they slowly added the monomer dropwise to the stirred initiator solution. The resulting polymerization was very similar to the living polymerization, that is,, the relative molecular mass increased linearly with monomer consumption. The number of macromolecules was comparable to the amount of active species, and the relative molecular mass distribution was narrow. Kennedy et al. believe that there are reversible chain transfer reactions and chain termination reactions in the polymerization, but under the conditions of slow dropping of monomers, these two reactions are limited at very low monomer concentrations, and it shows the characters of living polymerization. However, there is a fundamental difference between this polymerization and living polymerization, so it is called quasiliving polymerization.

In quasiliving polymerization, the active polymeric chain will lose its activity by the chain transfer reaction and become a dormant seed, but this transfer is reversible, and the dormant seed can be rapidly transformed into active centers, and this transition must be rapidly reversible, that is, the apparent rate of chain transfer reaction and chain termination reaction is zero. The dormant molecular chain is a special covalently stable form of active growth center and is in rapid equilibrium with the active center of the chain. This equilibrium is called quasiactivity equilibrium, and the relative molecular mass distribution of the polymer depends on the rate of equilibrium exchange. Since the decrease of the monomer concentration drives the quasiplanar move toward the formation of the active center, the reversible chain transfer reaction can be suppressed by slowly adding the monomer.

Many styrene derivatives, alkyl vinyl ethers and isobutylenes can undergo quasiactive cationic polymerization. The general requirement for active cationic polymerization is that there are chain transfer reactions and chain termination reactions in the polymerization, but it is required that the kinetic chain termination reaction, the proton elimination reaction, and the bimolecular chain transfer reaction to the monomers are all reversible. The terminated molecular chain can still reinitiate monomer polymerization.

1.3.5 Cationic living polymerization applications

1.3.5.1 Synthesis of terminal functional polymers

1. Functional group initiator method
Compounds formed by vinyl ether with a specific functional group X and the same equimolar amount of HI can form an initiator system with ZnI_2 to initiate the

cationic living polymerization of an alkyl vinyl ether, and the polymer has a functional group X on its first segment. The reaction process is shown in Fig. 1.23.

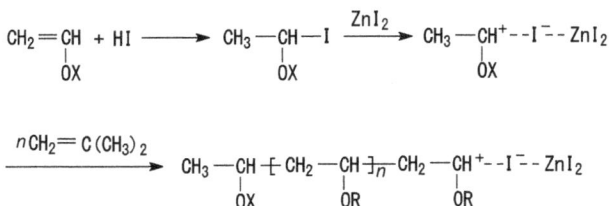

Fig. 1.23: Preparation of terminal functional group polymer by functional group initiator method.

2. The end cap method

The end cap method is used to terminate the cationic living chain with Nu–Y, a nucleophile with a specific functional group. The end of the final product is introduced into the functional group Y. The reaction is as shown in Fig. 1.24.

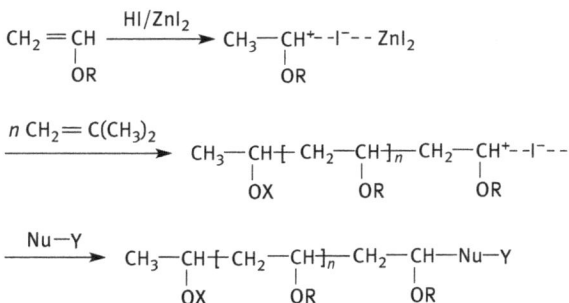

Fig. 1.24: Preparation of terminal functional group polymer by end cap method.

If functional initiator method and cap method are used together, the polymer obtained from the functional initiator is terminated with Nu–Y, a telechelic polymer having the same or different functional groups at both ends can be obtained.

In recent years, 1,1-diphenylethylene (DPE) has been often used as a capping agent for cationic living polymerization, because it does not polymerize itself and does not react with Lewis acid. DPE reacts with the very unstable polyisobutylene living chain and converts it to a stable carbocation with two benzene rings. After that, we select an appropriate terminator to control the termination reaction. A series of terminal functional group polymers can be prepared. The reaction is shown in Fig. 1.25.

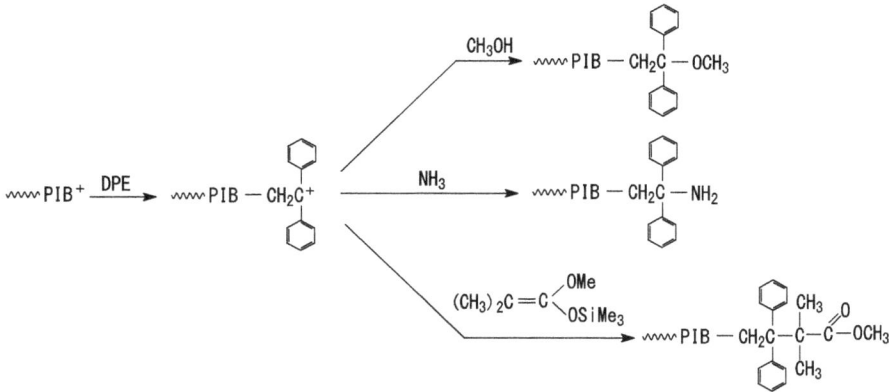

Fig. 1.25: Schematic diagram for DPE used as a capping agent for cationic living polymerization.

3. Inifer method

Inifer is a substance that can both initiate and transfer, and it was first discovered by Kennedy in 1979. It is both an initiator, a control initiator, and a transfer agent which causes directed chain transfer of the active growing chain to produce a polymer with a specific chain end functional group. Therefore, the terminal functional group polymer can also be prepared by cationic polymerization by this substance. For example, cumyl chloride acts as an inifer when cumyl chloride and BCl_3 are used to initiate the polymerization of isobutylene, as illustrated in Fig. 1.26.

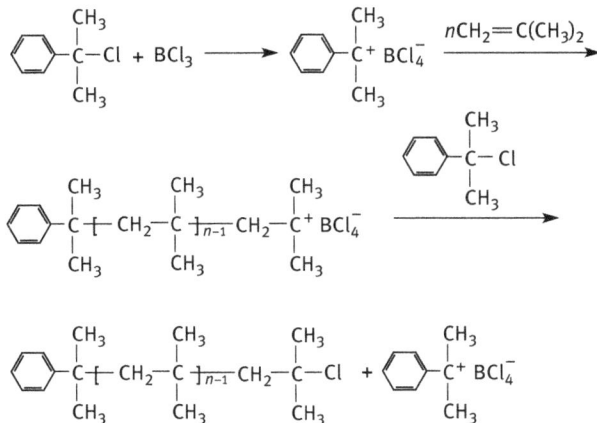

Fig. 1.26: Polymerization of isobutylene initiated by BCl_3.

A number of linear and star-shaped chlorine-terminated telechelic polyisobutylenes have been synthesized by this inifer method. It is also possible to continue the reaction with chlorine end groups to further synthesize block copolymers or convert chlorine end groups to other functional groups. The method has the advantages of high product functional rate, simple reaction steps, and the like, and is an effective method for preparing a block copolymer of a telechelic polymer in the field of cationic polymerization.

1.3.5.2 Synthesis of telechelic polymers

The use of compounds such as acetyl chloride (CH_3COCl) or chloromethyl ether (CH_3OCH_2Cl) as initiators, in combination with coinitiators, can initiate the cationic polymerization of styrene carbon cations. Part of the structure of the initiator can be introduced into the chain end of the polystyrene. Inspired by this, if a double-end initiator is used to initiate polymerization of the monomers, part of the structure of the initiator can be introduced into both ends of the polymer to form a telechelic polymer. The preparation process of a telechelic polymer is shown in Fig. 1.27.

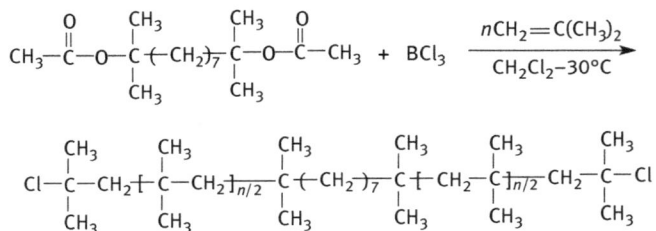

Fig. 1.27: Schematic diagram of preparing a telechelic polymer.

The polymer can also be further hydroxylated to form hydroxyl-terminated polyisobutylene, which can be used as a binder with good aging resistance and high temperature resistance and can also be used as an intermediate for preparing novel polyurethanes. In addition to the hydroxyl group and chlorine atom, telechelic polymers with a carboxyl group, an amino group, and an isocyanate at the molecular chain end have been prepared by a cationic living polymerization.

1.3.5.3 Synthesis of macromonomer

If isobutene is initiated by initiator with a double bond, a macromonomer containing a double bond at the chain end can be synthesized, as shown in Fig. 1.28(a).

Initiator with a halogen atom can also produce a macromonomer with a styrene structure as an end group, as shown in Fig. 1.28(b).

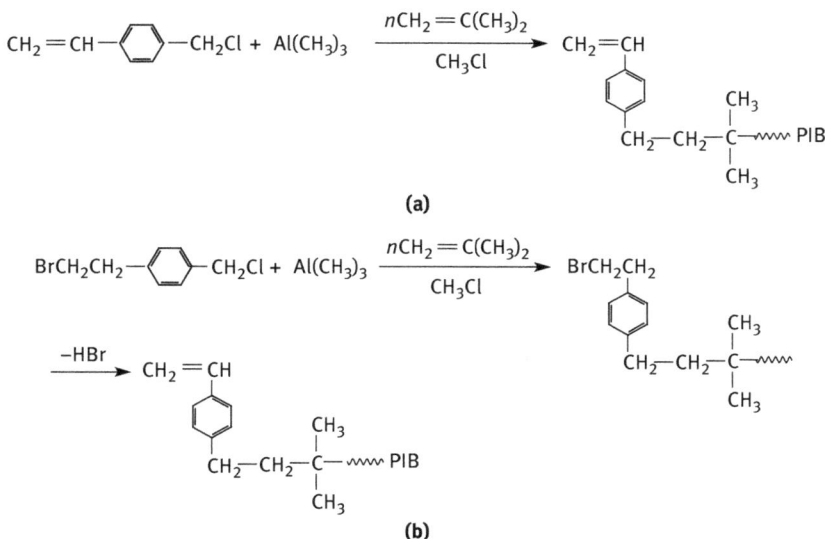

Fig. 1.28: Schematic diagram of preparing macromolecule monomer. (a) introduction of double bond by initiator containing double bond, and (b) introduction of double bond by dehydrohalogenation.

These macromonomers can undergo other polymerizations, such as anionic polymerization and radical polymerization, to prepare graft or comb copolymers.

1.3.5.4 Synthesis of block copolymers

Like anionic living polymerization, cationic living polymerization can also be used to prepare block copolymers. At present, there are two methods to prepare block copolymers by cationic living polymerization: macroinitiator method and sequential addition method.

1. Macroinitiator method

In carbocationic living polymerization, the R group of the initiator RX enters the polymer chain. If the initiator only has one functional group, the R group enters the head of the polymer chain; if the initiator is difunctional or polyfunctional material, the R group enters the center of the polymer chain. Therefore, controlling the initiation reaction means that it is possible to design telechelic polymers, linear block copolymers, and star polymers according to the structure and properties of the initiator.

The first block copolymer obtained by cationic living polymerization is a diblock copolymer of isobutylene and styrene, and its key technology is to prepare a macroinitiator. For example, the preparation of diblock copolymers of isobutylene and styrene includes the following steps, as shown in Fig. 1.29.

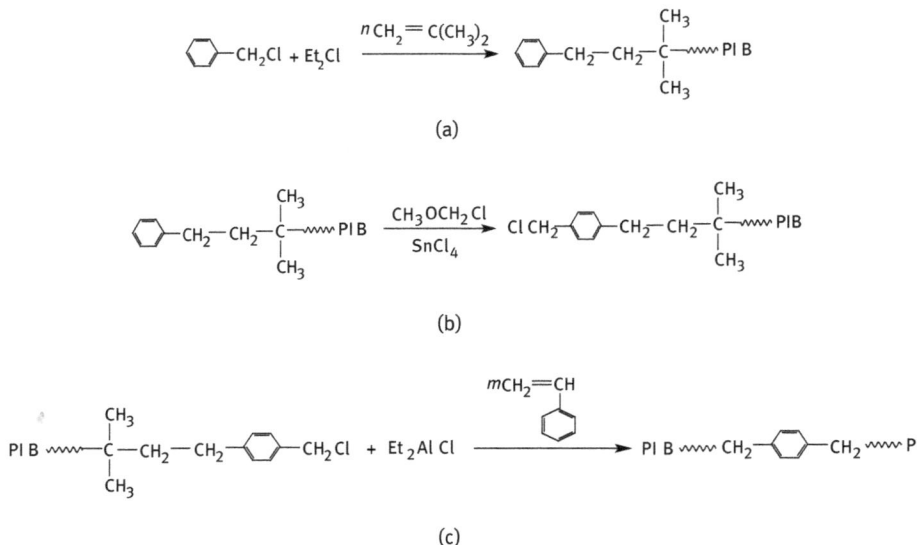

(a)

(b)

(c)

Fig. 1.29: Schematic diagram of diblock copolymers prepared by macromolecular initiator method. (a) synthesis of benzyl-terminated PIB, (b) benzyl end group chloromethylation, and (c) initiate styrene polymerization to get diblock copolymers.

Using the different activities of the functional groups in the initiator, macroinitiators and block copolymers can be obtained. For example, the styrene is initiated by initiators of different active sites first. Since the chlorine atom is more active than the bromine atom, the bromine atoms are retained, and the bromine-terminated polystyrene can be obtained. Then, polymerization of isobutylene was initiated with this macroinitiator to obtain a PS-b-PIB diblock copolymer. The reaction is shown in Fig. 1.30.

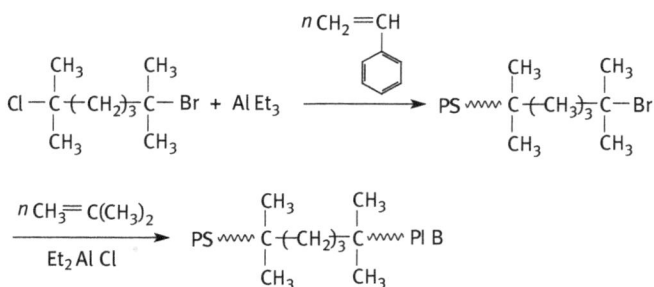

Fig. 1.30: Schematic diagram of preparation of diblock copolymers by using initiator activity difference.

If we use the bifunctional initiator to initiate the cationic living polymerization of isobutylene and styrene, the triblock copolymer can be easily prepared.

2. Sequential addition method

Similar to anionic living polymerization, block copolymers can also be obtained by cationic living polymerization through sequential addition method. For example, when the polymerization of isobutylene is initiated by $TiCl_4$/cumyl chloride in dichloromethane/ n-hexane, the relative molecular mass increases linearly with the monomer conversion rate, the relative molecular mass distribution is lower than 2.0, and the number of polymer active chains is basically constant, if we add monomer slowly. If we continue to add the second and third monomers to the system, the polymerization would not stop and eventually the diblock and triblock can be obtained. For example, Kennedy et al. had succeeded in preparing PS-b-PIB-b-PS triblock thermoplastic elastomers with a tensile strength of 20–22 MPa and a glass transition temperature of 98 °C.

The use of 1,1-DPE to convert the unstable polyisobutylene active chain ends to stable carbon cations with two benzene rings is described in the previous section. On this basis, we can adjust the concentration of the Lewis acid to initiate the polymerization of highly active monomers to synthesize high percentage block copolymers. For example, the polymerization of isobutylene can be initiated by cumyl chloride/carbon tetrachloride/pyridine at −80 °C. After all monomers are completely reacted, add DPE to terminate the active chain and the stable diphenylalkyl carbocation will generate. After that, excess titanate is added to invalidate the carbon tetrachloride in the system. Tin tetrabromide is then added to initiate the polymerization of α-methyl styrene to generate PIB-b-PαMS block copolymer. With the same method, triblock copolymers like PS-b-PIB-b-PS, PIBVE-b-PIB-b-PIBVE have been synthesized.

In addition to DPE, furan derivatives can also be capped with polyisobutylene active chains for chain-end design and synthesis of block copolymers.

In a block copolymer of hexadecyl vinyl ether (CVE) and methyl vinyl ether (MVE) initiated by sequential addition with HI/I_2, the PCVE segment is a hydrophobic polymer with long chain alkyl groups, while the PMVE segment is hydrophilic polymer. Therefore, this is an amphiphilic block copolymer with many special uses.

1.3.5.5 Synthesis of graft copolymers

Cationic living polymerization can be used to synthesize graft polymers that cannot be obtained by other reactions. For example, the chlorine atom on the polychloroprene or polyvinyl chloride chain is the active initiator of cationic polymerization. So, they are actually polyfunctional cationic polymerization initiators. Graft copolymers with PIB branch or α-polymethylstyrene can be prepared with Et_2AlCl or $AlEt_3$ as coinitiators.

Using polyvinyl acetate (PVAc) obtained by radical polymerization as a macroinitiator, $EtAlCl_2$ or $TiCl_4$ as coinitiator to graft isobutylene to PVAc, a graft copolymer PVAc-g-PIB can be obtained. The main chain of this is a polar polymer, and the branched chain is a nonpolar polymer. Therefore, it is an amphiphilic graft copolymer and is a valuable functional polymer material.

1.4 Other ionic living polymerization

1.4.1 Coordination anionic living polymerization

Although complex anion polymerization of α-olefins can synthesize a copolymer, it is not easy to carry out living polymerization because of easy chain dehydrogenation due to dehydrogenation during the polymerization. It was recently discovered that vanadium acetoacetate/alkylaluminum chloride can initiate propylene at −70 °C and obtain structural regular polypropylene with a narrow molecular weight distribution ($\bar{M}_w/\bar{M}_n = 1.05\text{~}1.12$). Obviously, this is a kind of coordination anionic living polymerization.

A typical example of coordinating anionic living polymerization is the polymerization of propylene initiated with soluble vanadium bodies. At present, the vanadium compound initiators that can be used for active polymerization of propylene are mainly four kinds of complexes of +3 vanadium ions (Fig. 1.31), and the coinitiator is AlR$_2$X (R = ethyl, propyl, butyl, etc.; X = Cl, Br, etc.).

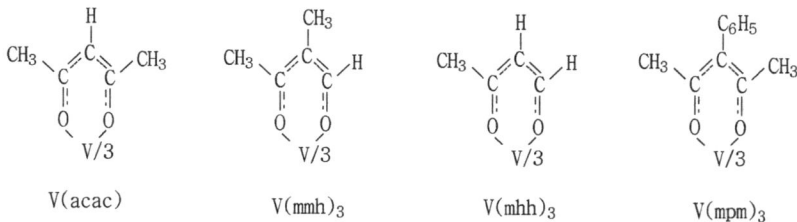

Fig. 1.31: Four vanadium compound initiators for living polymerization of propylene.

Table 1.11 is the initiating activity of various vanadium initiators mentioned previously for propylene and leads to the stereoregularity of the product. As shown from the data, the V(mmh)$_3$/AlEt$_2$Cl system has the highest polymer activity and can generate 700 g PP per gram of vanadium per hour at −40 °C. The random copolymerization of propylene and ethylene in this system also showed good activity characteristics (900 g polymer/(h · gV)) and formed a very narrow molecular weight EP rubber with very high polymerization characteristics ($\bar{M}_w/\bar{M}_n = 1.1$). In the active polymerization with this initiating system, the concentration of active centers can be maintained at 1.0 ± 0.1 mol/mol V, indicating that all vanadium ions act as active centers. From the table, it can be seen that a variety of initiator systems can produce polymers with

Tab. 1.11: The effect of vanadium initiator on propylene initiating activity and regularity of products.

Initiation system	Temperature (°C)	Polymerization activity/(gPP/ (h · gV))	Active species concentration/ (mol/mol V)	$\overline{M}_w/\overline{M}_n$	Percentage of syntactic structure (%)
V(mmh)₃/ AlEt₂Cl	−70	105	1.0	1.2	79
V(mmh)₃/ AlEt₂Cl	−60	206	1.0	1.2	79
V(mmh)₃/ AlEt₂Cl	−40	700	1.0	1.3	77
V(mhh)₃/ AlEt₂Cl	−70	17	0.1	1.2	79
V(mpm)₃/ AlEt₂Cl	−70	0.3	0.005	1.1	–
V(acac)₃/ AlEt₂Cl	−78	10	0.08	1.1	81
V(acac)₃/ AlEt₂Cl	−70	15	0.08	1.1	80
V(acac)₃/ AlEt₂Br	−78	1	0.06	1.2	66

higher syndiotactic content, but polymer activity and initiator efficiency are significantly affected by the initiation of system species, and the polymerization temperature also has a great influence.

In the coordinative anionic living polymerization of propylene, the growth reaction is performed by a coordination polymerization mechanism including the coordination and insertion to active species, as shown in Fig. 1.32.

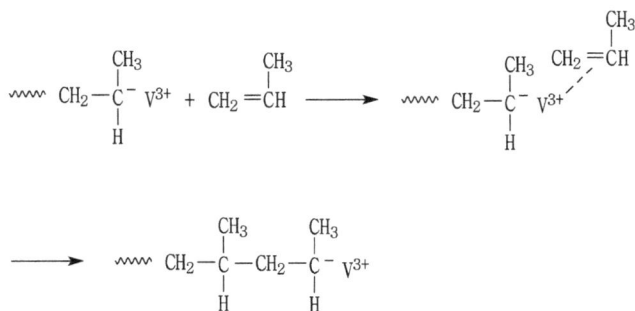

Fig. 1.32: Growth reaction of coordination anionic living polymerization of propylene.

At higher temperatures, active V^{3+} can be deoxidized to inactive V^{2+}, thus ending polymerization.

In the active anionic polymerization process of propylene, adding a small amount of ethylene will cause a high rated random copolymerization of ethylene–propylene. When the ethylene reaction is completed, the active chain began to add propylene again. Finally, PP-b-EP-b-PP block copolymer with narrow relative molecular mass distribution ($\overline{M}_w/\overline{M}_n = 1.1$).

1.4.2 Living ionic ring-opening polymerization

Active ring-opening polymerization is a developing study area, and it is of great significance as a living polymerization of vinyls.

1.4.2.1 Ring-opening polymerization of cyclosiloxanes

Trimethylcyclotrisiloxane (D_3) can carry out anionic ring-opening polymerization with BuLi and can also carry out cationic ring-opening polymerization with trifluoromethanesulfonic acid (CF_3SO_3 H). Polydimethylsiloxane (PDMS) has advantages like softness, water resistance, high ozone permeability, high stability, low surface tension, and excellent biocompatibility, and has received attention as a block material. There are two methods to prepare the block copolymers of PDMS and PMMA by anionic ring-opening polymerization with D_3. The first one is to prepare a dianion, such as a reaction product of benzophenone and potassium, as shown in Fig. 1.33. In which, oxygen anion first initiates D_3, and then the carbon anion initiates MMA. Another method is to prepare hydroxyl-terminated PMMA. The terminal hydroxyl group is then converted to an oxyanion to initiate D_3.

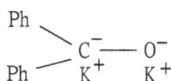

Fig. 1.33: A product of the reaction of diphenylketone with metal potassium.

Through the living ring-opening polymerization of cyclosiloxanes, we can also prepare multiblock polymers with epoxy resins and PDMS soft segments. The method is to prepare a product of the reaction of PMDS and aminoethyl-2,6-diazepane with carboxyl groups at both ends, as shown in Fig. 1.34. Then reactants react with epoxy resin to obtain multiblock copolymers.

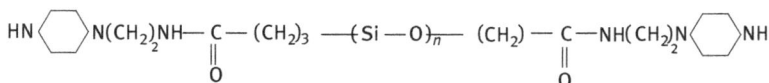

Fig. 1.34: A product of the reaction of PMDS and aminoethyl-2,6-diazohexane with carboxyl groups at both ends.

PDMS with carboxyl groups at both ends can be obtained by ring-opening polymerization of octamethylcyclotetrasiloxane (D_4) with CF_3SO_3 H as initiator.

1.4.2.2 Ring-opening polymerization of cyclic ethers

The cyclic ether mainly refers to ethylene oxide, propylene oxide, tetrahydrofuran, and so on, and their polymers are important raw materials for polyurethane preparation.

Both ethylene oxide and propylene oxide are three-membered rings and ring opening is easy, so both anionic and cationic polymerization can be carried out. Tetraphenylporphyrin/alkylaluminum chloride (TPPAlCl) can undergo anionic ring-opening polymerization of ethylene oxide and propylene oxide. TPPAlCl can also be used for the alternate living copolymerization of epoxides with CO_2 to produce polycarbonates and alternative copolymerization of epoxides and anhydrides to produce polyesters. In addition, TPPAlCl also enables the living polymerization of polar monomers such as MMA. For example, $Ph_3C^+SbF_6$ can initiate the polymerization of tetrahydrofuran at −58 °C, and the polymer relative molecular mass distribution is 1.04. The first-order dynamic curves shown in Fig. 1.35 and the conversion–polymerization curves shown in Fig. 1.36 clearly demonstrate the living characteristics of tetrahydrofuran carbon cation polymerization.

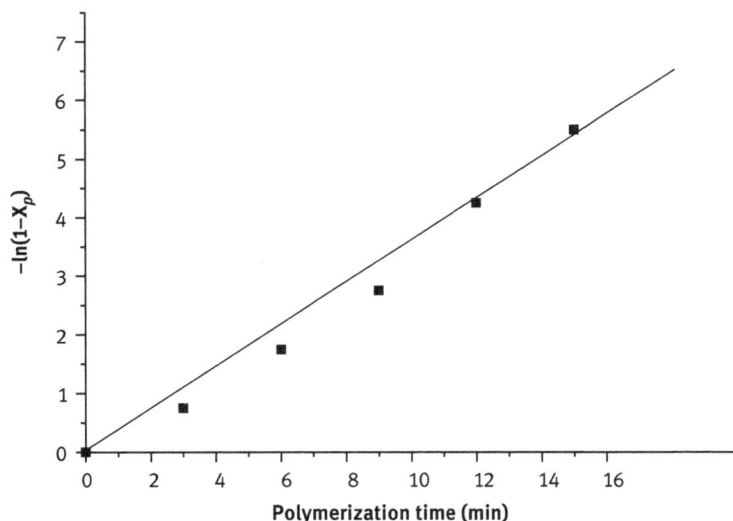

Fig. 1.35: First-order dynamic curve of the bulk polymerization of tetrahydrofuran initiated with IBC/AgSbF$_6$ at room temperature.

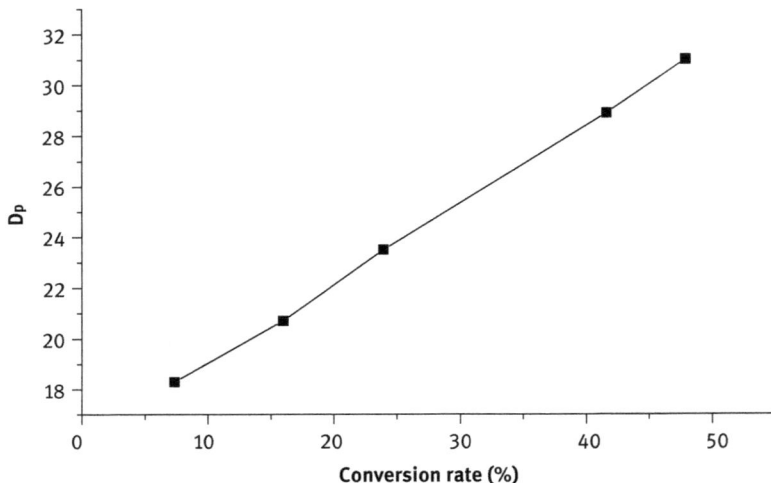

Fig. 1.36: The curve between conversion rate and degree of polymerization bulk polymerization of tetrahydrofuran initiated with IBC/AgSbF$_6$ at room temperature.

Superacids such as CF$_3$SO$_3$H and CF$_3$SO$_3$R, acyl halides/silver perchlorate can also initiate cationic living polymerization of tetrahydrofuran. For example, the process on how acyl halide/silver perchlorate initiate tetrahydrofuran is shown in Fig. 1.37.

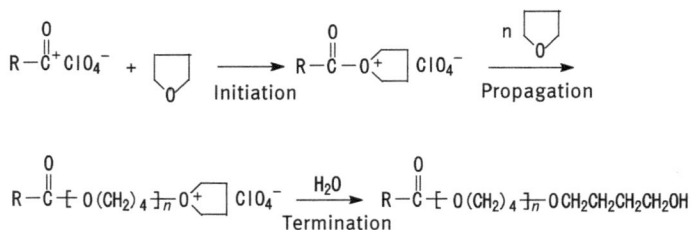

Fig. 1.37: Living polymerization of tetrahydrofuran initiated by silver perchlorate and acyl halide.

References

[1]   Szwarc, M, Levy, M, Milkovich, R. J Am Chem Soc. 1956;78:2656.
[2]   Szwarc, M. Carbanion, Living polymers, and Electron Transfer Processes. New York: Wiley-Interscience 1968.
[3]   Bywater, S, Worsfold, D. J Can J Chem. 1962;40: l56.4.
[4]   Hsieh, HL, Quirk, RP. Anionic Polymerization: Principles and Practical Applications. New York/Basel/Hong Kong: Marcel Dekkerfg Inc., 1996.
[5]   Aoshima, S, et al. Macromolecules. 1985;18:2097.

[6]   Aida, T, Inoue, S. Macromolecules. 1981;14(5):1166.

[7]   Fayt, R, Forte, R, Teyssie, P. Macromolecules. 1987;20(6):1442.

[8]   Reetz, MT. Angew Chem. 1988;100(7):1026.

[9]   Raynal, S. Eur Polym J. 1986;17(22):559.

[10]   Shi, ZQ, Hong, LG, Hu, CP, et al. J East China Univ Sci Technol. 1989;15(1):72.

[11]   Hsieh, HL, Quirk, RP. Anionic Polymerization: Principles and Practical Applications. New York: Marcel Dekker, Inc., 1996.

[12]   Xue, LB, Jin, GT. Theory and Application of Anionic Polymerization. Beijing: China Friendship Publishing Company, 1990.

[13]   Miyamoto, M, Sawamoto, M, Higashimura, T. Macromolecules. 1984;17(3):265–268.

[14]   Kojima, K, Sawamoto, M, et al. Macromolecules. 1987;20:1.

[15]   Kojima, K, Sawamoto, M, et al. Macromolecules. 1989;22:1009.

[16]   Kojima, K, Sawamoto, M, et al. Polymer Bull. 1990;23(2):149.

[17]   Kojima, K, Sawamoto, M, et al. Macromolecules. 1991;24(10):2658.

[18]   Kojima, K, Sawamoto, M, et al. Macromol, Symp. 1994;85:33.

[19]   Sawamoto, M, Okamoto, C, Higashimura, T. Macromolecules. 1987;20:2639.

[20]   Sawamoto, M, Fujimori, J, et al. Macromolecules. 1989;22(5):916.

[21]   Sawamoto, M, Shohi, H, Sawamoto, H, et al. J Macromol Sci, Pure Appl Chem. 1994;A31:1609.

[22]   Sawamoto, M, Higashimura, T. Macromol Chem, Macromol Symp. 1992;54–55:44.

[23]   Sawamoto, M. Prog Polym Sci. 1991;16:111.

[24]   Higashimura, T, Miyamoto, M, Sawamoto, M. Macromolecules. 1985;18:611.

[25]   Higashimura, T, Aoshima, S, Sawamoto, M Macromol Chem Macromol Symp. 1988;13:457.

[26]   Higashimura, T, Kojima, K, Sawamoto, M. Polymer Bull. 1988;19:7.

[27]   Higashimura, T, Kojima, K, Sawamoto, M. Macromol Chem. 1989;15:127.

[28]   Hishimora, T, Aoshima, S, Higashimura, T. Macromolecules. 1989;22:3877.

[29]   Aoshima, S, Iwasa, S, et al. Macromolecules. 1989;22(10):3877.

[30]   Aoshima, S, Iwasa, S, et al. J Polym Sci, Part A: Polym Chem. 1990;28(5):1137.

[31]   Aoshima, S, Iwasa, S, et al. Eur Polym J. 1994;30(8):912.

[32]   Shohi, H, Sawamoto, M, et al. Macromol Chem. 1992;193(7):1783.

[33]   Shohi, H, Sawamoto, M, et al. Macromol Chem. 1992;193(8):2027.

[34]   Storey, RF, Chrsholn, BJ. Macromolecules. 1993;26(25):6727.

[35]   Perce, V, Lee, M. Polymer. 1991;32(15):2862.

[36]   Perce, V, Lee, M. Polymer Bull. 1991;25(1):123.

[37]   Perce, V, Lee, M. Polymer Bull. 1992;33(4):703.

[38]   Perce, V, Lee, M. Macromolecules. 1991;24(17):4963.

[39]   Marsalka, S, Sawamoto, M Macromolecules. 1991;24(9):2309.

[40]   Faust, R, Kennedy, JP. Polym Bull. 1986;15(4):317.

[41]   Faust, R, Kennedy, JP. Polym Bull. 1986;15(5):411.

[42]   Faust, R, Kennedy, JP. J Polym Sci, Polym Chem. 1987;25(7):1847.

[43]   Zsuga, M, Kelen, T. Polymer Bull. 1986;16(3):285.

[44]   Kennedy, JP. Macromol Chem, Macromol Symp. 1994;85:70.

[45]   Mishra, MK, Kennedy, JP. J Macromol Sci Chem. 1987;A24:933.

[46]   Mishra, MK, Chen, CC, Kennedy, JP. Polymer Bull. 1989;20:413.

[47]   Kaszas, G, Pusakas, J, Kennedy, JP. Macromol Chem, Soc Div Polym Chem. 1990;31(1):466.

[48]   Crivello, JV, Lee, JL, Conlon, DA. Macromol Chem, Macromol Symp. 1988;13(14):145.

[49]   Uno, H, Takata, T, Endo, T Chem Lett. 1988;935.

[50]   Marek, M, Pecka, J, Halaska, V. Macromol Chem, Macromol Symp. 1988;13(14):433.

[51]   Nguyen, HA, Kennedy, JP. Polymer Bull. 1983;9:506.

[52]   Haque, HA, Biswas, M. Polymer. 1985;26:22.

[53] Gong, MS, Hall, HKJ. Macromolecules. 1986;19:3011.

[54] Roudet, J, Gandini, A. Makromol Chem, Rapid Commun. 1989;10:277.

[55] Sun, XM, Liu, YR, Wu, G, et al. Papers Collection of the National Conference on Polymer Academic Papers 1997;1:a55.

[56] Lu, J, Liang, H, Zhang, RJ, et al. Papers Collection of the National Conference on Polymer Academic Papers 1997;1:a348.

[57] Johnson, AF. J Polym Sci, Polym Symp. 1976;56:211.

[58] Sigwalt, P. Macromol Chem, Macromol Symp. 1991;47:179.

[59] Sigwalt, P, Polton, A, Tardi, M. J Macromol Sci, Pure Appl Chem. 1994;A31:953.

[60] Matyjaszewski, K, Sigwalt, P. Polym Int. 1994;35:1.

[61] Leleu, JM, Bernardo, V, Polton, A, Tardi, M, Sigwalt, P. Polym Int. 1995;37:219.

[62] Faust, R, Fehervari, A, Kennedy, JP. J Macromol Sci, Chem. 1982;A18:1209.

[63] Kennedy, JP, Marechal, E. Carbocationic polymerization, New York: Wiley Interscience, 1982:453.

[64] Peng, YX. Living polymerization, apparent living polymerization and quasi living polymerization, Chin J Chem Edu (in Chinese). 1998;(6):8–10, 3.

[65] Zhang, HM, Hou, YX. Living Polymerization. Beijing: Sinopec Press, 1998.

Exercises

1. Please complete the following blank to review anionic polymerization.

Questions	Answers
Reaction mechanism	
Active center	
Monomer	
Initiator	
Chain growth mode	
Chain termination mode	
Polymerization temperature	
Solvent	
Aggregation implementation method	
Polymerization inhibitor	

2. Which is the sufficient criteria of living polymerization?
    A) Linear dynamics
    B) Molecular weight increases linearly with conversion
    C) Product molecular weight distribution is narrow
    D) None of the above

3. Anionic polymerization in the presence of polar additives is ( ) than in nonpolar environments.
    A) Slower
    B) Faster
    C) The same
    D) Unsure

4. How to calculate the theoretical molecular weight of anionic polymerization?
    A) Monomer feed weight/butyllithium molar amount
    B) Monomer initial concentration/initiator concentration
    C) Monomer moles/initiator concentration
    D) Monomer mass/initiator molar concentration

5. Which of the following monomers are not suitable for anionic polymerization?
    A) Styrene
    B) Ethylene
    C) 1,3-cyclohexadiene
    D) Methacrylate

6. Please answer characteristics of anionic living polymerization?

7. The polymerization of styrene is initiated by RLi, RK, RNa in DME, and other conditions are the same, rank the polymerization rate of different initiators. Change the solvent to cyclohexane and rank the polymerization rate.

8. 100 styrene molecules are initiated with 500 n-BuLi molecules, assuming no termination and all initiator and monomers were consumed. Please calculate the number average molecular weight $(\overline{M}_n)$ of the living polystyrene.

9. How to control the polymerization rate and the degree of polymerization of anionic living polymerization?

Chapter 2
Group transfer polymerization

2.1 Introduction

Group transfer polymerization, a new living polymerization technique, was first reported in 1983 by O. W. Webster et al. (DuPont, USA). At the time, Webster announced in the name of the research group at the 186th National Conference of the American Chemical Society that they discovered a new method of chain polymerization, namely "group transfer polymerization." It is the fifth kind of chain polymerization technology except for free radical, cationic, anionic, and coordination anionic polymerization. It has been greatly appreciated and highly valued by the polymer academia worldwide. It has been another significant new polymerization technique after Ziegler and Natta have found the orientation polymerization of olefins with a coordination catalyst in the 1950s and Szwarc invented the anionic living polymerization.

The group transfer polymerization is a compound in which an α,β-unsaturated ester, a ketone, an amide, or a nitrile is used as a monomer, and a compound having a group such as a silicon, an anthracene or a tin alkyl group is used as an initiator, and an anionic type or Lewis acid type compound is used as a catalyst, using a suitable organic solvent as a solvent, by coordinating the silicon, ruthenium, and tin atoms of the initiator with the initiator, exciting silicon, germanium, and tin atoms to make them bond with oxygen or nitrogen atoms. The atoms combine to form a covalent bond, and the double bond in the monomer completes the addition reaction with the double bond in the initiator, and the silicon, germanium, tin, and alkyl groups move to the end to form an "active" compound. The above process was repeated to obtain the corresponding polymer.

In fact, group transfer polymerization is like other chain polymerization reactions and can be divided into the following three elementary reactions.

2.1.1 Chain initiation reaction

Webster et al. used a small amount of dimethyl ketene methyltrimethylsilyl acetal (MTS) as an initiator and a large amount of methyl methacrylate (MMA) monomer in an anionic catalyst (HF_2^-) to carry out this addition reaction. The reaction occurs as followings.

In the reaction, the trimethyl silane on the initiator is transferred to the carbonyl group of MMA, and the negatively charged α carbon atom on the double bond reacts with a positively charged double bond α carbon atom of the monomer, resulting in a trimethylsilyl group and a double bond on the end groups of the newly formed intermediate I.

https://doi.org/10.1515/9783110597097-002

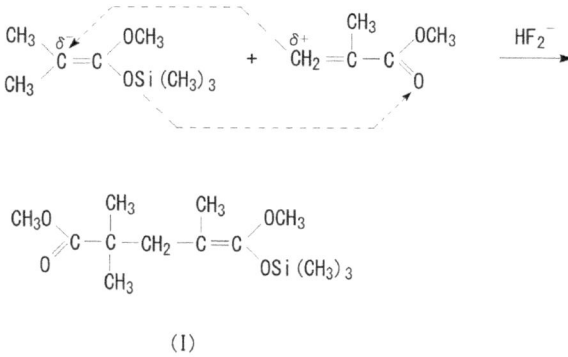

(I)

Fig. 2.1: Schematic diagram of the initiation reaction of the group transfer polymerization chain.

2.1.2 Chain growth reaction

Obviously, one end of the abovementioned addition product I still has a structure like MTS and can further undergo an addition reaction with the carbonyl oxygen atom of MMA. This process can be repeated until all the monomers have been consumed and finally the polymer is obtained. Therefore, the actual process of group transfer polymerization is that the active trimethyl siloxane group is first transferred from the initiator MTS to the addition product I, and then continuously transferred to the MMA monomer, so that the molecule continues to grow, the group transfer polymerization reaction is thus named. The growth reaction process can be expressed as follows:

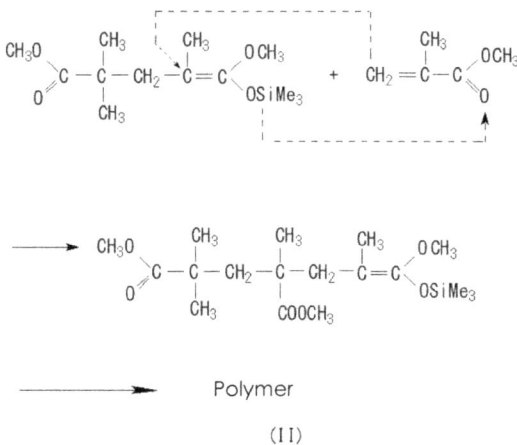

Polymer

(II)

Fig. 2.2: Schematic diagram of group transfer polymerization chain growth reaction.

2.1.3 Chain termination reaction

It can be seen from the living polymer (II) that before the addition of the terminator, the growing polymer chain contains a trimethylsilyl end group which can continue to add to the remaining same monomer or different monomers, thus it is a living polymer chain. Like anionic polymerization, such a living chain can also be killed by artificially adding a substance which reacts with a terminal group, that is, a chain termination reaction. For example, when methanol is used as a terminator, the following reaction occurs:

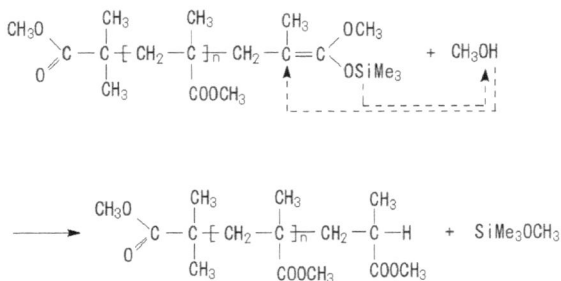

Fig. 2.3: Schematic diagram of termination reaction of group transfer polymerization chain.

As with anionic living polymerization, if there are impurities, such as active hydrogen (protons), which may react with the active center in the polymerization system, the active chain will be terminated. Therefore, the polymerization system is generally required to be very pure. Since the group transfer polymerization technique is the same as the anionic polymerization, it belongs to the category of "living polymerization." Therefore, the polymerization system is relatively stable at room temperature, and has the ability to continuously add when the corresponding monomer is added after several days of storage. In addition, the initiation rate of the initiator is greater than or equal to the chain growth rate, so all the activated active centers will simultaneously undergo a chain growth reaction, thereby obtaining a polymer having a "Poisson" distribution with a relatively narrow molecular mass distribution, generally $D = 1.03$ to 1.2. At the same time, the degree of polymerization of the product can be controlled by the molar concentration ratio of both the monomer and the initiator (DP = [M]/[I]).

When the molecular weight is between 1,000 and 20,000, the degree of polymerization of the product and its distribution can be controlled relatively accurately, but when a polymer with a higher degree of polymerization is prepared, it is more difficult to control the narrow distribution. Because the amount of initiator is small and is easily interfered by impurities in the system. However, when high-purity monomers, reagents, and solvents are used, polymers having a number average molecular weight of up to 100,000 to 200,000 have also been produced.

In fact, the origin of group transfer polymerization began in 1979. At that time, Webster and others were looking for a fluoride as a catalyst to prepare a new acrylate polymer. As the fluoride storage container was broken, the tidal air entered the tank and reacted with it to form a difluoride. This discovery led to the birth of a new technology, the group transfer polymerization, because they found that the difluoride has a remarkable catalytic ability for the polymerization of MMA.

Later, Webster et al. used the initiator, catalyst, solvent, monomer range, reaction mechanism, synthetic block copolymer, terminal functional group polymer, and conversion rate and time in the polymerization process. A series of detailed studies have been carried out on the relationship between the concentration of the initiator, the concentration of the catalyst, and the concentration of the monomer, and great achievements have been made. Of course, the current group transfer polymerization technology is still not very mature, and its reaction mechanism, reaction conditions, and monomer range have yet to be further explored.

Although the discovery of group transfer polymerization (GPT) technology has certain contingency, it is the inevitable result of people's continuous exploration and research in the field of active polymerization in recent decades. GPT has developed rapidly in a short period of time, and it has more advantages than the usual polymerization methods in controlling the relative molecular mass, relative molecular mass distribution, end group functionalization, and reaction conditions of the polymer. Thereby it has added a new method for the design of polymer. In practical applications, the use of this technology to produce automotive topcoats, synthetic liquid crystal polymers, and some special polymers, such as block, telechelic polymer materials, has been successful. It is expected that the group transfer polymerization technology will make more progress soon.

2.2 Characteristics of group transfer polymerization

2.2.1 Monomer

The main monomers of the anionic polymerization are monoolefins (such as styrene) and conjugated diene (such as butadiene) compounds, that is, nonpolar monomers. The anionic polymerization of a polar monomer easily causes a side reaction, thereby deactivating the polymerization system, an anionic polymerization such as MMA often does not obtain a living chain. In contrast, polar monomers are well suited for group transfer polymerization techniques. At present, the group transfer polymerization technology is limited to α,β-unsaturated esters, ketones, nitriles, and disubstituted amides. Such monomers can be represented by the formula $H_2C=CR'$ wherein $R' = H$, CH_3, $X = COOR$, $CONH_2$, COR, and CN, wherein R is an alkyl group.

The most studied monomers in the current group transfer polymerization are MMA and ethyl acrylate (EA). Because MMA is the most active, the research is deeper.

For some specific structural monomers, the use of group transfer polymerization techniques to obtain their corresponding polymers has special significance. For example, if the monomer 1 shown in Fig. 2.4 is polymerized by radical polymerization or other chain polymerization method, a crosslinking reaction generally occurs. The use of group transfer polymerization technology can smoothly eliminate the possibility of cross-linking, because according to the transfer law of trimethylsiloxy group, the $-CH_2CH=CH_2$ group on the monomer is not polymerized. Therefore, it is expected that these monomers can be prepared into an elastomer or a photosensitive polymer by a group transfer polymerization technique.

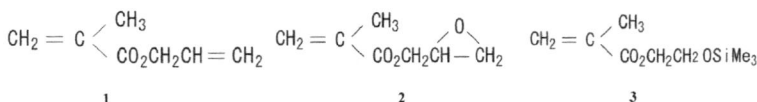

$$CH_2 = C \underset{CO_2CH_2CH=CH_2}{\overset{CH_3}{<}} \quad CH_2 = C \underset{CO_2CH_2CH-CH_2}{\overset{CH_3 \quad O}{<}} \quad CH_2 = C \underset{CO_2CH_2CH_2\,OS\,i\,Me_3}{\overset{CH_3}{<}}$$

1   2   3

Fig. 2.4: Group transfer polymerization monomer with specific structure.

Due to the unique polymerization mechanism of GPT, many monomers having sensitivity to other polymerization methods can be polymerized by a group transfer polymerization method, thereby leaving these groups unchanged. For example, the epoxy methacrylate 2 is polymerized by a group transfer polymerization technique below 0 °C to leave the epoxy group unreacted, and the product can be used as an epoxy resin. If anionic polymerization is employed, both the double bond, the epoxy group, and the carbonyl group may react to complicate the product.

Further, a polymer containing a pendant group CH_2CH_2OH can be obtained by a group transfer polymerization reaction like the following siloxy group-containing acrylate monomer 3. Therefore, a polymer having a specific functional group can be obtained very easily by a group transfer polymerization method.

2.2.2 Initiator

The initiators that have been found to be useful for group transfer polymerization are generally classified into the following categories:

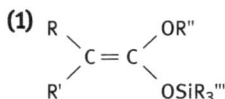

(1)
$$R \underset{R'}{\overset{}{>}} C = C \underset{OSiR_3'''}{\overset{OR''}{<}}$$

This is by far the best type of group transfer polymerization initiator, the most widely used of which is compound 4, expressed as MTS. In general, the larger the R'' group on Si, the smaller the reaction rate; and the R'' on the OR'' group can be greatly changed, thereby serving as a special functional group introduced into the polymer terminal. The route, for example, by using 5 as an initiator, can introduce a terminal carboxyl group to the polymer. The terminal $COOCH_2CH_2OH$ can be introduced by using 6 as an initiator.

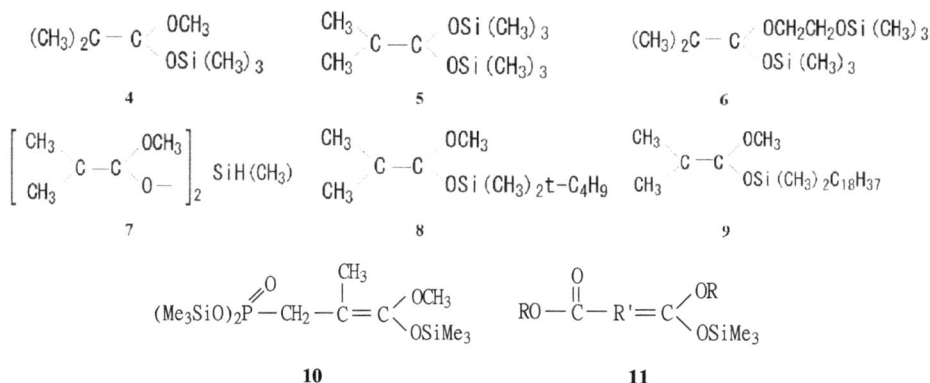

Fig. 2.5: MTS series initiator.

Such initiators tend to be isomerized in the polymerization system.

When R is CH_3, the isomerization rate does not affect the polymerization process, and when R is H atom, the isomerization rate is enough to affect the polymerization process and stop the polymerization reaction.

In addition, the size of the substituent groups on the silicon atom also has a large influence on the initiation rate. In general, the smaller the volume of the substituent on the silicon atom, the better the controllability of the polymerization process. For example, the initiator 7 is used to initiate the MMA polymerization, and a product having a good molecular weight control and a narrow molecular weight distribution ($\bar{X}_w/\bar{X}_n = 1.10$) can be obtained. The rate of MMA initiated by initiator 8 is very slow. The relative molecular mass of the product is higher than the theoretical value, and the relative molecular mass distribution is also wide, which has largely deviated from the living polymerization.

If the substituent group on the silicon atom is long, but the steric hindrance is small (such as compound 9), the relative molecular mass of the product still has a good control effect, but the relative molecular mass distribution is wide. This may be because the rate of exchange of the catalyst between the active molecules is less than the rate of growth.

(2)

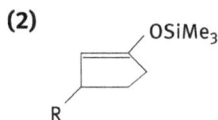

The initiation principle of such an initiator is basically the same as that of the first type, and therefore will not be described again.

(3) R$_3$SiX

The X in such an initiator may be CN, SCH$_3$, CH$_2$COOEt, and so on, and the initiation principle thereof is shown in Fig. 2.6. Such initiators often have a significant induction period during the initiation of MMA polymerization. The initiation process is carried out by addition to MMA, and the controllability of the relative molecular mass of the product is poor. The bifunctional initiator (CH$_3$)$_3$SiSCH$_2$CH$_2$SSi(CH$_3$)$_3$ containing −SSi(CH$_3$)$_3$ also belongs to this type of initiator and can be used to prepare bifunctional end group polymers.

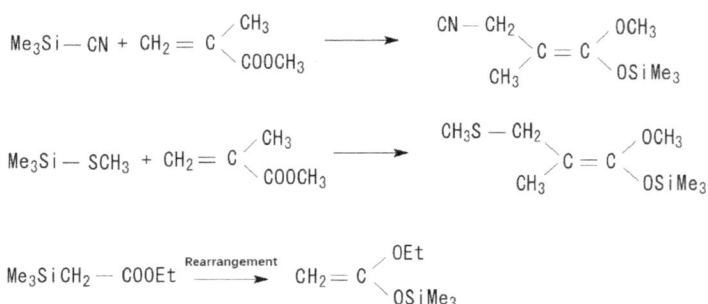

Fig. 2.6: The initiation principle of R$_3$SiX series initiator.

(4) P(OSiMe$_3$)$_3$

Such an initiator must first be reacted with MMA at 110 to 120 °C to form 10 (Fig. 2.5) and then initiate addition polymerization. Similar phosphorus-containing compound initiators are (Et$_2$O)$_2$POSiMe$_3$, (Me$_2$N)$_2$POSiMe$_3$, and the like. Such initiators generally have good controllable properties for the polymerization process.

Pan et al. synthesized a batch of compounds with the structure 11 (Fig. 2.5) as initiators for group transfer polymerization. The synthesis method is simple and the activity is also high. They can be attributed to the first category.

As an initiator for group transfer polymerization, it is easily decomposed by an active hydrogen-containing compound because it contains a relatively active R$_3$M−C bond or an R$_3$M−O bond. Therefore, as with the operation and requirements of anionic polymerization, the presence of proton-containing compounds such as water, alcohol, acid, and so on must be avoided throughout the reaction system. All instruments, equipment, and reagents must undergo strict drying pretreatment and then pumped. Exhaust air and high-purity nitrogen are aerated or vacuumed.

2.2.3 Catalyst

The group transfer polymerization differs from the anionic polymerization in that it is generally carried out with the addition of a catalyst. The catalyst is mainly divided into two major categories: anionic and Lewis acid.

(1) Anionic

The anions of such catalysts include HF_2^-, CN^- and $F_2Si(CH_3)_3^-$, and the most common and most effective anion catalyst is $[(CH_3)_2N]_2SHF_2$ (abbreviated as $TASHF_2$). Other anionic catalysts are generally prepared as quaternary ammonium salts, such as Bu_4NF and Bu_4NCN, or as a salt of $[(CH_3)_2N]S^+$ as a cation (TAS^+X^-), which is readily soluble in organic solvents and therefore has larger activity.

The catalytic mechanism of the anionic catalyst is generally considered to be the process in which the initiator forms an intermediate state of the supervalent silicon under the action of the catalyst to activate the initiator.

As with other types of polymerization, the amount of catalyst has a large effect on the group transfer polymerization. Generally, the amount of the anion catalyst is 0.01–0.1% by mole of the initiator, so that the reaction can be carried out at a certain speed, but there is often an induction period, and the latter is shortened as the catalyst increases the proportion of the initiator, so the actual amount is generally 1% to 5% by mole of the initiator.

For example, $TASHF_2(R)$ has a higher catalytic efficiency for group transfer polymerization and is generally used in an amount of 0.1% by mole based on the initiator. However, for initiators such as $P(OSiMe_3)_3$, the amount of $TASHF_2$ must be as high as 4–11% by mole. This is due to the coordination between the catalyst and the P atom resulting in an increase in the amount of catalyst.

KHF_2 can also be used as a catalyst for group transfer polymerization, but it has a low solubility in a general organic solvent and requires polymerization in a polar solvent such as acetonitrile or dimethylformamide. It was found that the combination of KHF_2 and 18-crown ether-6 promoted the polymerization of MMA, which resulted in high-conversion and low-dispersion polymethyl methacrylate (PMMA).

(2) Lewis acid type

Commonly used catalysts of this type are zinc halide inorganics such as $ZnCl_2$, $ZnBr_2$, ZnI_2, and the like. In addition, there are also aluminum alkyl organics such as R_2AlCl and $(RAlOR)_2O$. The former is used in an amount of 10–20% by mole of the monomer to completely convert the monomer, and the latter is used in an amount of 10–20% by mole based on the initiator. In summary, the Lewis acid type catalyst is several times more abundant for the group transfer polymerization than the anionic type. ZnX_2 works well at room temperature, while the aluminum alkyl catalyst has a competitive decomposition reaction at room temperature and, therefore, generally needs to be used at low temperatures (−78 °C).

The catalytic mechanism of the Lewis acid type catalyst is that the Lewis acid coordinates with the carbonyl group in the monomer to activate the monomer, making the initiator more prone to nucleophilic attack.

When using Lewis acid as a catalyst, the use of an electron-rich solvent should be avoided, and a nonpolar solvent such as an alkyl halide or an aromatic hydrocarbon is usually used.

2.2.4 Solvent

Group transfer polymerization as with some other polymerization types, can be performed by bulk polymerization. However, since the reaction of bulk polymerization is very rapid, a large amount of reaction heat is difficult to be eliminated, resulting in uncontrollable polymerization and even explosion. For example, when MMA is bulk polymerized at room temperature, the temperature of the system can rise to 84 °C in 15 min. Pan et al. also found that when acrylonitrile is subjected to bulk polymerization by group transfer polymerization technology, explosion polymerization occurs. Therefore, in order to facilitate the control of the polymerization rate, general group transfer polymerization is carried out in a solvent. Another role of using solvents is to facilitate the accurate formulation of the initiator and catalyst for accurate metering purposes.

Depending on the type of catalyst, two different types of solvents can be used accordingly. When an anionic catalyst is selected for group transfer polymerization, it is generally preferred to use a solvent for the electron donor. The most common ones are THF and CH_3CN, and $CH_3O\text{-}CH_2CH_2\text{-}OCH_3$, $CH_3CH_2O\text{-}CH_2CH_2\text{-}OCH_2CH_3$, and the like. When a Lewis acid type catalyst is used, the solvent of the electron donor should be avoided because it is easily complexed with the Lewis acid, thereby hindering the coordination of the catalyst with the carbon-based oxygen of the monomer. Therefore, in this case, halogenated alkanes and aromatic hydrocarbons such as toluene, CH_3Cl, and $C_1CH_2CH_2Cl$ are generally used.

2.2.5 Reaction temperature

When the group transfer polymerization technique is employed, the reaction temperature can be carried out in a very wide range depending on the kind of the monomer, for example, between −100 and + 150 °C, but a suitable reaction temperature is 0–50 °C. For MMA, it is best at room temperature, which is convenient in operation and application. For acrylates, the reaction is preferably at 0 °C or lower because of their relatively large activity. Since the (meth)acrylates react relatively quickly, it is usually controlled by gradually adding monomers.

In addition, according to the literature, the difference between the catalyst and the reaction temperature has a certain influence on the stereostructure of the polymerization product. PMMA was prepared by group transfer polymerization at room temperature with nucleophilic HF_2^- as a catalyst. The product contained nearly equal amounts of syndiotactic and random structure polymers. When polymerized at −78 °C, syndiotactic polymer was mainly obtained. If isothermal group transfer polymerization of MMA is carried out using THF as a solvent, the stereocomponent of the product at different temperatures is determined by nuclear magnetic resonance spectroscopy, and it is found that the isotactic PMMA content is always low, and the reaction temperature is 60 °C. When 60 °C drops to −90 °C, the syndiotactic PMMA increases from 50% to over 80%. When Lewis acid is used as the catalyst, the syndiotactic PMMA in the product is much higher than that obtained with the anionic catalyst. At most temperatures, the ratio of syndiotactic and random PMMA is close to 2:1. Further, it was found that the stereoregularity of the polymer obtained by the group transfer polymerization is independent of the polarity of the solvent used, which is different from the anionic polymerization which is related to the solvent used and has little relationship with temperature.

2.3 Mechanism of group transfer polymerization

The reaction mechanism of group transfer polymerization varies depending on the catalyst. When an anionic catalyst is employed, Webster et al. suggest that a possible mechanism for silyl transfer in group transfer polymerization is by the formation of an intermediate of supervalent silicon under the action of a nucleophilic catalyst. That is, the nucleophilic anion Nu^- is first coordinated to the silicon atom on the active end group of the initiator or the living polymer to activate the silicon atom, and then the activated silicon atom is bonded to the carbon-based oxygen atom in the monomer to form a hexacoordinate silicon, the intermediate transition state. Subsequently, trimethyl silane forms a covalent bond with the carbon-based oxygen atom of the monomer, so that the double bond of the initiator completes the addition reaction with the double bond in the monomer, the catalyst Nu is extruded, and the monomer is formed at the front end of the bond. The carbon–carbon single bond, $Si(CH_3)_3$, moves to the end of the chain to form the living polymer. The reaction equation is shown in Fig. 2.7.

It can be seen from the above process that in the presence of the HF_2-catalyst, the initiator reacts with the double bond of the monomer, and the trialkylsilyl group on the initiator moves to the carbon-based oxygen atom of the monomer to form an intermediate product. The intermediate product continues to react with another monomer unit, and the above process is continuously repeated to cause the molecular chain to grow, and finally, the H^+ is decomposed, and the trialkylsilyl group is detached to obtain a polymer.

The core of the above mechanism is the formation of an intermediate transition state of hexacoordinated silicon, which Farmham and Sogah consider to be in the association mechanism. To this end, they prepared a stable pentacoordinate silyl enol as an initiator to confirm the method by first reacting MTS with BuLi to form

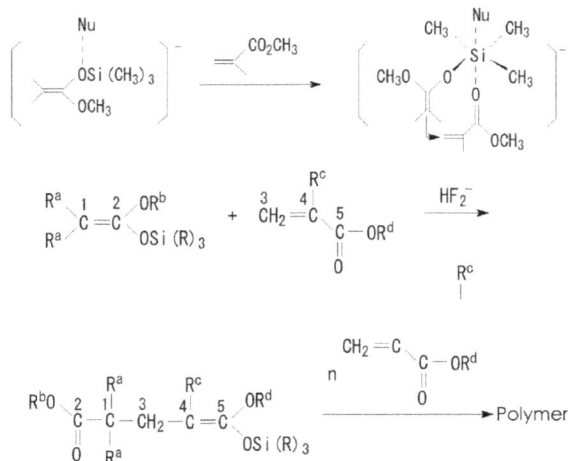

Fig. 2.7: Schematic diagram of group transfer polymerization mechanism using anionic catalyst.

lithium enolate, which is formed by the lithium enolate. A similar structure in the intermediate transition state of silicon, the group transfer polymerization of MMA can be carried out at 25–65 °C without additional catalyst, and a polymer having a relatively narrow dispersion is obtained, which is a previously proposed group transfer polymerization. This is evidenced by the transfer of intermediate transition states of supervaluated silicon.

When Lewis acid is used as the catalyst, the reaction mechanism of the group transfer polymerization may be caused by the catalyst first coordinating with the carbon-based oxygen atom of the monomer to form a complex to activate the monomer and then undergoing nucleophilic reaction by the initiator. Recently, China's Pan Ronghua and others found in the laboratory that when the Lewis acid and the solvent were first mixed in the monomer for a period, and then the initiator was added, the reaction immediately occurred without an induction period. If the initiator, the solvent, and the catalyst are mixed for a while, then the monomer is added, a certain induction period is produced, and the reaction proceeds slowly, which also confirms that the catalyst may react after the monomer is activated.

The stereoregularity of polymers synthesized by group transfer polymerization is generally only related to the catalyst used and the polymerization temperature. When the MMA group transfer polymerization is carried out by the Lewis acid catalyst, the ratio of the syndiotactic to the atactic structure of the obtained PMMA is almost 2:1, and when an anionic catalyst is used, the ratio of the two is almost 1:1. However, no matter what kind of catalyst is used, the content of the isotactic body is small. When the polymerization temperature rises, the ratio of isotactic and atactic structures increases, while the proportion of syndiotactic decreases. In addition, studies have shown that the stereoregularity of the group transfer polymerization product is also related to the size of the ester group in the monomer used, and the ratio of the syndiotactic body decreases as the ester group increases. For example, the ratio of syndiotactic poly (t-butyl methacrylate) at 20 °C is lower than that of PMMA.

2.4 Application of group transfer polymerization

Group transfer polymerization has many similar characteristics to anionic polymerization, so it is not difficult to synthesize standard samples of narrow relative molecular mass distribution by using this technique, and preparation of random and block copolymers and functionalized telechelic polymers are used. The following is a brief introduction to the application of group transfer polymerization.

2.4.1 Synthesis of narrow relative molecular mass distribution homopolymer

Table 2.1 lists the results of group transfer polymerization of several monomers under different catalysts. It can be seen from the table that the dispersion of most polymers is very small, generally between $D = 1.03$ and 1.20, and only a few are close to 2. It can also be seen from the data in the table that when ZnX_2 is used as the catalyst, the product obtained by using ZnI_2 has the smallest D value and the best control of the relative molecular mass; while with $ZnCl_2$, the control effect is the worst. In addition, most of the number average molecular weight is close to the theoretical value. As for some data with large differences, it may be related to the purity of polymerization systems such as reagents and solvents. In summary, controlling the relative molecular mass of the polymer and preparing a narrow distribution sample by using the difference in the molar ratio of the initiator to the monomer amount is one of the advantages of the group transfer polymerization. Therefore, it has been proved that it is feasible to use a group transfer polymerization technique to prepare a sample of a monodisperse polymer having a predetermined relative molecular mass.

2.4.2 Synthesis of copolymer

2.4.2.1 Synthesis of random copolymers
The group transfer polymerization of vinyl benzyl methacrylate/methyl methacrylate (VBM/MMA) is carried out by using $(CH_3)_2C=C(OCH_3)[OSi(CH_3)_3]$ as an initiator, and the resulting copolymer is active polymer with PMMA as the main chain and styrene as a side chain. The presence of styrene side chains is important for further formation of polymer networks or interpenetrating networks. Table 2–1 shows the experimental results (Table 2.2).

According to the principle of group transfer polymerization, during the polymerization, the double bond in the MMA can be opened, and the double bond in the styrene

Tab. 2.1: Examples of group transfer polymerization.

Monomer	Initiator	Catalyst	Solvent	Polymer			
				$\bar{M}_n$	$\bar{M}_w$	$\bar{M}_w/\bar{M}_n$	理论$\bar{M}_n$
Methyl methacrylate	MTS	TASF$_2$SiMe$_3$	THF, −78	1,120	1,750	1.56	2,040
		TASN$_3$	CH$_3$CN	3,000	3,100	1.03	3,700
		TAS 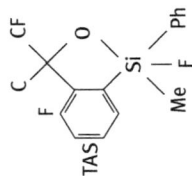	CH$_3$CN	1,700	1,900	1.11	2,000
		ZnBr$_2$	ClCH$_2$CH$_2$Cl	6,020	7,240	1.20	3,400
Methyl methacrylate	Me$_3$SiCN	TASCN	CH$_3$CN	800	800	1.03	1,000
Methyl methacrylate		KHF$_2$	CH$_3$CN	18,000	21,400	1.18	20,200
Ethyl acrylate		ZnI$_2$	CH$_2$Cl$_2$	3,300	3,400	1.03	3,360
Methyl methacrylate 57% Butyl methacrylate 32% Allyl methacrylate 32%	MTS	TASHF$_2$	THF	3,800	4,060	1.07	4,060
Methyl methacrylate 58% Butyl methacrylate 17% Glycidyl methacrylate 32%	MTS	TASHF$_2$	THF	3,010	4,290	1.10	4,092

N,N-Dimethylformamide	Me₃SiCH₂CO₂Et	TASF₂SiMe₃	THF	1,400	2,000	1.43	2,260
Methyl ketene	MTS	TASF₂SiMe₃	CH₃CN/THF	490	944	1.93	800
Methyl methacrylate	(OSiMe₃ structure)	TASF₂SiMe₃	CH₃CN/THF	2,100	2,400	1.14	2,000
Methyl methacrylate 90% Lauryl methacrylate 10%	MTS	TASF2SiMe3	THF	6,540	7,470	1.14	7,000
Methyl methacrylate 75% 2-Ethylhexyl methacrylate 25%	MTS	TASHF2	THF	41,500	54,200	1.30	46,300
Ethyl acrylate	MTS	ZnBr2	ClCH2CH2Cl	17,000	26,600	1.57	10,100
Ethyl acrylate	MTS	ZnCl₂	C₆H₅CH₃	5,800	11,300	1.96	4,100
Ethyl acrylate		Et₂AlCl	CH₂Cl₂	2,340	2,740	1.17	2,100
Ethyl acrylate	MTS	i-Bu₂AlCl	CH₂Cl₂	2,370	2,520	1.06	2,660
Ethyl acrylate	MTS	(i-Bu₂Al)₂O	C₆H₅CH₃	1,380	1,580	1.19	2,100
Methyl methacrylate		Et₂AlCl	CH₂Cl₂	1,800	3,300	1.83	1,030

Tab. 2.2: Results of group transfer polymerization of VBM/MMA.

Serial number	VBM (mol)	MMA (mol)	Initiator (mol)	Yield (%)	Theoretical $\overline{M}_n$	Actual		$\overline{M}_w/\overline{M}_n$	T_g	Polymerization condition
						$\overline{M}_n$	$\overline{M}_w$			
1	0.987	0	0.067	55	2,980	3,720	30,900	8.3	56	In ice water, 1.5 h
2	0.705	0.692	0.135	46	1,550	6,110	17,230	2.8	74	In ice water, 1.5 h
3	0.504	1.439	0.076	61	3,340	8,330	21,300	2.6	75	Room temperature, 2.5 h
4	0.196	2.683	0.076	40	4,090	8,040	12,400	1.6	131	Room temperature, 1.5 h

cannot be opened, so that the styrene double bond in the monomer can be retained. When the temperature is high, some of the styrene may participate in the thermal polymerization, so the polymerization should be carried out at a lower temperature.

2.4.2.2 Synthesis of block copolymers

The characteristics of the living polymer can be formed by the group transfer polymerization technique. As with the anionic polymerization, the block copolymer can be conveniently prepared by sequentially adding different monomers (sequential feeding method). For example, the same type of monomers (acrylates/acrylates or methacrylates/methacrylates) having little difference in activity can be used, and the second monomer can be added after the first monomer is reacted. The reaction is continued to form an AB-type block copolymer. If a difunctional initiator is employed, a triblock copolymer can be formed, and the structure of the difunctional initiator is as follows:

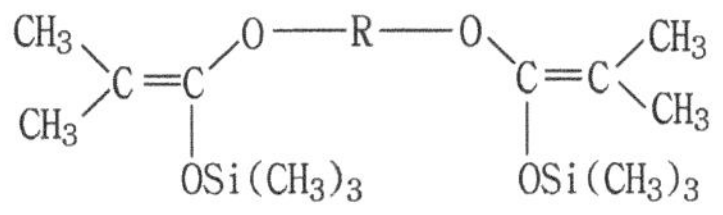

Chemical structure 1

The combination of group transfer polymerization with other polymerization methods is also an effective method for preparing block copolymers. A variety of block copolymers have been prepared by this method. For example, aldol–GTP (described later) in combination with GTP is used to prepare a polyvinyl alcohol–polymethyl methacrylate block copolymer. The synthesis process is as shown in Fig. 2.8. First, a polymethyl methacrylate active polymer is produced by a GTP method. Second, the silicon–oxygen side group-containing polymer is prepared by the aldol–GTP method. Third, the above two polymer chains are coupled and treated with methanol to convert the $OSi(CH_3)_2Bu$

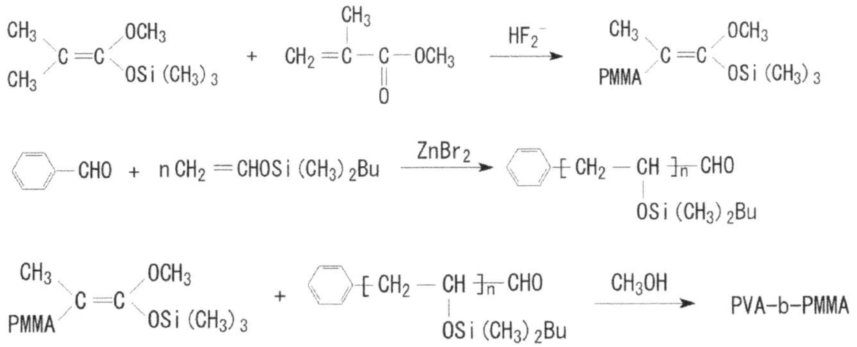

Fig. 2.8: Aldol–GTP combined with GTP to prepare block copolymer.

group on the copolymer into a hydroxyl group; A PVA-b-PMMA block copolymer was obtained.

Similarly, a PMMA-b-PS block copolymer was synthesized by combination with GTP and atom transfer radical polymerization. The specific method is to terminate the PMMA active chain prepared by the GTP method with bromine to obtain a bromine end group PMMA. Then, CuBr/bpy is used as an initiation system, and styrene is used as a monomer for atom transfer radical polymerization to obtain a PMMA-b-PS block copolymer, as shown in Fig. 2.9.

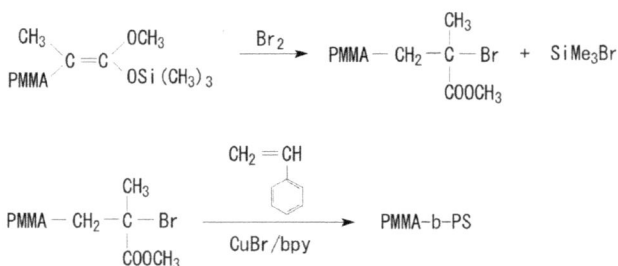

Fig. 2.9: Preparation of block copolymer by GPC and atom transfer radical polymerization.

Block and graft copolymers of PMMA and polydimethylsiloxane can also be synthesized by combination of group transfer polymerization and anionic living polymerization. Further, a polyfunctional monomer, such as $CH(CH_2OOCCH=CH_2)_3$, $C(CH_2OOCCH=CH_2)_4$, is first subjected to group transfer polymerization to obtain a polyactive center polymer, which is then converted into polymerization by other mechanisms. A star polymer can be prepared by polymerization of other monomers.

It should be noted that when the block copolymer is prepared by sequential feeding method, if nonidentical monomers (acrylates/methacrylates) are selected and the difference in activity is large, the low activity singles should be polymerized first. After the active chain is formed, a more active monomer is added. If MMA is copolymerized with EA, it should be converted into active chain A.

Its terminal double bond contains two alkyl groups, which is more active than the active chain B containing only one alkyl group.

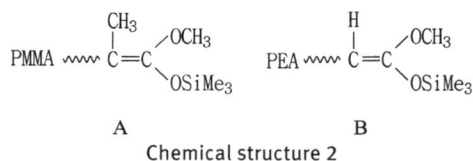

A   B

Chemical structure 2

Therefore, it is necessary to first polymerize into A by MMA before adding EA, so that all active chains will simultaneously produce a product with uniform chain length. Conversely, if B is first synthesized and then MMA is added, once the reaction begins, the same chain ends as A is formed. Due to the high reactivity at the end of the chain, a product of relatively high molecular mass is quickly formed, and most of the B will remain in the polymerization system without increasing or growing, so that a pure block copolymer cannot be obtained. The relative molecular mass is also uneven. The principle of this sequential addition of block copolymers is completely similar to other living polymerizations.

If the first monomer is added after obtaining the AB-type block copolymer as described above, an ABA-type triblock copolymer can be produced. It is, of course, also possible to add a third monomer to form an ABC type triblock copolymer.

2.4.2.3 Synthesis of graft copolymer

There have been many successful examples of the preparation of graft copolymers by combining group transfer polymerization with other living polymerizations. The graft copolymer obtained by this method has the characteristics of clear structure and controllable branch length, and is, therefore, one of the important methods for synthesizing graft copolymer. For example, Pan et al. used a reactive polystyrene prepared by anionic living polymerization, combined with polymethyl methacrylate prepared by group transfer polymerization, to obtain PS-g-PMMA graft copolymer with better control of relative molecular mass of main chain and branch. The specific process is as follows:

First, polystyrene with narrow molecular weight distribution was prepared by anionic living polymerization method, and active PMMA was prepared by group transfer polymerization. Polystyrene is chloromethylated with sulfuryl chloride in the presence of a catalyst anhydrous zinc chloride powder to give poly(p-chloromethyl)styrene (PCMS). Finally, PCMS was coupled with active PMMA in the following manner to obtain a graft copolymer.

2.4.3 Synthesis of telechelic polymers

By means of group transfer polymerization technology, the telechelic polymer can be prepared by anionic polymerization, and its functional group can reach the theoretical value, which is superior to other synthetic methods. Taking MMA polymerization as an example, initiator 1 initiates MMA to form intermediate 2, and then this intermediate is reacted with a proton-containing solvent or a suitable alkylating agent, and then treated with Bu$_4$NF, a polymer containing 100% terminal hydroxyl group can be obtained quantitatively. If $Br_2/TiCl_4$ or p-dibromomethylbenzene/H^+ is used as a terminator and then treated with Bu_4NF, a difunctional telechelic polymer can be obtained. The reaction formulas can be expressed as shown in Fig. 2.10.

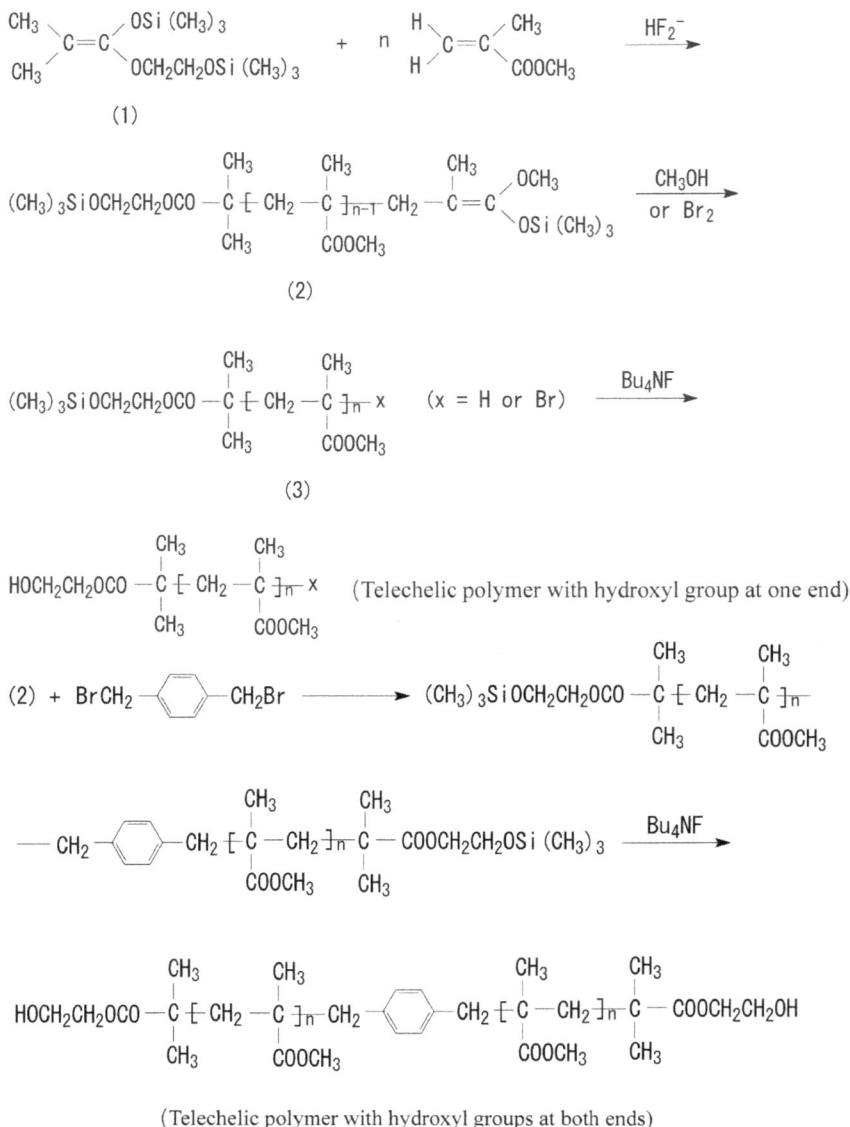

Fig. 2.10: Synthesis of telechelic polymer.

2.5 New progress in group transfer polymerization

Since Webster and others announced the discovery of group transfer polymerization technology, they have done a lot of work in addition to their own in-depth study of the reaction mechanism of this new polymerization method, the search for new initiators, the expansion of applicable monomers, and development applications. It has also aroused the interest of other polymer scientists in this field. After extensive exploration, many new developments have been made.

2.5.1 Initiator or catalyst

Sogah found that benzaldehyde could be used as an initiator, $ZnBr_2$ or $Bu_2A_1C_1$ used as a catalyst, and silyl vinyl ether used as a monomer. Through the group transfer polymerization technology, continuous aldol condensation was achieved and polyvinyl alcohol was directly synthesized. Unlike the general group transfer polymerization, in this polymerization process, $-SiR_3$ starts from the transfer of the monomer to the initiator, and as a result, the end of the living polymer chain continues to contain the aldehyde group, thus continuing aldehyde transfer, eventually forming a high polymer, as shown in Fig. 2.11.

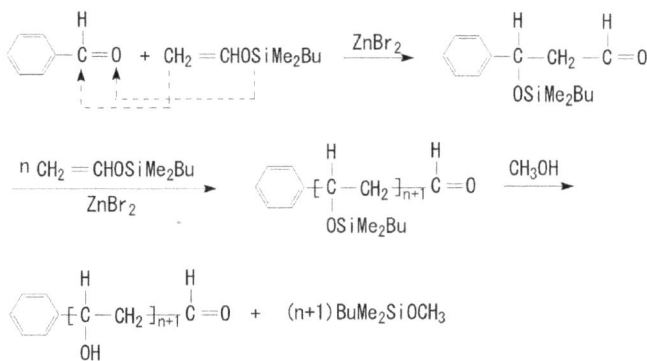

Fig. 2.11: Schematic diagram of a1dol-group transfer polymerization process.

The polymerization is characterized in that not only the initiator is benzaldehyde, but also the amount of the catalyst is only $10^{-4}\%$ to $10^{-2}\%$ by mole of the monomer, which is enough for the reaction to proceed smoothly.

Since the above reaction is caused by the aldol condensation, the transfer of the $-SiR_3$ group occurs, so it is called a1dol–GTP. According to research, the reaction mechanism of this group transfer polymerization may be that the initiator first activates by coordination with the catalyst, forms an intermediate with the monomer, then decomposes the catalyst and completes the following transfer and addition reactions, as shown in Fig. 2.12.

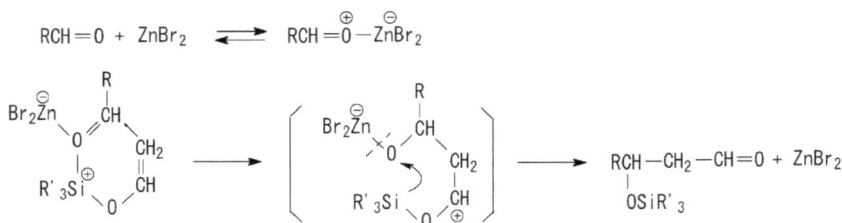

Fig. 2.12: Schematic diagram of the initiation mechanism of a1dol-group transfer polymerization.

In addition to aromatic aldehyde initiators, fatty aldehydes or ketones can also be used as initiators. If an aromatic dialdehyde is used as the initiator, the reaction proceeds to both ends like the bifunctional initiator in the anionic polymerization. Besides aldehydes, electrophilic benzyl halides and acetals $[C_6H_5CH(OCH_3)_2]$ can also be used as initiators. When 1,4-di(bromomethyl)benzene is used as the initiator, a polymerization product like that of terephthalaldehyde can be obtained, and the active chain has an aldehyde group at both ends, as shown in Fig. 2.13.

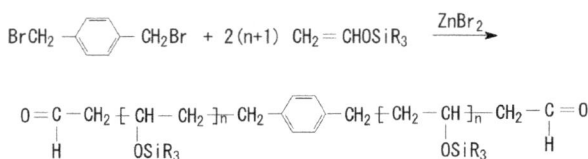

Fig. 2.13: A1dol-group transfer polymerization initiated by benzyl halide.

Recently, Zou et al. used a silyl enol initiator of diethyl malonate, such as ethyl 3-ethoxy-3-trimethylsiloxyacrylate $[CH_3CH_2O_2CCH = C(OCH_2CH_3)OSiMe_3]$ and nucleophilic-type tetraethylammonium dihydrogen fluoride $[(CH_2CH_2)_4NHF_2]$ to carry out the group transfer polymerization of MMA, BA, EA, and MA to obtain a polymer which was nearly monodisperse and substantially conformed to the designed relative molecular mass.

2.5.2 Monomer

In recent years, progress has also been made in the polymerization of certain specific monomers using group transfer polymerization techniques. For example, Pugh utilizes the following methacrylates and acrylates containing a series of special groups such as carbazole or a benzene ring as a monomer for group transfer polymerization, and many of the obtained polymers are smectic liquid crystal polymers, as shown in Fig. 2.14.

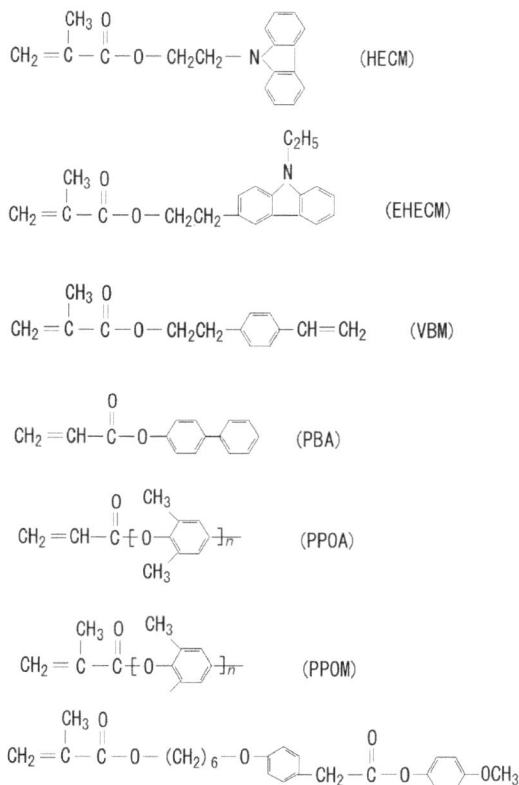

Fig. 2.14: Methacrylate and acrylate containing special groups which can be polymerized by GTP.

2.5.3 Block copolymer

The following PMMA–PVA block copolymers can be synthesized by the action of an active PMMA having an end containing $-OSiR_3$ and another active polymer having an aldehyde group at the end. This is a new method of synthesizing block copolymers by which different amounts of PVA can be introduced to control the hydrophilicity of the product, as shown in Fig. 2.15.

In addition, Xu et al. recently started with the basic theoretical research, using MTS as initiator and ZnI_2 as catalyst to carry out group transfer polymerization in dichloroethane solvent, respectively, the concentration of catalyst is greater than

$$\sim\!\!\!\sim\!\!\!\sim -\!\!\!\left[CH_2-\underset{\underset{OSiR_3}{|}}{CH}\right]_n\!\!\!-CH_2\overset{\overset{H}{|}}{C}=O \ + \ \sim\!\!\!\sim\!\!\!\sim -\!\!\!\left[CH_2-\underset{\underset{COOCH_3}{|}}{\overset{\overset{CH_3}{|}}{C}}\right]_n\!\!\!-CH_2-\overset{\overset{CH_3}{|}}{C}=C\overset{OMe}{\underset{OSiR_3}{\diagdown}}$$

$$\xrightarrow{HF_2^-} \sim\!\!\!\sim\!\!\!\sim -\!\!\!\left[CH_2-\underset{\underset{OSiR_3}{|}}{CH}\right]_{n+1}\!\!\!\!\!\left[CH_2-\underset{\underset{COOCH_3}{|}}{\overset{\overset{CH_3}{|}}{C}}\right]_{n+1}\!\!\!\!\sim\!\!\!\sim\!\!\!\sim \quad \xrightarrow[CH_3OH]{Bu_4NF}$$

$$\sim\!\!\!\sim\!\!\!\sim -\!\!\!\left[CH_2-\underset{\underset{OH}{|}}{CH}\right]_{n+1}\!\!\!\!\!\left[CH_2-\underset{\underset{COOCH_3}{|}}{\overset{\overset{CH_3}{|}}{C}}\right]_{n+1}\!\!\!\!\sim\!\!\!\sim\!\!\!\sim$$

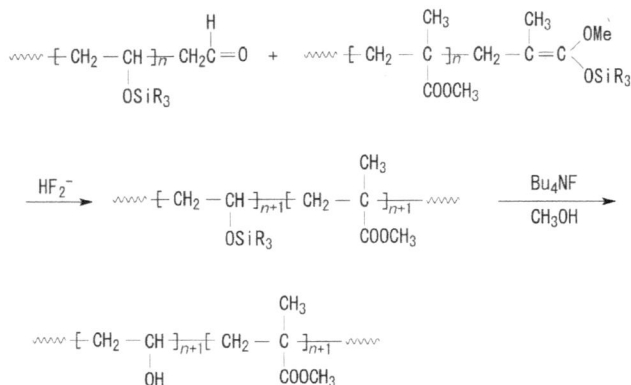

Fig. 2.15: Preparation of PMMA--PVA block copolymer by a1dol–GTP combined with GTP.

10% of the monomer dosage (molar) and less than 40% by mole of the monomer used to carry out the polymerization kinetics of EA at 0 °C and above 0 °C, the reaction curve was determined, the effect of temperature on the polymerization reaction was investigated, the polymerization product and the relative molecular mass distribution was measured. Some new conclusions have been obtained.

2.5.4 Product development

In industrial preparation, group transfer polymerization is first used in acrylate-based automotive coatings. The polymer obtained by the group transfer polymerization technique has a relatively uniform molecular mass and a solid content of up to 60%, while the general free radical polymerization generally has a solid content of only 20%; and at the same time, since there is no terminally reacted monomer, it is volatilized during finishing. The amount is small, reducing environmental pollution. In addition, the acrylate-based automotive facemask coating synthesized by the group transfer polymerization can be cured at 82 °C, and it is possible to cure at room temperature, while usually the automotive overcoat paint needs to be cured at 116–127 °C.

Group transfer polymerization has great advantages in molecular structure control. It can obtain a certain molecular weight of polymer by this method, and it is also possible to prepare biopolymer materials, which has important theoretical and practical value.

DuPont has developed automotive topcoat coatings using a group transfer polymerization process. It is also planned to produce photosensitive resins by this method to prepare semiconductor wafers, to produce coatings for optical fibers, thermoplastic elastomers, and composites that may replace metals. The application prospect of group transfer polymerization is very broad.

References

[1]   Webster, OW, Hertler, WR, Sogah, DY, Farnham, WB, RajanBabu, TV. J Am Chem Soc.
       1983;105(17):5706–5708.
[2]   Webster, OW, Hertler, WR, Sogah, DY, et al. Polym Prepr. 1983;24(2):52.
[3]   Webster, OW. Living polymers and process for their preparation. United States patent
       4,417,034. 1983.
[4]   Webster, OW, Hertler, WR, Sogah, DY, et al. J Macromol Sci Chem. 1984;A21(8/9):943.
[5]   Webster, OW. Polym Prepr. 1986;27(1):161.
[6]   Suenaga, J, Sutherlin, DM, Stille, JK. Macromolecules. 1984;17(12):2913–2916.
[7]   Sogah, DY, Webster, OW. Polym Prepr. 1983;24(2):54.
[8]   Sogah, DY, Hertler, WR, Webster, OW. Polym Prepr. 1984;25(2):3.
[9]   Sogah, DY. Polym Prepr. 1986;27(1):163.
[10]   Sogah, DY, Hertler, WR, Webster, OW, Cohen, GM. Macromolecules. 1987;20(7):1473–1488.
[11]   Farnham, WB, Sogah, DY. Polym Prepr. 1986;27(1):167.
[12]   Farnham, WB, Sogah, DY. Process for preparing living polymers. United States patent
       4414372, 1983.
[13]   Farnham, WB, Tris (dialkylamino) sulfonium bifluoride catalysts. United States patent
       4598161, 1986.
[14]   Hertler, WR. Polym Prepr. 1986;27(1):165.
[15]   Hertler, WR, Sogah, DY, Webster, OW. Macromolecules. 1984;17(7):1415–1417.
[16]   Hertler, WR. Macromolecules. 1987;20(12):2976–2982.
[17]   Hertler, WR, Sogah, DY, Boettcher, FP. Macromolecules. 1990;23(5):1264–1268.
[18]   Bandermann, F, Speikamp, H-D. Macromol Rapid Commun. 1985;6(5):335–339.
[19]   Pudgh, CR, Percec, V. Polym Prepr. 1985;190:110–116.
[20]   Pugh, C, Percec, V. Polym Bull. 1985;14(2):109–116.
[21]   Reetz, MT, Ostarek, R, Piejko, K-E, Arlt, D, Bömmer, B.. Angew Chem. 1986;98(12):1116–1118.
[22]   Willi, K, Webster, OW. Macromol Rapid Commun. 1986;7(1):5–13.
[23]   Eastmond, GC, Grigor, J. Macromol Rapid Commun. 1986;7(6):375–379.
[24]   Xia, Y, Yang, ZJ. Proceedings of polymer academic paper Symposium (in Chinese) 1985; 60726.
[25]   Zou, YS, Pan, RH. Acta Polym Sin (in Chinese). 1988;4:301–304.
[26]   Pan, RH, Xia, HP. Journal of Xiamen University (in Chinese). 1984;26:429.
[27]   Pan, RH, Xia, HP. Chin J Appl Chem (in Chinese). 1987;4(3):1–10.
[28]   Dai, LZ, Zou, YS, Yao, QQ, et al. Chin J Appl Chem (in Chinese). 1997;14(5):23–27.
[29]   Dai, LZ, Fu, X, Zhou, SK, et al. Chin J Appl Chem (in Chinese). 1999;16(2):14.

Exercises

1. **What is the group transfer polymerization?**
    The group transfer polymerization is a compound in which an α,β-unsaturated ester, a ketone,
    an amide, or a nitrile is used as a monomer, and a compound having a group such as a silicon,
    an anthracene or a tin alkyl group is used as an initiator, and an anionic type or Lewis acid
    type compound is used as a catalyst, using a suitable organic solvent as a solvent, by coordi-
    nating the silicon, ruthenium, and tin atoms of the initiator with the initiator, exciting silicon,
    germanium, and tin atoms to make them bond with oxygen or nitrogen atoms. The atoms com-
    bine to form a covalent bond, and the double bond in the monomer completes the addition

reaction with the double bond in the initiator, and the silicon, germanium, tin, and alkyl groups move to the end to form an "active" compound.

2. **What are the steps of group transfer polymerization?**
   (1) Chain initiation
   (2) Chain growth
   (3) Chain termination

3. **Please list some advantages of GTP.**
   (1) There was no obvious termination reaction and the product was "active polymer." Polymers with various structures can be obtained by further treatment.
   (2) Polymers with special functional groups at the end of the initiator can be easily synthesized by changing the functional groups at the end of the initiator.
   (3) Polymerization can be carried out in a wide temperature range, for example, -100–150 °C.
   (4) Polymers with narrow molecular weight distribution can be obtained.

4. **What are the applications of group transfer polymerization?**
   (1) Synthesis of homopolymers with narrow molecular weight distribution.
   (2) Produce random, block, graft copolymer.
   (3) Synthesis of telechelic polymer.

5. **Please list some catalysts used in GTP.**

   (1)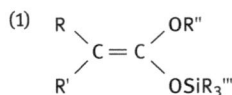

   (2)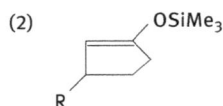

   (3) R_3SiX
   (4) $P(OSiMe_3)_3$

Chapter 3
Living controlled polymerization of free radical

3.1 Overview

Since the advent of the first synthetic polymer in the early twentieth century, polymer chemists have been working tirelessly to find ways to prepare polymer materials with specified structures and molecular sizes. In the early 1950s, the emergence of coordination polymerization technology opened up a new field of stereoregular polymerization, leading to the birth of a number of high-quality new materials such as linear polyethylene and stereoregular polypropylene. This has led to the development of the entire petrochemical industry. In the mid-1950s, the development of anionic living polymerization technology provided the possibility of molecular design, synthetic structure and combined controllable polymers of polymer materials. Living polymerization has been one of the most active research directions in the field of polymer chemistry for nearly 50 years.

Regardless of the chemical nature of the reactive species, any linked polymerization reaction contains three major elementary reactions: chain initiation, chain extension, and chain termination Due to the chain termination reaction (including irreversible chain transfer reaction), the conventional chain polymerization method generally cannot control the structure and size of the polymer molecule, and has a wide relative molecular mass distribution.

In 1956, American scientist Szwarc et al. reported a cross-aged discovery that there is no styrene anion polymerization initiated by sodium naphthalate under anhydrous, anaerobic, impurity-free, low-temperature conditions, and tetrahydrofuran as solvent. The reaction was terminated and transferred, and the obtained polymer solution was stored under low temperature and high vacuum for several months, and the active species concentration remained unchanged. If styrene is further added, the polymerization can be continued to obtain a higher molecular weight polystyrene. When a second monomer such as butadiene is added, a pure butadiene–styrene–butadiene triblock copolymer can be obtained. Based on this finding, Szwarc et al. first proposed the concept of living polymerization.

There are four characteristics that distinguish the living polymerization from the traditional polymerization reaction: (1) the initiation reaction rate is much larger than the growth reaction rate, and there is no chain termination and chain transfer reaction. All the polymer chains grow at the same time, and the number of growth chains is always the same. Molecular mass distribution is very narrow ($\bar{M}_w/\bar{M}_n \leq 1.1$). (2) The relative molecular mass of the polymer is proportional to the ratio of the concentration of the consumed monomer to the initial concentration of the initiator. (3) The relative molecular mass of the polymer increases linearly with the conversion rate, and the degree of polymerization of the obtained polymer can be controlled by

https://doi.org/10.1515/9783110597097-003

controlling the amount of monomer and initiator charged. (4) When the conversion rate of the first monomer reaches 100%, another monomer is added to synthesize a block copolymer having a predetermined structure.

After Szwarc proposed living polymerization for more than 40 years, living polymerization has developed into one of the most academic and industrial applications in the field of polymer chemistry. The most important significance is that living polymerization provides a means for polymer chemists to synthesize structures and control molecules with a relative molecular mass without the traditional polymerization methods. After decades of efforts, a series of living polymerization systems suitable for polymerization of different monomers have been successfully developed, such as living anionic polymerization, living cationic polymerization, active ring-opening polymerization, active ring-opening disproportionation polymerization, group transfer polymerization, anionic polymerization, and metal-free anionic polymerization. The dream of molecular design of polymer materials for many years has become a reality. However, practice has shown that although these active polymerizations have been developed to produce some structurally controllable polymers, there are not many large-scale industrial productions. The main problem is that their reaction conditions are generally harsh, and the reaction process is also complicated, resulting in high industrialization costs of products. At the same time, the existing living polymerization technology has a narrow coverage of monomers, mainly monomers such as styrene and (meth)acrylate, which makes the molecular structure less configurable, thus greatly limiting the living polymerization technology application in the field of polymer materials.

Based on the development and dilemma of living polymerization, polymer chemists naturally associate with free radical polymerization. The traditional free radical polymerization has the advantages of wide monomer, various synthetic processes, simple operation, low industrialization cost, and so on, and also allows various functional groups to be carried on the monomer, and can be used as a polymerization medium by using a proton-containing solvent and water. Part of the monomer is copolymerized and so on. Currently, about 70% of polymer materials are derived from free radical polymerization. However, the radical polymerization has a primitive reaction or a side reaction that contradicts the living polymerization, such as a coupling termination reaction of a radical, a disproportionation termination reaction, and a chain transfer reaction, which makes the polymerization reaction process difficult to control. Therefore, the realization of free radical living polymerization or controlled polymerization has been a subject of interest.

In the development of "active" radical polymerization, early work dates back to the 1989 initiation–transfer–termination method (iniferter method) used by Otsu, Turner, and Blevins. The most influential contribution is the use of the concept of stable free radicals. In 1993, researchers at Xerox Canada first reported high-temperature (120 °C) bulk polymerization of styrene initiated by TEMPO (2,2,6,6-tetramethyl-1-piperidine oxide)/benzoyl peroxide (BPO) initiation system. This is the first living radical polymerization system ever. Georges et al. used this method to prove that the relative molecular mass of the growing polymer chain increases with increasing conversion rate when the temperature is increased, and the relative molecular mass distribution of the polymer decreases to 1.10–1.30, which is much lower than the theoretical calculation. The limit is 1.5. These findings mark the fact that the living radical polymerization process has been achieved.

In 1995, Professor Matyjaszewski of Carnegie Mellon University in the United States and Dr. Wang Jinshan, a Chinese scholar in the United States, successfully discovered atom transfer radical polymerization (ATRP) and realized the active (controlled) polymerization of free radicals based on years of research on living polymerization. As soon as this technology is reported, scientists in various countries engaged in active polymerization research have given extremely high evaluations. It is considered to be an important discovery in the field of polymer synthesis chemistry for decades, and it is a historic breakthrough in the field of "active" radical polymerization.

3.2 Difficulties and solutions for free radical controllable polymerization

3.2.1 Difficulties in controlled polymerization of free radicals

Based on the fact that most commercially valuable polymer products are derived from free radical polymerization, it can be said that the research and development of free radical active (controlled) polymerization determines the direction, future and fate of the entire living polymerization research. Only the large-scale industrialization of activity and controlled radical polymerization can fully reflect the significance and role of living polymerization.

From the viewpoint of polymerization mechanism, the radical polymerization process of vinyl monomers includes the following four elementary reactions:

Chain initiatin $\quad R-R' \longrightarrow R^{\cdot} + R'^{\cdot}$

Chain growth $\quad R^{\cdot} + (n+1)M \longrightarrow R\text{-}[M\text{-}]_n\text{-}M^{\cdot}$

Chain end $\quad 2\,R\text{-}[M\text{-}]_n\text{-}M^{\cdot} \longrightarrow R\text{-}[M\text{-}]_n\text{-}M-H + R\text{-}[M\text{-}]_n\text{-}M'$

$\qquad\qquad\qquad\qquad \longrightarrow R\text{-}[M\text{-}]_n\text{-}M-M\text{-}[M\text{-}]_n\text{-}R$

Chain transfer $\quad R\text{-}[M\text{-}]_n\text{-}M^{\cdot} + HSR \longrightarrow R\text{-}[M\text{-}]_n\text{-}M-H + SR^{\cdot}$

First, the initiator decomposes to generate free radicals; in turn, the free radicals attack the unsaturated C = C double bonds of the monomers, and successive additions cause the chains to grow; the growth of free radicals reacts (coupling or disproportionation) leading to chain termination. If the free radical reacts with the chain transfer agent, chain transfer occurs, forming an inactive macromolecule and inducing the formation of a new polymer chain. In general, the free radicals in the free radical polymerization system may be a transition state with a short lifetime, and the reaction rate is extremely fast. The relative molecular mass of the polymer is randomly determined by the competitive reaction between chain growth, chain termination, and chain transfer.

In ionic polymerization, the growth of carbon anions or carbocations does not react with each other due to electrostatic repulsion. However, free radicals strongly exhibit a tendency to terminate the reaction by coupling or disproportionation, and the termination reaction rate constant is close to the diffusion control rate constant $(k_t = 10^7 - 10^9 \ m^{-1} \cdot s^{-1})$, which is five orders of magnitude higher than the corresponding growth reaction rate constant $(k_p = 10^2 - 10^4 \ m^{-1} \cdot s^{-1})$. In addition, the slow decomposition of classical free radical initiators $(k_d = 10^{-6} - 10^{-4} \cdot s^{-1})$ often leads to incomplete initiation. These kinetic factors (slow initiation, fast growth, rapid termination, and easy transfer) determine the uncontrollability of traditional free radical polymerization.

In addition, from the perspective of free radical polymerization kinetics, the rate of initiator decomposition is closely related to the chemical bond dissociation in the initiator molecule, and the dissociation energy is a function of temperature. Increasing the temperature can increase the decomposition rate of the initiator. At the same time, it accelerates the chain growth reaction rate and leads to an increase in side reactions such as chain transfer. Thus, research on living radical polymerization has focused on controlling chain growth, which stabilizes free radicals.

3.2.2 Countermeasures for free radical controllable polymerization

From the knowledge of polymer chemistry, the ratio of chain termination rate to chain growth rate can be expressed by the following equation:

$$\frac{R_t}{R_p} = \frac{k_t[P\cdot]}{k_p[M]} \tag{3.1}$$

where R_p, R_t, k_p, k_t, [p·], [M] are growth rate, termination rate, growth rate constant, termination rate constant, free radical transient concentration, and monomer transient concentration, respectively.

It is not difficult to see from formula (3.1) that the smaller the k_t/k_p value, the smaller the influence of the chain termination reaction on the entire polymerization reaction. Usually k_t/k_p is 10^4–10^5. Therefore, the chain termination reaction has a great influence on the polymerization process. In addition, R_t/R_p also depends on the ratio of free radical concentration to monomer concentration. For example, in free radical bulk polymerization, $[M]_0$ is 1–10 mol/L, which is difficult to change under normal conditions. It can be seen that reducing the R_t/R_p value should be achieved mainly by reducing the instantaneous radical concentration in the system. Assuming a monomer concentration of 1 mol/L in the system:

$$\frac{R_t}{R_p} \approx 10^4 - 10^5 [P \cdot]$$ (3.2)

Of course, the concentration of free radical active species cannot be reduced without limitation. In general, $[p\cdot]$ is around 10^{-8} mol/L, and the polymerization rate is still considerable. At such free radical concentrations, $R_t/R_p = 10^{-3}$–10^{-4}, R_t is negligible relative to R_p. On the other hand, the decrease in the concentration of the radicals necessarily reduces the rate of polymerization. However, since the chain growth reaction activation energy is higher than the chain termination reaction activation energy, increasing the polymerization temperature not only increases the polymerization rate (increased k_p), but also effectively lowers the ratio k_t/k_p and inhibits the termination reaction. For this reason, "living" radical polymerization should generally be carried out at higher temperatures.

In practice, to enable free radical polymerization as a controlled polymerization, a low and constant concentration of free radicals must be present in the polymerization system. For the concentration of free radicals, the termination reaction is a kinetic secondary reaction, and the growth reaction is a kinetic first-order reaction. To maintain a considerable rate of polymerization (free radical concentration cannot be too low), but also to ensure that the active species do not occur in the reaction process (elimination of chain termination and chain transfer reaction), there are two problems to be solved. One is how to control the concentration of such low reactive species from the beginning of the polymerization until the end of the reaction; the second is how to avoid the excessive polymerization degree of the polymer $(DP_n = [M]_0/[P] = 1/10^{-8} = 10^8)$ obtained by polymerization at such a low concentration of the reactive species that cannot be designed. This is a contradiction. In order to solve this contradiction, polymer chemists are inspired by living cationic polymerization to introduce the concept of reversible chain termination reaction and chain transfer reaction into free radical polymerization between active species and dormant species (temporarily inactivated active species). Establishing a rapid exchange response, that is, establishing a reversible equilibrium response, successfully achieved the unity of opposites of the above contradictions, as shown in the following equation:

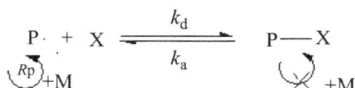

The basic principle of this consideration is: if a quantity of artificially controlled reactant X is added to the polymerization system, the reactant X cannot initiate polymerization of the monomer, but can react with the radical P· to cause a passivation
reaction. A "dormant species" P–X that does not initiate polymerization of the
monomer is produced. The dormant species can be split into growth free radicals
P· and X under the experimental conditions. The concentration of free radical active
species present in such a system will depend on three parameters: the concentration of reactant X, the passivation rate constant k_d, and the activation rate constant
k_a. Among them, the concentration of the reactant X can be artificially controlled,
which solves the first problem raised above. Studies have shown that if the conversion rate of the passivation reaction and the activation reaction is fast enough (not
less than the continuous growth rate), then the relative molecular mass of the polymer will not be determined by P· but by the concentration of P–X in the case of a
very low concentration of active species, that is:

$$\overline{DP_n} = \frac{[M]_0}{[P-X]} \times d \tag{3.3}$$

Where in d is the monomer conversion. This solves the second problem raised above.

It can be seen that the rapid equilibrium reaction by means of X not only controls the radical concentration to be low, but also controls the relative molecular
mass of the product. Therefore, controlled radical polymerization becomes possible. However, the above method merely changes the concentration of the radical
active center without changing the nature of the reaction, and is thus a controllable polymerization and not a living polymerization in the true sense. In order to
distinguish it from the true living polymerization, such macroscopic polymerization processes similar to living polymerization are generally referred to as active/
controlled polymerization. The above method is easy to understand conceptually,
but it is not so easy in practice. To put this idea into practice, the majority of scientists have conducted a long-term research and exploration. Three ways to achieve the
above goals have been found:
(1) Reactive growth of free radicals and stable free radicals to form dormant cova-
    lent compounds:

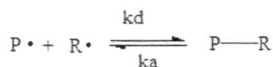

$$P\cdot + R\cdot \underset{k_a}{\overset{k_d}{\rightleftharpoons}} P\text{—}R$$

where k_d and k_a are the reaction rate constants for activation and deactivation, respectively. The stable free radical R· compounds that have been found so far mainly
include nitroxide compounds (such as TEMPO), dithiocarbamate, tritylmethyl and
diphenylmethyl derivative, a transition metal compound (e.g., an alkyl porphyrin
cobalt, a copper halide/dipyridyl complex), and so on.

(2) Growth of free radicals and nonradical materials reversibly form dormant and persistent free radicals:

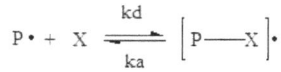

$$P\bullet + X \underset{ka}{\overset{kd}{\rightleftharpoons}} \left[P\text{---}X \right]\bullet$$

X is usually an organometallic compound that reacts with increasing free radicals to form relatively stable, highly coordinated free radicals, such as alkyl aluminum–TEMPO complexes.

(3) Reversible deuteration transfer between growth free radicals and chain transfer agents

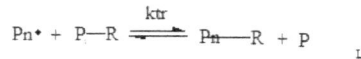

$$Pn\bullet + P\text{---}R \overset{ktr}{\rightleftharpoons} Pn\text{---}R + P$$

The ideal chain transfer agent should have a high chain transfer constant k_{tr}, such as commonly used alkyl iodides, and dithioesters.

After decades of efforts, research on the controllable polymerization of free radical activity has made a breakthrough. In 1993, researchers at Xerox Canada first proposed high-temperature (120 °C) bulk polymerization of styrene initiated by TEMPO radical/BPO. This is the first living radical polymerization system ever. However, TEMPO does not polymerize other types of monomers other than styrene. In addition, TEMPO is expensive and difficult to industrialize.

In 1994, Wayland et al. used tetrakis(trimethylphenyl) porphyrin-2,2′-dimethylpropyl cobalt [TEM-CO-CH2(CH3)3] to initiate the polymerization of methyl acrylate and found acrylic acid. The relative molecular mass of the methyl ester has a linear relationship with the monomer conversion rate, and its relative molecular mass distribution is narrow ($M_w/M_n = 1.10$–1.21), so it is a living radical polymerization. However, this system does not allow other types of monomers to be polymerized, and the price is also very expensive, and it is difficult to have an industrial future.

In 1995, Matyjaszewski and Wang Jinshan used a heterogeneous system consisting of 1-PECl/CuCl/bpy to initiate the polymerization of styrene and acrylate. The relationship between the monomer conversion and the relationship between time and the relative molecular mass and monomer of the polymer were found. The conversion rate is linear and close to the theoretical value, while the relative molecular mass distribution of the polymer is very narrow (DP = $\triangle[M]/[I]$), so the polymerization process exhibits "active characteristics" ($M_w/M_n \leq 1.5$). This is the sensation of polymer chemistry, which is the most important ATRP in the field of living polymerization. In this chapter, several more important controlled activities of free radical activity can be described in detail.

3.3 Controlled free radical polymerization of metal-free initiated systems

3.3.1 Priming transfer-terminating method (iniferter method)

In 1982, Japanese scholar Otsu and others proposed the concept of initiating a transfer terminator and successfully applied it to free radical polymerization to bring free radical activity/controlled polymerization into a new historical development period.

From the characteristics of living polymerization and the reaction mechanism of radical polymerization, the key to achieving free radical activity/controlled polymerization is how to prevent inactive polymer chains from being caused by chain termination reaction and chain transfer reaction during polymerization. The initiator has a high activity for increasing the free radical to the chain transfer reaction of the initiator itself, or the radical generated by the decomposition of the initiator is highly active to the initiator's own chain transfer reaction, or is decomposed by the initiator. A part of the free radical is prone to the termination reaction with the chain radical, and the radical polymerization process of the vinyl monomer can be represented by the following formula:

$$R—R' + nM \longrightarrow R—[M]n—R'$$

According to the above reaction mechanism, the radical polymerization can be simply regarded as a continuous insertion reaction of a monomer molecule into an R–R' bond in an initiator molecule, and a structural feature of the polymerization product is a fragment having an initiator at both ends. Otsu et al. have thus come to the conclusion that if a suitable initiator satisfying the above conditions can be found, a monofunctional or difunctional polymer can be easily synthesized by radical polymerization, thereby achieving the purpose of polymer structure design. Since the initiator set, transfer, and termination functions are integrated, it is called an initiation–transfer terminator.

3.3.1.1 The basic type of initiation–transfer terminator

Initiation–transfer terminator refers to a class of compounds that simultaneously act to initiate, transfer, and terminate during free radical polymerization. Studies have found that many of the compounds that can be used as initiation–transfer terminators are generally classified into two types: thermal decomposition and photolysis. There are few reports on the thermally induced transfer terminator in the literature. Except that the triphenylmethylazobenzene and tetraethylthiuram are azo bond and S–S bond, respectively, the other are the symmetric six-substituted ethane of C–C bond. Among the class of compounds, the 1,2-disubstituted

R = H , X = Y = CN , OC_6H_5 , $OSi(CH_3)$

R = OCH_3 , X = Y = CN

R = H , X = H, Y = C_6H_5

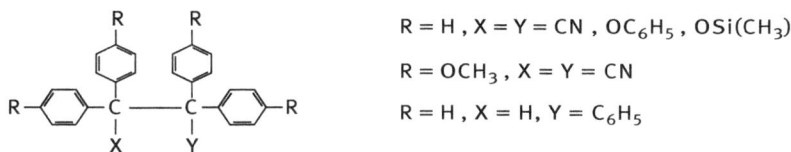

Fig. 3.1: General formula of 1,2-disubstituted tetraphenylethane derivatives. Reprinted with permission from [30]. Copyright Chinese Academy of Sciences, 2002.

tetraphenylethane derivative is discussed, and its general formula is shown in Fig. 3.1, including tetraphenylsuccinonitrile (TPSTN), tetrakis(p-methoxy)phenylbutanenitrile, pentaphenylethane (PPE), 1,1,2,2,-tetraphenyl-1,2-diphenoxyethane, and 1,1,2,2-tetraphenyl-1,2-trimethylsiloxyethane.

It is found that these symmetric carbon–carbon bond thermally initiated transfer terminators induce the polymerization of polar monomer methyl methacrylate (MMA) to be actively controlled polymerization, and the activity sequence of the initiator is PPE > TPPSTN > TPSTN. The obtained poly methyl methacrylate (PMMA) can be used as a macroinitiator to initiate polymerization of a second monomer such as styrene to prepare a PMMA-b-PS copolymer, but the block efficiency is relatively low. However, for the polymerization that initiates the nonpolar monomer St, their action is like that of the conventional radical polymerization initiator, and there is no characteristic of living polymerization. It is believed that when the 1,2-disubstituted tetraphenylethane derivative initiates the polymerization of styrene, the ω end of the obtained polymer is a penta-substituted C–C bond, and the bond energy is relatively high, and cannot be decomposed when heated, and is a dead-end polymerization. When the MMA polymerization is initiated, the ω end of the obtained polymer is a six-substituted C–C bond, and the bond energy is low, and it can be reversibly decomposed when heated. Living radical polymerization can be achieved.

Photoinitiated transfer terminator mainly refers to a compound containing a diethyldithiocarbamoyloxy (DC) group. Relatively speaking, it has many types, such as *N,N*-diethyldithiocarbamate (BDC), bis(*N,N*-diethyldithiocarbamic acid)-*p*-benzene, dimethyl ester, benzyl *N*-ethyldithiocarbamate, bis(*N*-ethyl and thiocarbamic acid) terephthalate, 2-*N,N*-diethyl ethyldithiocarbamoyloxyisobutyrate, 2-M,M-diethyldithiocarbamoyloxypropionate, and *N,N*-diethyl (4-vinyl)decyl thiocarbamate (VBDC) or the like. These photoinitiated transfer terminators are mostly used to initiate living polymerization of vinyl monomers to prepare end group functionalized polymers and block and graft copolymers. A significant advantage of photoinitiated transfer terminators is the large number of polymerizable monomers, especially the active polymerization of monomers such as vinyl acetate and isoprene, which are not possible or difficult with other current living radical polymerization processes. Figure 3.2 shows the structural formula of a commonly used photoinitiated transfer terminator.

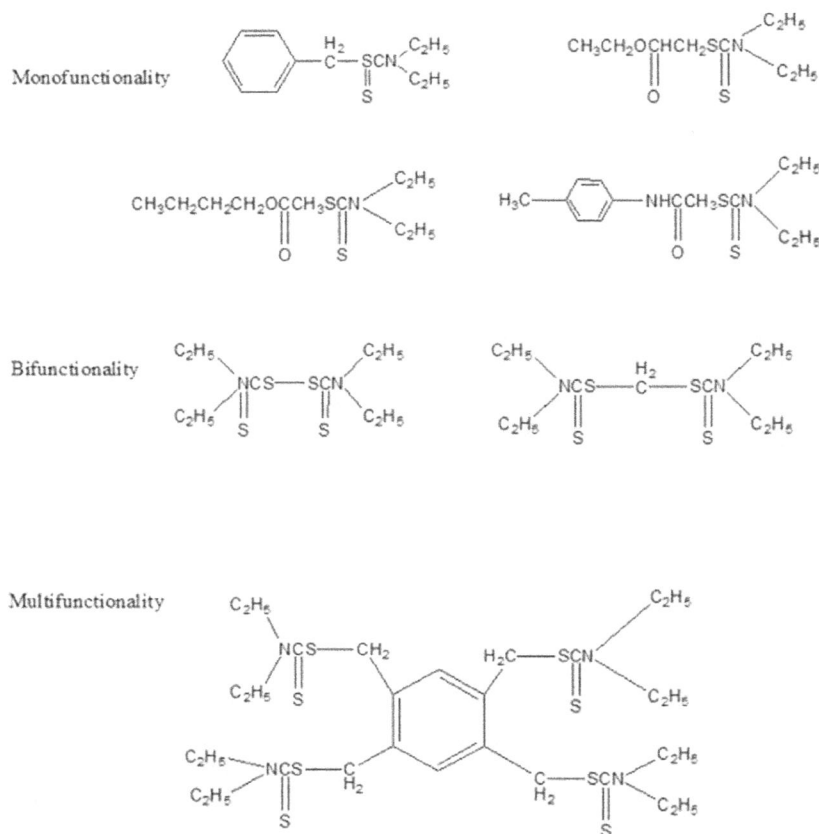

Fig. 3.2: Common photoinitiated transfer terminator structure.

In recent years, in-depth research conducted in the field of initiating transfer terminators, highly active thermally initiated transfer terminators and photoinitiated transfer terminators, polymerizable photoinitiated transfer terminators, and multifunctional initiation transfer terminators have been synthesized. They were investigated to initiate the living behavior of vinyl monomers and to prepare corresponding copolymers.

As mentioned in the earlier description, the initiation transfer terminator can be divided into two types: symmetrical and asymmetric. Symmetrical initiation of the same free radicals produced by the decomposition of the terminator can be initiated and terminated:

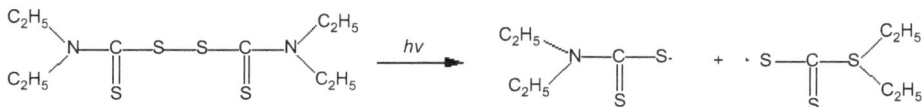

The decomposition of the asymmetric initiation transfer terminator can be carried out in two different ways, resulting in different free radicals as follows:

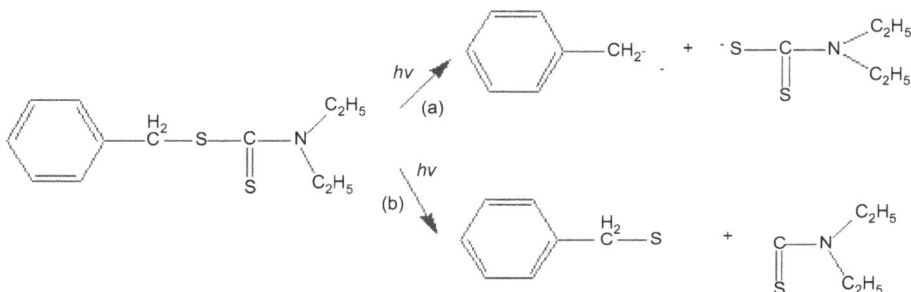

From the route (a), a benzyl radical and a $(C_2H_5)_2NC(S)S \cdot$ radical can be produced, the former being used for initiating polymerization and the latter for terminating polymerization. Both $PhCH_2S \cdot$ and $(C_2H_5)_2NCS \cdot$ free radicals can be produced by route (b), and neither can initiate polymerization. In general, the C–SR bond is weaker than the S–R bond, so decomposition is mainly performed by the first method. However, if the energy given to the system is high, decomposition of pathway (b) may also occur.

3.3.1.2 C–C bond type high-activity thermal initiation transfer terminator

Due to the small number of thermally initiated transfer terminators reported in the literature, the activity is low and the living polymerization of the polar monomer MMA can only be achieved at higher temperatures (>80 °C). The polymerization of the thermally initiated transfer terminator to the nonpolar monomer St is still a conventional free radical polymerization, inactive polymerization feature. Therefore, it is a meaningful work to synthesize and design a novel structure of highly active thermally initiated transfer terminator in order to achieve active polymerization of MMA at a lower temperature and to expand the range of active polymerization monomers.

Two C–C bonded thermally initiated transfer terminators have been studied: diethyl 2,3-dicyano-2,3-diphenylsuccinate (DCDPST) and 2,3-dicyano-2,3-di-(p-tolyl)-diethylsuccinate (DCDTS) initiate the polymerization of vinyl monomers. These two compounds can be used as a conventional radical polymerization initiator to initiate St polymerization. Similarly, the living radical polymerization of MMA can be carried out at 60–100 °C. The obtained PMMA was confirmed by ^{1}H NMR analysis, and the end group was a fragment derived from the initiator DCDPST. This polymer can be used as a macromolecular initiation transfer terminator for chain extension reaction and block copolymerization to obtain the corresponding chain extension product and block copolymer. In addition, the use of DCDPST for the first time achieved the living radical polymerization of the nonpolar monomer St in the field of small-molecule thermally initiated transfer terminators. Figure 3.3 shows the relationship between the relative molecular mass of the obtained polymer and the monomer conversion rate at different temperatures. When the polymerization is carried out at 70, 85, and 100 °C. The number average molecular weight of the obtained polymers increased linearly with the increase of monomer conversion rate (relative molecular mass distribution around 1.50), indicating that DCDPST initiated St bulk polymerization at 70, 85, and 100 °C.

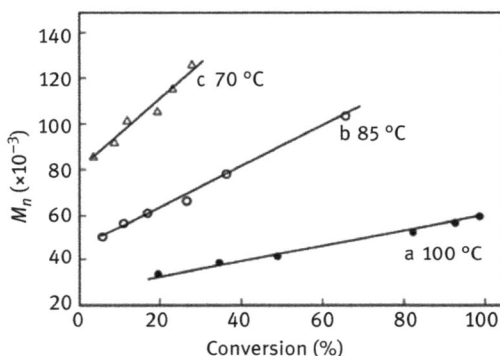

Polymerization condition: [St] = 8.7 mol/L; [DCDPST] = 4.4 × 10^{-2} mol/L

Fig. 3.3: Relationship between Mn and conversion of St (styrene) bulk polymerization with DCDPST as transfer termination agent. Reprinted with permission from [30]. Copyright Chinese Academy of Sciences, 2002.

It is shown in Fig. 3.3 that the linear relationship between the number average molecular mass of the polymer and the monomer conversion rate does not pass through the origin, which has an intercept and the intercept decreases as the reflected temperature increases. It is caused by the fact that the growth chain free radicals and the reversible termination of the free radicals do not reach the dynamic equilibrium in the initial stage of the reaction. At the same time, it can be seen that at higher reaction temperature, the faster equilibrium is established. A similar phenomenon occurs for the polymerization of MMA. Similarly, the obtained PS was confirmed by ^{1}H NMR analysis that the end group was a fragment derived from the initiator DCDPST, which was used as a macromolecular initiation transfer terminator to carry out St chain extension reaction and initiate MMA polymerization to prepare PS-b-PMMA copolymer that both have been successful.

The mechanism by which DCDPST initiates the polymerization of MMA and St is like that of other thermally initiated transfer terminators reported in the literature. Taking St polymerization as an example, as shown in FIg. 3.4, in the initiation phase, DCDPST is split into two identical radicals B, which can initiate St

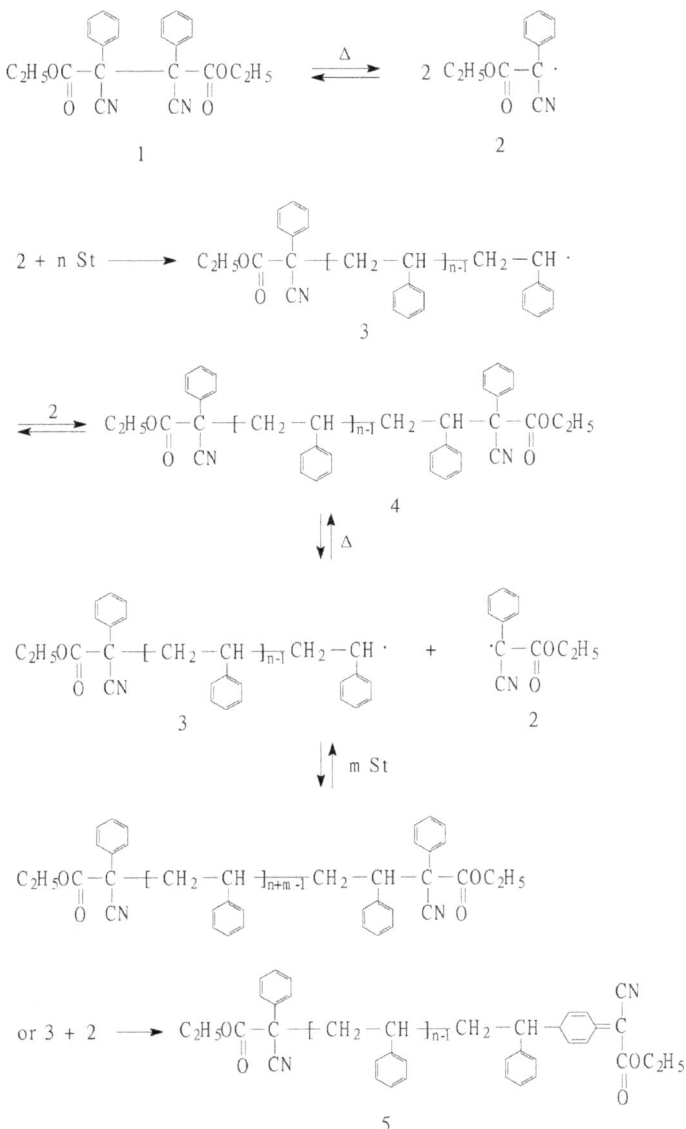

Fig. 3.4: Structure of a novel photoinitiated transfer terminator. Reprinted with permission from [30]. Copyright Chinese Academy of Sciences, 2002" should be revised as" Fig. 3.4: Mechanism of St polymerization initiated by DCDPST.

polymerization and reversibly terminate the growth chain radicals to form α and ω. The polymer D with the initiator fragment at the end and the C–C bond of the pentasubstituent group at the ω end of D can still be reversibly decomposed into the growth chain radical C and the radical B under the action of heat. Free radical C continues to initiate monomer polymerization, while the activity of free radical B is relatively low, which is mainly involved in the formation of dormant species D by primary free radicals, thus repeatedly forming a dynamic balance between growth chain free radicals and dormant species, and achieving living polymerization. The side reaction that may occur during the polymerization process is that the free radical B may undergo an isomerization reaction to form a radical on the benzene ring. This radical terminates the growth of the chain radical to form a dead-end polymer E, which can no longer be reversibly decomposed. Initiating polymerization is partly responsible for the wide distribution of relative molecular masses:

DCDTS can also realize the polymerization of MMA and St. It has active polymerization characteristics in bulk polymerization, the initiating activity of DCDTS is larger than DCDPST, and the polymerization mechanism of the two is consistent. Compared with the tetraphenyl-substituted ethane derivative-based thermal initiation transfer terminator, DCDPST and DCDTS have higher activity, which not only achieve the active polymerization of MMA at lower temperatures, but also the first thermal initiation of transfer delay agent in small molecules. The field achieves the living polymerization of St. The reason for its higher activity is mainly due to the change of the structure of the initiator, which is reflected in two aspects: first, the structure of the primary free gene polyphenyl ring is destroyed by introducing electron-withdrawing cyano and ester functional groups. The stability resulting from the presence of the conjugation increases the initiating activity and easily initiates polymerization of the vinyl monomer; at the same time, due to its electron-withdrawing action, the six (or five) substituted CC of the end of the formed polymer is lowered. The electron cloud density on the bond reduces the strength of the chemical bond, making it easy to break when heated, and not only achieves the living polymerization of MMA but also easily realizes the living polymerization of St. Second, due to the small number of phenyl substituents on the DCDPST and DCDTS structures, there is only one phenyl group per primary radical, which reduces the probability of generating dead-end polymer E during polymerization, thus controlling the polymerization process compared to tetraphenylene. The ethane derivative is better.

Another reason why DCDPST and DCDTS can initiate St-active polymerization is reflected in the initiation stage. Due to the electron-withdrawing action of the functional groups of the cyano group and the ester group, the primary radical generated is weakly positively charged, while the monomer St is double. The bond is weakly negatively charged, so the resulting primary radical is more likely to cause St to carry out living polymerization.

3.3.1.3 Novel photoinitiated transfer terminator

Several new photoinitiated transfer terminators containing DC groups have been reported in the literature: 2-N,N-diethyldithiocarbamoyloxyacetyl-p-toluidine (TDCA), 2-N,N-diethyldithiocarbamoyloxyacetate and butyl ester [(EDCA) and (BDCA)] and 2-N,N-diethyldithiocarbamoyloxyacetate (BzDCA), and their structures are shown in Fig. 3.5.

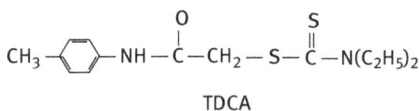

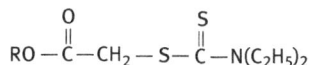

$$CH_3-\langle\bigcirc\rangle-NH-\overset{O}{\overset{|}{C}}-CH_2-S-\overset{S}{\overset{||}{C}}-N(C_2H_5)_2$$

TDCA

$$RO-\overset{O}{\overset{||}{C}}-CH_2-S-\overset{S}{\overset{||}{C}}-N(C_2H_5)_2$$

R = C₂H₅   EDCA

R = C₄H₉   BDCA

R = CH₂C₆H₅   BzDCA

Fig. 3.5: Structure of a novel photoinitiated transfer terminator. Reprinted with permission from [30]. Copyright Chinese Academy of Sciences, 2002.

It was found that when these novel photoinitiated transfer terminators were used for styrene polymerization, the relative molecular mass of the obtained polymer increased with reaction time and conversion rate (Tab. 3.1), and the obtained PS contained DC end groups, which could be further expanded: chain reaction and block copolymerization. It is indicated that the polymerization has the characteristics of living radical polymerization, but the relative molecular mass distribution is wide.

These radicals generated by photolysis of the transfer terminator were detected by 2-methyl-2-nitrosopropane (MNP) capture and electron spin resonance (ESR) spectroscopy to detect free radicals and DC sulfur center radicals at the methyl carbon center attached to the carbonyl group. It is stated that during the polymerization, these photoinitiated transfer terminators are carried out according to the mechanism shown by the following reaction formula, which is consistent with the polymerization mechanism of other photoinitiated transfer terminators containing DC groups reported by Ostu.

3.3.1.4 Polymerizable photoinitiated transfer terminator

The polymerizable photoinitiated transfer terminator is a class of double bond-containing compounds having a DC group and capable of undergoing radical polymerization. Little is known about such systems. Otsu has reported that N,N-diethyldithiocarbamic acid p-vinyl oxime ester (VBDC) can be copolymerized with a second monomer, followed by graft copolymerization under light or light-induced others. The monomers are polymerized and then copolymerized with a vinyl monomer. Recently, two new types of photoinitiated transfer terminators containing methacrylate-type and allyl ester-type polymerizable double bonds have been designed and synthesized. They are 2-*N*,*N*-diethyldithioamino formyloxyacetic acid-β-methacryloyloxyethyl ester (MAEDCA) and 2-*N*,*N*-diethyldithiocarbamoyloxyacetic acid allyl ester (ADCA). This initiates the polymerization of MMA and St, and the product is the corresponding macromonomer. The obtained macromonomer is copolymerized with the second monomer to synthesize a comb-like graft copolymer. It is also possible to carry out copolymerization of the polymerizable photoinitiated transfer terminator with MMA, and then photoinitiated graft polymerization using a DC group on the side chain to obtain a comb-like graft copolymer. Taking MAEDCA as an initiator, this process is shown in the following reaction formula:

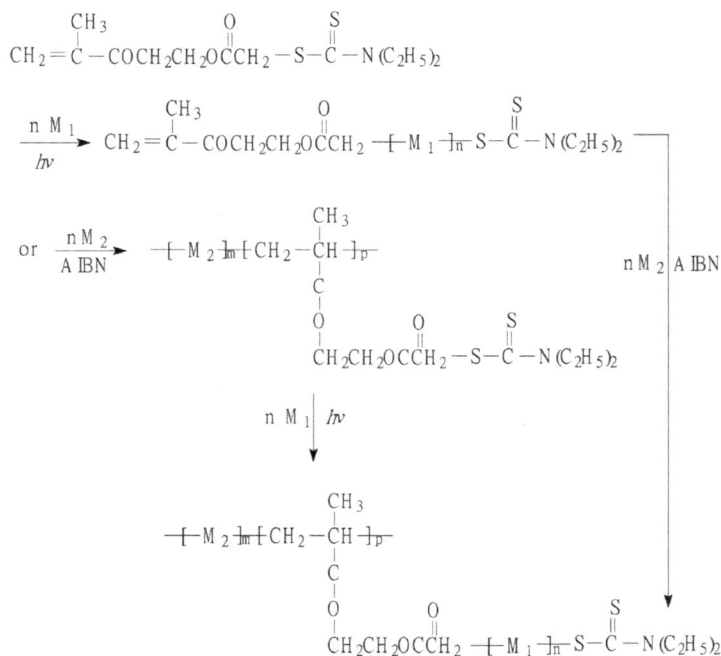

$$
\begin{array}{c}
\underset{CH_3}{\underset{|}{}} \quad \underset{O}{\overset{O}{\parallel}} \quad \underset{S}{\overset{S}{\parallel}} \\
CH_2{=}C{-}COCH_2CH_2OCCH_2{-}S{-}C{-}N(C_2H_5)_2
\end{array}
$$

$$
\xrightarrow[h\nu]{n\,M_1} \quad
CH_2{=}\underset{|}{\overset{CH_3}{C}}{-}COCH_2CH_2O\overset{O}{\overset{\parallel}{C}}CH_2 {-\!\!\!\!+}M_1{+\!\!\!\!-}_{\!n}S{-}\overset{S}{\overset{\parallel}{C}}{-}N(C_2H_5)_2
$$

or $\xrightarrow[\text{A IBN}]{n\,M_2}$
$$
{-\!\!\!+}M_2{+\!\!\!-}_{\!m}{+}CH_2{-}\underset{\underset{\displaystyle CH_2CH_2OCCH_2{-}S{-}\overset{S}{\overset{\parallel}{C}}{-}N(C_2H_5)_2}{\overset{\displaystyle C}{\overset{\displaystyle |}{\overset{\displaystyle O}{\overset{\displaystyle |}{}}}}}}{\overset{CH_3}{\underset{\displaystyle |}{C}}}H{+\!\!\!-}_{\!p}
$$

$n\,M_2$ | A IBN

$n\,M_1$ | $h\nu$

$$
{-\!\!\!+}M_2{+\!\!\!-}_{\!m}{+}CH_2{-}CH{+\!\!\!-}_{\!p}
$$

$$
CH_2CH_2OCCH_2 {-\!\!\!+}M_1{+\!\!\!-}_{\!n}S{-}\overset{S}{\overset{\parallel}{C}}{-}N(C_2H_5)_2
$$

It is found that MAEDCA initiates the polymerization of MMA and St into a living radical polymerization process. Although the macromonomers produced at high conversion rate will participate in the copolymerization, the ratio of monomer to initiator can be adjusted and the polymerization can be controlled. Conversion rate, a series of macromonomers with well-defined structures, and controlled molecular weights were prepared. When a macromonomer is copolymerized with a second monomer to prepare a graft copolymer, the relative molecular mass and concentration of the macromonomer used have a certain influence on the polymerization process. The

Tab. 3.1: St bulk polymerization using a novel photoinitiated transfer terminator.

Photoinitiated transfer terminator	t (h)	Mn ($\times 10^{-3}$)	M_w/M_n	Number of end-group EtNCSS
TDCA	3	5.57	2.77	1.04
	6	5.81	2.78	1.05
	9	6.29	2.78	1.09
	12	7.66	2.72	0.95
	15	8.21	2.79	0.96
BDCA	3	6.97	2.64	1.00
	6	7.02	2.68	1.02
	9	7.92	2.55	1.00
	12	8.96	2.71	1.00
	15	9.67	2.66	0.98
EDCA	3	6.62	2.75	0.97
	6	7.24	2.68	0.99
	9	8.80	2.51	1.01
	12	9.42	2.68	1.00
	15	9.56	2.69	0.99

Note: ① polymerization condition: 30 °C, [St] = 8.74 mol/L, [TDCA] = 0.1 mol/L; ② polymerization condition: 30 °C, [St] = 8.74 mol/L, [BDCA] = [EDTD] = 0.05 mol/L. Reprinted with permission from [30]. Copyright Chinese Academy of Sciences, 2002.

ADCA-initiated MMA and St polymerization are both living radical polymerization processes, and the products are also well-defined macromonomers.

MAEDCA can be copolymerized with MMA in any ratio, and the resulting random copolymer side chain contains a DC group. It has been found that the use of this random copolymer as a macroinitiator to initiate St polymerization is also a living polymerization process. Therefore, the length of the graft chain can be controlled by controlling the conversion rate of St. At the same time, the position of the grafting point is also fixed. This is a convenient method for preparing a well-defined graft copolymer.

3.3.1.5 New multifunctional trigger transfer terminator

The thermally initiated transfer terminator and the photoinitiated transfer terminator can, respectively, initiate different radicals for living radical polymerization and have respective advantages. Design a new compound by combining a hexa-substituted ethane-type CC bond and a DC group into one molecule: 2,3-dicyano-2,3-di(*p*-diethyldithiocarbamoyl) diethyloxyphenyl phenyl succinate (DDDCS), which combines the advantages of a photoinitiated transfer terminator and a thermally initiated transfer terminator, is a multifunctional initiation transfer terminator. The synthetic route of DDDCS is as follows:

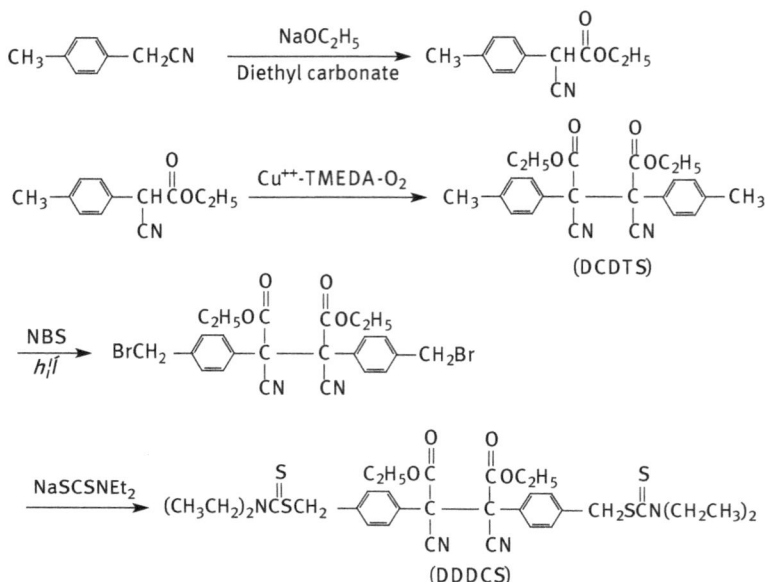

$$CH_3-\!\!\langle\!\bigcirc\!\rangle\!-CH_2CN \xrightarrow[\text{Diethyl carbonate}]{NaOC_2H_5} CH_3-\!\!\langle\!\bigcirc\!\rangle\!-\underset{\underset{CN}{|}}{CH}\overset{\overset{O}{\|}}{C}OC_2H_5$$

$$CH_3-\!\!\langle\!\bigcirc\!\rangle\!-\underset{\underset{CN}{|}}{CH}\overset{\overset{O}{\|}}{C}OC_2H_5 \xrightarrow{Cu^{++}\text{-TMEDA-}O_2} CH_3-\!\!\langle\!\bigcirc\!\rangle\!-\underset{\underset{CN}{|}}{\overset{\overset{O}{\|}}{\underset{}{C_2H_5O}C}}\!\!-\!\!\underset{\underset{CN}{|}}{\overset{\overset{O}{\|}}{C}OC_2H_5}\!\!-\!\!\langle\!\bigcirc\!\rangle\!-CH_3$$

(DCDTS)

$$\xrightarrow[h,l]{NBS} BrCH_2-\!\!\langle\!\bigcirc\!\rangle\!-\underset{\underset{CN}{|}}{\overset{\overset{O}{\|}}{\underset{}{C_2H_5O}C}}\!\!-\!\!\underset{\underset{CN}{|}}{\overset{\overset{O}{\|}}{C}OC_2H_5}\!\!-\!\!\langle\!\bigcirc\!\rangle\!-CH_2Br$$

$$\xrightarrow{NaSCSNEt_2} (CH_3CH_2)_2N\overset{\overset{S}{\|}}{C}SCH_2-\!\!\langle\!\bigcirc\!\rangle\!-\underset{\underset{CN}{|}}{\overset{\overset{O}{\|}}{\underset{}{C_2H_5O}C}}\!\!-\!\!\underset{\underset{CN}{|}}{\overset{\overset{O}{\|}}{C}OC_2H_5}\!\!-\!\!\langle\!\bigcirc\!\rangle\!-CH_2S\overset{\overset{S}{\|}}{C}N(CH_2CH_3)_2$$

(DDDCS)

TMEDA: N,N,N',N'-Tetramethylethylenediamine

Among them are TMEDA (tetramethylethylenediamine), NBS (*N*-bromosuccinimide), NaSCSNEt₂ (sodium *N,N*-diethylaminodithioformate).

DDDCS can initiate monomer polymerization under heating and ultraviolet (UV) light, respectively. Figures 3.6 and 3.7 show the results of St polymerization initiated by DDDCS heating without UV light and with UV light at room temperature. In both cases, the number average molecular weight of the obtained polymer linearly increases as the monomer conversion increases, and is characterized by living radical polymerization. A similar conclusion was obtained for the polymerization that used MMA.

The end-group analysis of the obtained polymer by ESR technique showed that the DDDCS initiated the polymerization of the monomer through the reversible cleavage of the intermediate hexa-substituted ethane-type C–C bond, and the C–C bond at both ends remained unchanged. Thermally initiated transfer terminator application while irradiated by UV light at room temperature initiates polymerization of the monomer by reversible cleavage of the C–S bond at both ends, and the intermediate hexa-substituted ethane-type C–C bond remains unchanged, thus being a photoinduced transfer terminator.

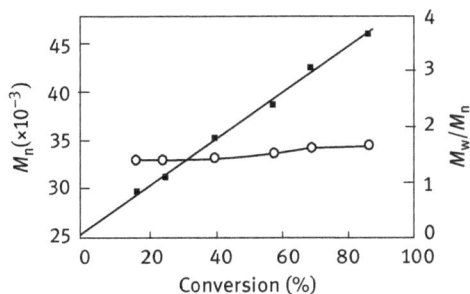

Fig. 3.6: Relationship between relative molecular mass and conversion of St thermal polymerization using DDDCS as the initiation of transfer terminator. Reprinted with permission from [30]. Copyright Chinese Academy of Sciences, 2002.

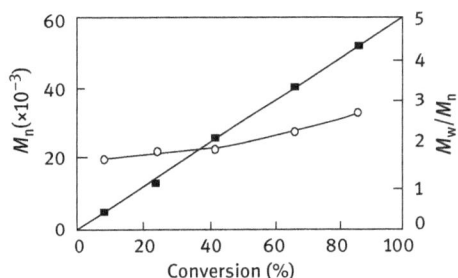

Fig. 3.7: Relationship between relative molecular mass and conversion of St photopolymerization using DDDCS as the initiation of transfer terminator. Reprinted with permission from [30]. Copyright Chinese Academy of Sciences, 2002.

Since DDDCS can initiate the living polymerization of vinyl monomers in both cases, it is possible to carry out the polymerization of MMA, St, isoprene, and vinyl acetate in the order of thermal decomposition (or reverse) after photodecomposition. A series of ABA-type triblock copolymers with controlled composition and chain length, in particular, the preparation of PVAc-b-PSt-b-PVAc triblock copolymers has a unique advantage because PVAc cannot be subjected to living radical polymerization by ATRP.

Figure 3.7 shows the results of VAC polymerization initiated by the thermally induced St polymerization product (PS-CCDCM) of DDDCS as a macromolecular photoinitiated transfer terminator. The gel permeation chromatography (GPC) curves of the obtained polymers are symmetric single peaks under different monomer conversion rates, and the GPC peaks shift to the direction of high relative molecular mass as the polymerization time increases, indicating that different molecular masses are generated, as shown in Fig. 3.8. PVAc-b-PS-b-PVAc block copolymer, the conversion of

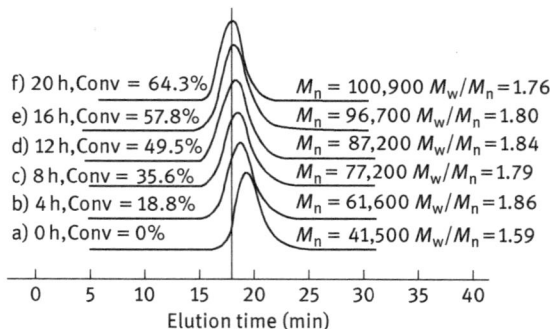

Polymerization condition: [Vac] = 10.8mol/L, [PSTt-CCDCM] = 1.6× 10^{-3} mol/L.

Fig. 3.8: GPC curve of PS-CCDCM and PVAc-b-PS-b-PVAc block copolymer. Reprinted with permission from [30]. Copyright Chinese Academy of Sciences, 2002.

Vac, and the relative molecular mass of the resulting block copolymer increased with increasing reaction time, and the relative molecular mass distribution of the block copolymer was less than 1.90. The resulting block copolymer was further character-ized by 1 H NMR and differential scanning calorimetry (DSC) analysis. Figure 3.9 shows the 1 H NMR spectrum of the composition change of the copolymer during polymerization, which is monitored by the proton peak on the benzene ring in the polystyrene segment, and the signal at the chemical shift of 4.9 ppm is polyacetic acid. The resonance absorption of the acetoxy group in the vinyl ester segment shows that the intensity of this signal increases as the reaction progresses, indi-cating that the content of polyvinyl acetate in the copolymer is continuously

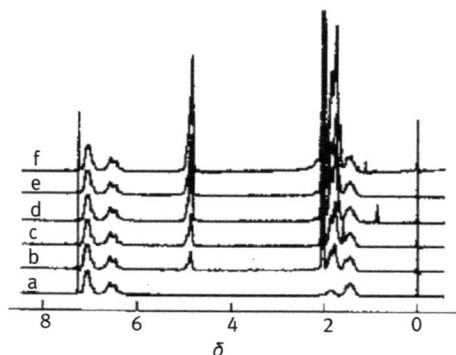

Concentration of VAc(%): a, 0; b, 34.7; c, 43.2; d, 56.3; e, 60.1; f, 72.3.

Fig. 3.9: 1 H NMR spectrum of PS-CCDCM and PVAc-b-PS-b-PVAc block copolymer. Reprinted with permission from [30]. Copyright Chinese Academy of Sciences, 2002.

increasing. Typical DSC curves for block copolymers obtained at different VAc conversions are shown in Fig. 3.10, where the transition at 42.5 °C is the glass transition temperature of the PVAc segment and the conversion to the PS segment at 105 °C (the glass transition temperature). At lower VAc conversions, the PVAc segment content of the block copolymer is lower than that of the PS segment, and its glass transition temperature is not as pronounced as that of the PS segment (Fig. 3.10 curve a). As the content of PVAc segment in the copolymer increases gradually (the content of PS segment decreases gradually), the glass transition temperature of the PVAc segment becomes more and more obvious, and the glass transition temperature of the PS segment becomes blurred until finally disappeared (Fig. 3.9 curve d). DSC data indicate that as the VAc conversion increases, the amount of PVAc component in the copolymer increases.

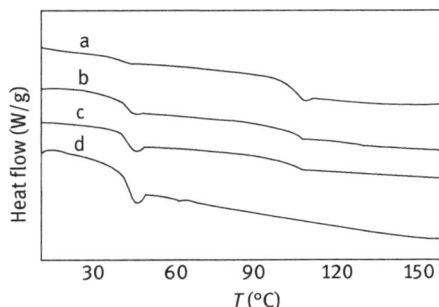

Concentration of VAc(%): a-18.8; b-49.5; c-57.8; d-64.3

Fig. 3.10: DSC curve of PVAc-b-PS-b-PVAc block copolymer. Reprinted with permission from [30]. Copyright Chinese Academy of Sciences, 2002.

The above results of GPC, 1 H NMR and DSC of block copolymers show that PS-CCDCM photoinitiated block copolymerization of vinyl acetate is a living radical polymerization process. The extent of the PVAc segment in the copolymer can be controlled by the conversion of VAc; and the length of the PSt segment is determined by the relative molecular mass of the macroinitiator used.

The PtBMA-b-PS-b-PtBMA block copolymer can also be prepared by initiating the polymerization of t-butyl methacrylate (t-BMA) using PS-CCDCM. Studies have shown that this polymerization is also a living polymerization. By controlling the conversion of t-BMA, a series of copolymers containing different PtBMA chain lengths can be prepared. After hydrolysis of the ester group of the block copolymer, an amphiphilic block copolymer having a polymethacrylic acid segment at both ends can be obtained, as shown in the following formula. This is a simple and effective method for preparing amphiphilic block copolymers containing polymethacrylic acid segments by free radical polymerization.

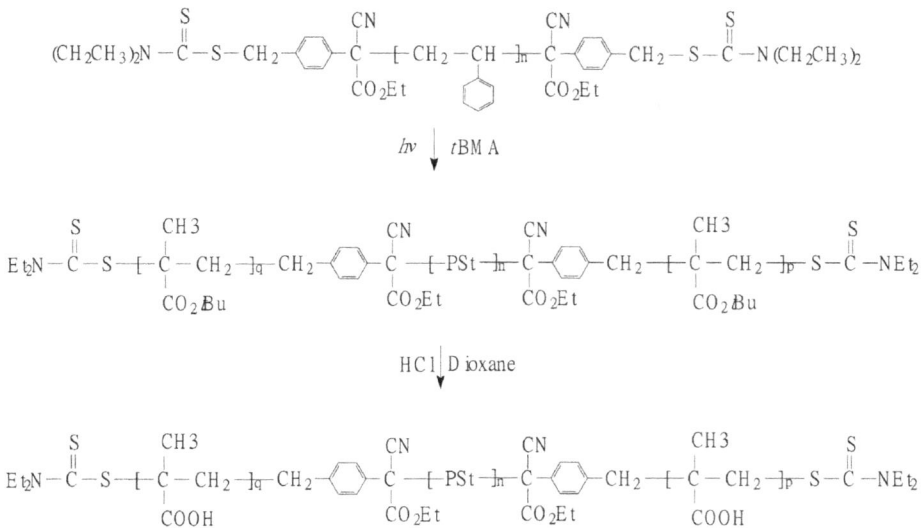

$$(CH_2CH_3)_2N-\overset{\overset{S}{\|}}{C}-S-CH_2-\!\!\langle\bigcirc\rangle\!\!-\overset{\overset{CN}{|}}{\underset{CO_2Et}{C}}-\!\!\{CH_2-CH\}_n\!\!-\overset{\overset{CN}{|}}{\underset{CO_2Et}{C}}-\!\!\langle\bigcirc\rangle\!\!-CH_2-S-\overset{\overset{S}{\|}}{C}-N(CH_2CH_3)_2$$

$$hv \downarrow tBMA$$

$$Et_2N-\overset{\overset{S}{\|}}{C}-S-\!\{\overset{\overset{CH_3}{|}}{\underset{CO_2Bu}{C}}-CH_2\}_q\!-CH_2-\!\!\langle\bigcirc\rangle\!\!-\overset{\overset{CN}{|}}{\underset{CO_2Et}{C}}-\!\{PSt\}_n\!-\overset{\overset{CN}{|}}{\underset{CO_2Et}{C}}-\!\!\langle\bigcirc\rangle\!\!-CH_2-\!\{\overset{\overset{CH_3}{|}}{\underset{CO_2Bu}{C}}-CH_2\}_p\!-S-\overset{\overset{S}{\|}}{C}-NEt_2$$

$$HCl \downarrow Dioxane$$

$$Et_2N-\overset{\overset{S}{\|}}{C}-S-\!\{\overset{\overset{CH_3}{|}}{\underset{COOH}{C}}-CH_2\}_q\!-CH_2-\!\!\langle\bigcirc\rangle\!\!-\overset{\overset{CN}{|}}{\underset{CO_2Et}{C}}-\!\{PSt\}_n\!-\overset{\overset{CN}{|}}{\underset{CO_2Et}{C}}-\!\!\langle\bigcirc\rangle\!\!-CH_2-\!\{\overset{\overset{CH_3}{|}}{\underset{COOH}{C}}-CH_2\}_p\!-S-\overset{\overset{S}{\|}}{C}-NEt_2$$

3.3.1.6 Macromolecular photoinitiated transfer terminator
Macromolecular photoinitiated transfer terminators are also an important research direction in the field of initiating transfer terminators, mainly for the preparation of block copolymers. Otsu has done a lot of work in this area and has a monograph. The polymer PS-DC obtained by photopolymerization of a small-molecule photoinitiated transfer terminator such as TDCA, EDCA and BDCA to carry out monomer St polymerization has a DC end group, and can also be used as a macroinitiator to initiate copolymerization of MMA, BA, and others under illumination. An AB-type block copolymer was obtained.

The above macromolecular photoinitiated transfer terminators are prepared by initiating polymerization of a monomer using a small-molecule photoinitiated transfer terminator, and each component in the block copolymer prepared is a vinyl polymer. In recent years, it has been reported that polyethylene polyethylene glycol (PEG) (I) and polytetrahydrofuran (PTHF)-I having a DC group at both ends are synthesized by a two-step reaction from PEG and polyoxytetramethylene (PTHF). Under UV light, the end group of the macroinitiator is broken to form an active center with carbon radicals at both ends, thereby initiating polymerization of monomers such as St, MMA, VAc, and BA to obtain an ABA-type block copolymer. The synthesis of macromolecular photoinitiated transfer terminators and the preparation of ABA block copolymers are as follows:

$$H\!-\!\!\left[O\!-\!\!\left(CH_2\right)_{\!n}\right]_{\!m}\!\!-OH \quad \xrightarrow{\ \ ClOCCH_2CL\ \ } \quad ClCH_2\overset{O}{\overset{\|}{C}}\!-\!\!\left[O\!-\!\!\left(CH_2\right)_{\!n}\right]_{\!m}\!\!-O\overset{O}{\overset{\|}{C}}CH_2Cl$$

$$\xrightarrow[\ \ C_2H_5OH\ \ ]{\ NaSSCN(CH_2CH_3)_2\ } \quad (C_2H_5)_2N\overset{S}{\overset{\|}{C}}S\,CH_2\overset{O}{\overset{\|}{C}}\!-\!\!\left[O\!-\!\!\left(CH_2\right)_{\!n}\right]_{\!m}\!\!-O\overset{O}{\overset{\|}{C}}CH_2S\overset{S}{\overset{\|}{C}}N(C_2H_5)_2$$

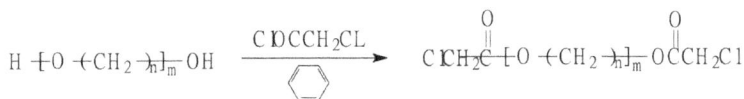

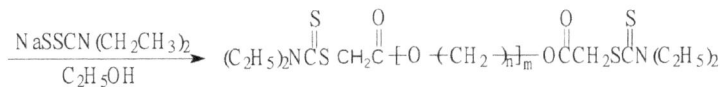

$n = 2 : PEG\text{-}I; \quad n = 4 : PTHF\text{-}I$

$$PEG\text{-}I\ or\ PTHF\text{-}I \quad \xrightarrow{\ hv\ } \quad \cdot CH_2\overset{O}{\overset{\|}{C}}\!-\!\!\left[O\!-\!\!\left(CH_2\right)_{\!n}\right]_{\!m}\!\!-O\overset{O}{\overset{\|}{C}}CH_2\cdot \ +\ 2\ \cdot S\overset{S}{\overset{\|}{C}}N(C_2H_5)_2$$

$$\Big\downarrow M$$

$$M_n\!-\!CH_2\overset{O}{\overset{\|}{C}}\!-\!\!\left[O\!-\!\!\left(CH_2\right)_{\!n}\right]_{\!m}\!\!-O\overset{O}{\overset{\|}{C}}CH_2\!-\!M_n$$

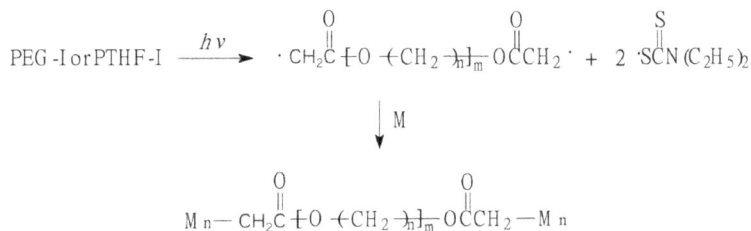

3.3.2 TEMPO initiation system

TEMPO is a free radical scavenger commonly used in organic chemistry. In the late 1970s, Australia's Rizarrdo and others first introduced TEMPO into a free radical po-lymerization system to capture growth chain free radicals to prepare acrylate oligomers. In 1993, in order to develop high-quality ink additives, researchers at Xerox Canada developed a high-temperature polymerization of styrene based on the research work of Rizarrdo et al. They found that the bulk polymerization of sty-rene initiated by TEMPO/BPO (1/1 molar ratio) as the initiation system at 120° C was a living polymerization. During the polymerization process, TEMPO is a stable free radical, which only undergoes a coupling reaction with a growing free radical to form a covalent bond, and this covalent bond can be decomposed to generate free radicals at high temperatures. Therefore, after TEMPO captures the growth of free radicals, it is not the true death of the active chain, but only temporarily inactivated and becomes a dormant species, as shown in the following formula:

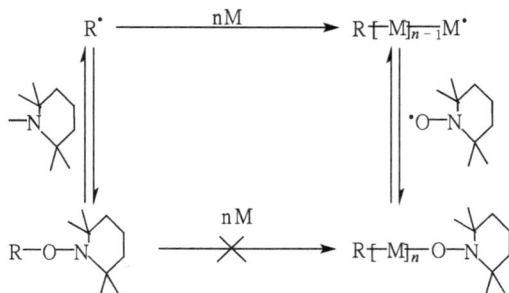

The TEMPO-controlled free radical living polymerization not only has the typical characteristics of controlled polymerization, but also avoids the harsh reaction conditions required for the anion active and the cationic living polymerization, thus causing the common interest of the polymer academia and industry. A large number of research results have emerged. For example, 4-hydroxy-2,2,6,6-tetramethyl oxynitride (HTEMPO) is esterified with methacryloyl chloride to obtain a nitroxide radical MTEMPO with a living double chain. The reaction formula is as follows:

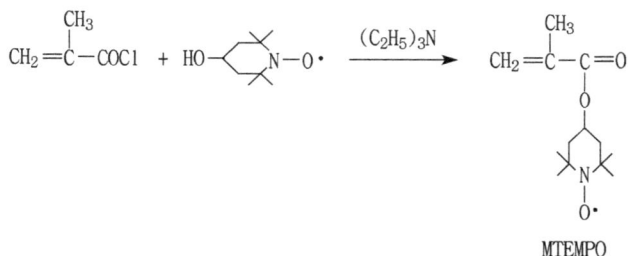

MTEMPO

MTEMPO has a dual function of capturing free radicals and participating in polymerization. After MTEMPO is polymerized onto the polymer chain, the free radical trapping ability of these TEMPOs is greatly reduced due to the shielding effect of the polymer chain conformation. The number of sleep chains is reduced, and the number of growth chains is increased, thereby accelerating the polymerization rate. Both theory and experiment show that the polymerization rate can be accelerated by about 2.5 times, while the relative molecular mass distribution remains basically unchanged. Using the characteristics of styrene monomer initiated by TEMPO radical, the above MTEMPO and methyl methacrylate were subjected to ATRP in the presence of bromoethylbenzene/copper chloride/bipyridine initiator system to obtain PMMA containing MTEMPO chain and then the macromolecule and the benzoyl peroxide constituting system, the graft copolymerization of styrene, finally obtain the PMMA-g-PS graft copolymer:

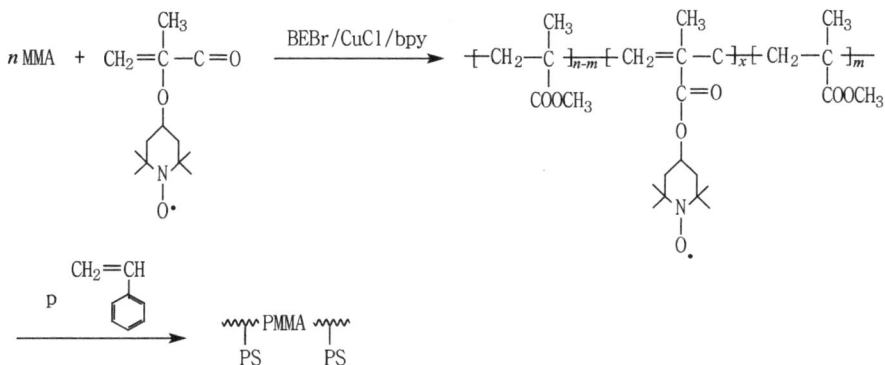

The living polymerization of styrene is carried out by using HTEMPO and (azodiiso-butyronitrile AIBN) as an initiation system, and polystyrene capped with HTEMPO can be prepared, which then reacted with C_{60}. A monosubstituted C_{60}-containing polystyrene can be obtained, and the reaction formula is as follows. The polymer can also be modified onto carbon nanotubes and graphene in a similar manner.

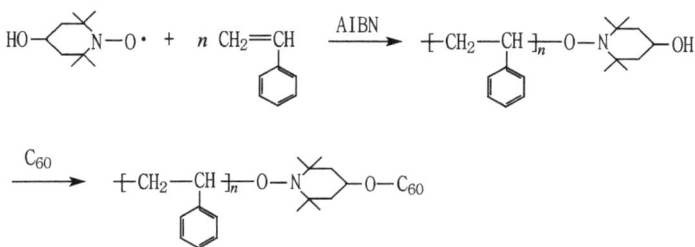

A disadvantage of the living polymerization initiated by the TEMPO initiating system is that the reaction rate is slower and the time required to achieve high conversion is longer. For example, styrene polymerization with TEMPO/BPO as the initiation system requires about 70 h at 125 °C, and the conversion rate can reach 90%. This long reaction time limits its industrial application. It has been found that the addition of a small amount of acidic material to the TEMPO initiation system accelerates the rate of polymerization of the system. In 1994, Georges et al. reported that in the presence of low concentrations of camphorsulfonic acid (CSA), the rate of polymerization of styrene was significantly improved, the conversion rate of 6 h was 90%, and the relative molecular mass distribution of the product was narrow ($M_w/M_n < 1.25$). And both the conversion rate and the relative molecular mass distribution increase with increasing CSA concentration. In 1995, Odell et al. found that the addition of a small amount of 2-fluoro-1-methylpyridinium salt of p-toluenesulfonic acid was more effective in increasing the polymerization rate of styrene than the addition of the same concentration of CSA. In 1997, Hawker et al. reported that a series of acylating agents, such as acetic anhydride and trifluoroacetic anhydride, also significantly improved the polymerization rate of the TEMPO initiating system. In recent years, it has been found that acetylacetone, acetoacetate, diethyl malonate, and others can significantly increase the rate of styrene polymerization initiated by the TEMPO system, and the relative molecular mass and relative molecular mass distribution are controllable. The mechanism of action of the above-mentioned speed-increasing agent is still not fully understood and further research is needed.

It is generally believed that the TEMPO initiating system is only suitable for the living polymerization of styrene and its derivatives, but not for polar monomers such as methyl methacrylate. However, recent studies have shown that methyl methacrylate with TEMPO/BPO as the initiation system in the presence of the speed-increasing agent trifluoroacetic anhydride, when the monomer conversion rate is less than 30%, a polymer with a relative narrow mass distribute ($M_w/M_n < 1.35$) and the relative molecular weight increasing linearly with the conversion rate can be obtained. As the conversion rate increases further, the relative molecular mass becomes uncontrollable. By 1 H NMR analysis of the model polymer, it was found that at higher conversion, the partial detachment of nitroxide radicals occurred, which caused the termination of the polymer chain double group, resulting in the uncontrollable molecular mass and distribution of the polymerization system. The reason why the nitroxide radicals are easily detached from the molecular chain is believed to be due to the weak electron-withdrawing effect of the ester group weakening the C–O bond, thereby weakening the connection between the nitroxide radical and the macromolecular chain.

Since the TEMPO initiating system is currently only suitable for the living polymerization of styrene and its derivatives, the range of molecular design of polymer materials through this system is very limited. In addition, the high price of TEMPO also determines that the industrial value of the system is not large. However, the free radical polymerization initiated by the TEMPO initiation system is the first free radical living polymerization. Its discovery has led polymer synthesis chemists to see the dawn of free radical living polymerization, and more firmly believe in the pursuit of free radical living polymerization.

3.3.3 Reversible addition-fragment chain transfer radical polymerization

3.3.3.1 Proposal of reversible addition-fracture chain transfer radical polymerization

The principle of the TEMPO initiation system leading to free radical living polymerization is the reversible chain termination of the growing chain radicals and the reversible addition-fragmentation chain transfer (RAFT) radical polymerization process to achieve the reversible chain transfer of the growing chain radicals.

It is well known that in classical free radical polymerization, irreversible chain transfer side reactions are one of the main factors leading to uncontrollable polymerization reactions. Reversible chain transfer can form dormant macromolecular chains and new priming active species. The establishment of this concept points the way for the study of active controlled radical polymerization. The key to putting this principle into practice is whether the ideal chain transfer agent A-X can be found as follows:

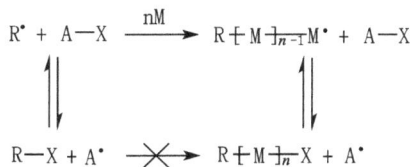

$$R^{\bullet} + A-X \xrightarrow{\ nM\ } R + M +_{\overline{n-1}} M^{\bullet} + A-X$$

$$R-X + A^{\bullet} \xrightarrow{\hspace{0.3em}\times\hspace{0.3em}} R + M +_{\overline{n}} X + A^{\bullet}$$

In 1995, Matyjaszewski et al. reported the free-radical polymerization of styrene and butyl acrylate (BA) with 1-iodoethylbenzene as the transfer agent and azobisisobutyronitrile as initiator. The system was found to have the following activity characteristics: ① the conversion rate is linear with time; ② the relative molecular mass of the polymer increases monotonically with the conversion rate; ③ after the first monomer is substantially consumed, the second monomer is added; and ④the polymerization can be continued to finally obtain a block copolymer. However, the experimental results show that the relative molecular mass distribution of the polymer is still relatively wide, which may be because the chain transfer constant of 1-iodoethylbenzene is small, resulting in a slow conversion rate between the active species and the dormant species.

In 1998, Rizzardo made a report on "Tailored Polymers by Free radical processes" at the 37th International Polymer Symposium, and proposed the concept of RAFT radical polymerization. The proposal of this polymerization mechanism has aroused strong reaction from experts and scholars. As described earlier, in the classical radical polymerization, the irreversible chain transfer side reaction is one of the main factors leading to uncontrollable polymerization. However, when the chain transfer constant and concentration of the chain transfer agent are sufficiently large, the chain transfer reaction is irreversible to reversible, and the polymerization behavior is also qualitatively transformed from uncontrollable to controllable. The discovery of RAFT reveals the dialectical development of nature from quantitative to qualitative. RAFT radical polymerization is the key to successful controlled radical polymerization, which is the discovery of a chain transfer agent dithioester (ZCS_2R) with a high chain transfer constant and a specific structure. Its chemical structure is shown in Fig. 3.11.

Table 3.2 lists some examples of polymer molecular design using the principle of RAFT radical polymerization. It can be seen from the table that a monofunctional, difunctional, and polyfunctional dithioester compound can be used as a chain transfer agent to successfully prepare a polymer having a complex molecular structure such as a block or a star.

Monofunctionality

$$S=C\overset{\displaystyle Z}{\underset{\displaystyle S}{\big|}}R$$

Z = ph, CH$_3$

R = C(CH$_3$)$_2$ph, CH(CH$_3$)ph, CH$_2$ph, CH$_2$phCH=CH$_2$
C(CH$_3$)$_2$CN, C(CH$_3$)(CN)CH$_2$CH$_2$CH$_2$OH,
C(CH$_3$)(CN)CH$_2$CH$_2$COOH, C(CH$_3$)(CN)CH$_2$CH$_2$COONa

Bifunctionality

Z—CS—C(CH$_3$)$_2$—⟨ph⟩—C(CH$_3$)$_2$—CS—Z

Multifunctionality

ZCS$_2$CH$_2$, ZCS$_2$CH$_2$ / CH$_2$CS$_2$Z, CH$_2$CS$_2$Z

ZCS$_2$CH$_2$, ZCS$_2$CH$_2$ / CH$_2$CS$_2$Z, CH$_2$CS$_2$Z, CH$_2$CS$_2$Z, CH$_2$CS$_2$Z

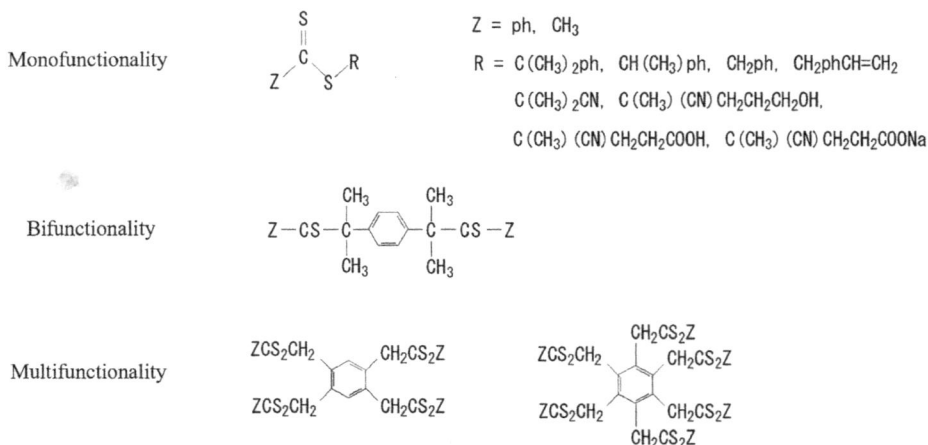

Fig. 3.11: Chemical structure of partial chain transfer agent dithioester.

3.3.3.2 Mechanism of reversible addition-fragment chain transfer radical polymerization

The mechanism of RAFT radical polymerization can be expressed by the following reaction formula:

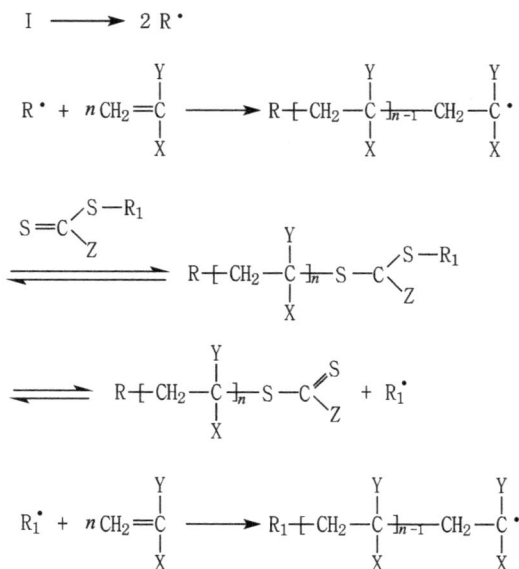

$$I \longrightarrow 2R^{\bullet}$$

$$R^{\bullet} + n\,CH_2=\overset{Y}{\underset{X}{C}} \longrightarrow R\text{+}CH_2-\overset{Y}{\underset{X}{C}}\text{+}_{n-1}CH_2-\overset{Y}{\underset{X}{C}}^{\bullet}$$

$$S=C\overset{S-R_1}{\underset{Z}{<}}$$

$$\rightleftharpoons R\text{+}CH_2-\overset{Y}{\underset{X}{C}}\text{+}_n S-C\overset{S-R_1}{\underset{Z}{<}}$$

$$\rightleftharpoons R\text{+}CH_2-\overset{Y}{\underset{X}{C}}\text{+}_n S-C\overset{S}{\underset{Z}{<}} + R_1^{\bullet}$$

$$R_1^{\bullet} + n\,CH_2=\overset{Y}{\underset{X}{C}} \longrightarrow R_1\text{+}CH_2-\overset{Y}{\underset{X}{C}}\text{+}_{n-1}CH_2-\overset{Y}{\underset{X}{C}}^{\bullet}$$

Tab. 3.2: Examples of structurally controllable polymers prepared by RAFT polymerization.

Product name	Conversion rate (%)	$\overline{M}_n$	$\overline{M}_w/\overline{M}_n$
PBA-b-PAA	8.3	52,400	1.19
PMMA-b-PSt	23.5	35,000	1.24
PHEMA-b-MMA-b-PHEMA	40.2	28,500	1.18
Star-shaped PST	72.1	80,000	1.67
Star-shaped P (BA-S)	71.4	82,500	2.16

The correctness of this mechanism can be confirmed by the analysis of nuclear magnetic resonance and UV-visible spectroscopy, and the presence of fragments of chain transfer agent molecules at the polymer chain ends.

It is not difficult to see from the above reaction mechanism that the two groups Z and R in the chain transfer agent play a crucial role. Wherein, Z should be a group capable of activating C = S addition to a radical, usually an aryl group or an alkyl group; and R should be an active radical leaving group, and the radical R · generated after breaking the bond should be effective. In reinitiation polymerization, the commonly used are cumene, cyanoisopropyl, and the like.

3.3.3.3 Comparison of RAFT with iniferter and TEMPO systems

The chain transfer agent used in the RAFT radical polymerization is essentially different from the initiation transfer terminator discussed earlier. Taking the initiation of the photoinitiated transfer terminator benzyl-*N,N*-diethyldithiocarbamate as an example, photodegradation of the CS bond produces a highly reactive carbon radical (which initiates polymerization of the monomer) and a low-activity sulfur radical (mainly terminated with chain radicals) and therefore different from the chain transfer process in RAFT radical polymerization:

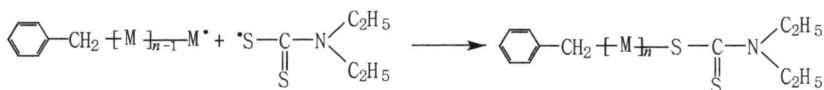

In the reaction mechanism, the RAFT radical polymerization and the free radicals of the TEMPO system are all derived from the thermal decomposition of classical initiators. The key step of RAFT is the reversible addition of free radicals to the chain transfer agent molecule C=S. The cleavage of S–R forms a new active species R· . The formation of free radicals only requires a specific structure of the chain transfer agent, independent of the monomer species. The TEMPO system involves the reversible chain termination, and the polymerization system has a large selectivity to monomers.

In the polymerization process, RAFT radical polymerization can be carried out at low temperatures, while TEMPO systems require higher temperatures. In addition, the RAFT radical polymerization process has no limitation of the polymerization method, and various polymerization processes such as bulk polymerization, solution polymerization, emulsion polymerization, and suspension polymerization at a lower temperature (50–70 °C) have been realized, and the polymerization of the TEMPO system is currently used for bulk and solution polymerization.

Similar to other free radical living polymerizations, RAFT radical polymerization also has limitations on the relative molecular mass of the polymer, because at a certain monomer concentration, to obtain a high relative molecular mass, the chain transfer agent must have reduced concentration. Obviously, this is at the expense of the controllable behavior of the aggregation process.

RAFT radical polymerization is not only suitable for common monomers such as styrene, (meth)acrylate, acrylonitrile (AN), and ethyl acetate but also for functional monomers such as acrylic acid, sodium styrene sulfonate, methacrylic acid-β-hydroxyethyl ester, and dimethylaminoethyl methacrylate. Obviously, RAFT radical polymerization inherits the main advantages of traditional free radical polymerization, and has strong molecular design ability. Therefore, it is one of the most industrialized prospects in active controlled radical polymerization. However, due to the complex preparation process of the chain transfer agent dithioester, including many steps of organic synthesis, the research on this system is not yet deep.

3.4 Atom transfer radical polymerization

3.4.1 Fundamental principle

The concept of ATRP is derived from transition metal catalyzed atom transfer radical addition (ATRA) in organic chemistry. ATRA is an effective method for forming C–C bonds in organic chemistry. The reaction process is shown in Fig. 3.12.

First, the reduced transition metal species M_t^n abstracts the halogen atom X from the organic halide R–X to form an oxidation state transition metal species M_t^{n+1} and a carbon radical R·; thereafter, the radical R· reacts with the alkene M to produce an intermediate radical R–M·. The intermediate radical is then reacted with the transition metal species in the oxidation state to obtain the target product R–M–X, and at

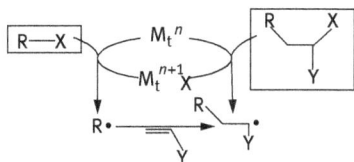

Fig. 3.12: Schematic diagram of atom transfer radical addition reaction. Reprinted with permission from [8]. Copyright American Chemical Society, 1995.

the same time, a reduced transition metal species M_t^n is produced, which in turn can react with the halide R–X to start a new round of oxidation–reduction cycle. This transition metal-catalyzed atom transfer reaction is highly efficient, and the yield of the adduct R–M–X is often greater than 90%. This fact indicates that the redox reaction of M_t^n/M_t^{n+1} can produce a low concentration of free radicals, thereby greatly inhibiting the termination reaction between free radicals. It can be assumed that if the macromolecular halide RMX has sufficient reactivity to M_t^n and the monomer is greatly excessive, then a series of ATRA reactions can control the free radical polymerization, as shown in Fig. 3.13.

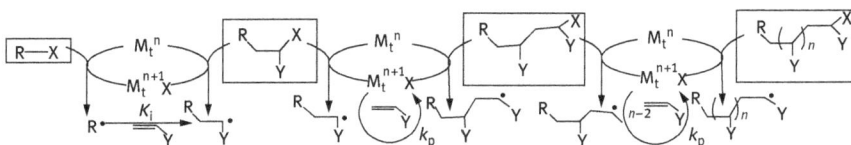

Fig. 3.13: Schematic diagram of continuous atom transfer radical addition reaction. Reprinted with permission from [8]. Copyright American Chemical Society, 1995.

In order to confirm this idea, Matyjaszwski and Dr. Wang Jinshan used α-chlorophenylethane as an initiator, a complex of cuprous chloride and 2,2′-bipyridine as a catalyst, and carried out benzene at 130 °C. Bulk polymerization of ethylene not only obtains polystyrene with narrow relative molecular mass distribution, but also the measured relative molecular mass of the polymer is very close to the theoretical calculation. When the second monomer methyl acrylate was added, block copolymerization was successfully achieved with significantly active polymerization characteristics. Accordingly, they proposed a new concept of ATRP.

Although ATRP is evolved from ATRA, the occurrence of an ATRA reaction is only a necessary and not enough condition for an ATRP reaction. Whether ATRA can be converted into ATRP depends not only on the influencing factors such as the reaction medium, reaction temperature, transition metal ions, and the nature of the ligand, but also on the molecular structure of the alkyl halide and the unsaturated compound. ATRAs research object is how the halogen atom can be smoothly added to the double bond, and whether the halogen atom in the additive can be successfully transferred is a problem solved by ATRP. Theory and practice have shown that the conjugation or induction effect in the molecular structure should be able to weaken the intensity of the α-position in C–X bond. This conclusion has not only become the principle of selecting ATRP initiators but also determines the range of monomers that ATRP is adapted to.

According to the concept proposed by Matyjaszwski and Wang Jinshan, the basic principle of typical ATRP is as follows:

$$R{-}X + M_t^n \rightleftharpoons R^{\cdot} + M_t^{n+1}X \qquad \text{Initiation}$$

$$\downarrow \; + M \qquad\qquad k_i\downarrow \; + M$$

$$R{-}M{-}X + M \rightleftharpoons R{-}M^{\cdot} + M_t^{n+1}X$$

$$M_n{-}X + M_t^n \rightleftharpoons M_n^{\cdot} + M_t^{n+1}X \qquad \text{Propagation}$$

$$\underset{k_P}{\overset{+M}{\nearrow}} \qquad\qquad \underset{k_P}{\overset{+M}{\nearrow}}$$

In the initiation phase, the transfer metal halide (salt) M_t^n in the low oxidation state extracts the halogen atom X from the organic halide R–X to form a metal halide M_t^{n+1}–X, which initiates the radical R· and is in a high oxidation state. The radical R· can initiate polymerization of the monomer to form a chain radical R–Mn·. R–Mn· can recapture the halogen atom from the highly oxidized metal complex M_t^{n+1}–X to form a passivation reaction, form R–Mn–X, and reduce the high-oxidation metal halide to a low oxidation state M_t^n. If R–Mn–X is the same as R–X (not always the same), it can react with M_t^n to form the corresponding R–Mn· and M_t^{n+1}–X. Meanwhile, if R–Mn· and M_t^{n+1}–X can be reversed when the passivation reaction occurs to form R–Mn–X and M_t^n, the radical polymerization reaction proceeds simultaneously with the reversible conversion equilibrium reaction of a radically active species with the organic macromolecular halide dormant species.

Since the reversible transfer in this polymerization reaction involves a cyclic atom transfer process in which a halogen atom is transferred from an organic halide to a metal halide and then from a metal halide to a radical, it is an atom transfer polymerization. At the same time, since the reactive species is a radical, it is called ATRP. ATRP is a catalytic process. The reversible transfer of catalysts M_t^n and M_t^{n+1}–X controls [Mn·], that is, R_t/R_p (controllability of polymerization process), while fast halogen atom conversion controls relative molecules. The mass and relative molecular mass distribution (controllability of the polymer structure) provides great convenience for artificially controlling the polymerization.

3.4.2 Atom transfer radical polymerization initiator

Since the publication of the first paper on ATRP in 1995, ATRP research has rapidly expanded to the world. The research content includes the following three aspects: new initiation and polymerization systems, polymer structure and material properties, and polymerization process and industrial product development. The research on the new trigger system is the most concerned.

Matyjaszwski et al.'s study of the initiation reaction system of ATRP showed that all alkyl halides containing a conjugated group at the α-position can initiate the ATRP reaction. Soon, Percec et al. successfully developed aryl-sulfonyl chloride initiators. The dissociation energy of the S–Cl bond in such an initiator is low, and the initiation efficiency is greater than that of the alkyl halide.

The typical ATRP initiators reported so far mainly include α-halophenyl compounds such as α-chlorophenylethane, α-bromophenylethane, benzyl chloride, and benzyl bromide; α-halocarbonyl compounds such as ethyl α-chloropropionate, ethyl α-bromopropionate, and ethyl α-bromoisobutyrate; α-halogenated cyano compounds such as α-chloroacetonitrile and α-chloropropionitrile; many halides such as carbon tetrachloride and chloroform. Further, substituted arylsulfonyl chlorides containing a weak S–Cl bond are effective initiators for styrene and (meth)acrylate monomers. Recent studies have found that halogenated alkane (such as dichloromethane, 1,2-dichloroethane) without a conjugated or inducing group in the molecular structure can also initiate methacrylic acid butyl under the catalysis of $FeCl_2$ $4H_2O/PPh_3$. The controlled polymerization of the ester broadens the range of initiator choices for ATRP.

3.4.3 Atom transfer radical polymerization catalyst and complexing agent

The catalyst of the first-generation ATRP technology initiation system was CuX (X = Cl, Br). Later, Sawamoto and Teyssie et al. used the complex of Ru and Ni as catalysts to carry out the ATRP reaction of MMA, which was successful. Later, an ATRP reaction using a ferrous halide as a catalyst was discovered. The successful research of these catalysts laid the foundation for the development of efficient and pollution-free initiation systems.

The complexing agent is an important component of the ATRP initiation system, which has the function of stabilizing the transition metal and increasing the solubility of the catalyst. The coordination agent first used by Matyjaszewski et al. is bipyridyl. The initiating system consisting of halogenated alkane and copper halide is a heterogeneous system. The dosage is large, the initiation efficiency is not high, and the relative molecular mass distribution of the product is also wide. Later, Matyjaszewski et al. used 2-pyridine aldehyde imine to replace the bipyridine in the first-generation ATRP technology, and Haddleton et al. used 2-pyridine acetal imide as the complexing agent to achieve the homogeneous reaction of ATRP. Cheng Guanglou et al. used phenanthroline for ATRP polymerization of monomers such as styrene and methyl methacrylate, which greatly improved the catalytic activity and selectivity of the catalyst copper halide.

The substituted bipyridines of the homogeneous system are more expensive and the polymerization rate is much slower than the heterogeneous system. Nowadays, cheap polyamines such as *N,N,N',N'',N''*-pentamethyldiethylenetriamine, and imines such as 2-pyridine acetal *n*-propylamine, the replacement of expensive bipyridyls by amino ethers such as bis(dimethylaminoethyl)ether has been reported to be as effective as the substituted bipyridine.

3.4.4 Atom transfer radical polymerization monomer

Compared to other living polymerizations, ATRP has the widest range of monomer options, which is perhaps the biggest attraction of ATRP. There are currently three types of monomers that can be reported to be polymerized by ATRP:
(1) Styrene and substituted styrene, such as *p*-fluorostyrene, *p*-chlorostyrene, *p*-bromostyrene, *p*-methylstyrene, *m*-methylstyrene, *p*-chloromethylstyrene, *m*-chloromethylstyrene, trifluoromethylstyrene, *m*-trifluoromethylstyrene, and *p*-tert-butylstyrene.
(2) (Meth)acrylate such as methyl (meth)acrylate, ethyl (meth)acrylate, *n*-butyl (meth)acrylate, t-butyl (meth)acrylate, isobornyl (meth)acrylate, 2-ethylhexyl (meth)acrylate, and dimethylaminoethyl (meth)acrylate.
(3) (Meth)acrylate with functional groups, such as 2-hydroxyethyl (meth) acrylate, propyl (meth) acrylate, glycidyl (meth)acrylate, and vinyl acrylate; special (methyl) acrylates such as 1,1-dihydroperfluorooctyl (meth)acrylate, pentafluorooctyl glycol (meth)acrylate, and (meth)acrylic acid-b-(*N*-ethyl-perfluorooctylsulfonyl)aminoethyl ester, and (meth)acrylic acid-2-perfluorodecenyloxyethyl ester; (meth)acrylonitrile; 4-vinylpyridine; or the like.

Up to now, the use of ATRP technology has not been possible for the polymerization of monomers such as olefin monomers, diene monomers, vinyl chloride, and vinyl acetate.

3.4.5 Reverse atom transfer radical polymerization

Although ATRP has powerful molecular design functions, it also has some fatal shortcomings. For example, the initiator of ATRP is usually an organic halide, which is highly toxic; the reduced transition metal compound in the catalyst is easily oxidized by oxygen in the air, which makes storage and experimental operations difficult; the catalytic system activity is not too high, and the amount is large. Metal salts as catalysts are not good for environmental protection. To this end, in recent years, an improved ATRP – reverse ATRP (RATRP) – technique has surfaced.

The RATRP technology uses a conventional free radical initiator (such as azobisisobutyronitrile and dibenzoyl peroxide) and a high-valence transition metal complex (such as $CuCl_2$ and $CuBr_2$) to form an initiation system. The reaction process can be as follows:

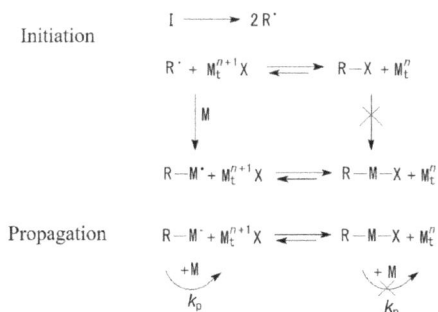

Initiation

$$I \longrightarrow 2R\cdot$$

$$R\cdot + M_t^{n+1}X \rightleftharpoons R-X + M_t^n$$

$$\downarrow M$$

$$R-M\cdot + M_t^{n+1}X \rightleftharpoons R-M-X + M_t^n$$

Propagation

$$R-M\cdot + M_t^{n+1}X \rightleftharpoons R-M-X + M_t^n$$

$$\underset{k_p}{\overset{+M}{\frown}} \qquad \underset{k_p}{\overset{+M}{\frown}}$$

Unlike the conventional ATRP, which first activates the dormant species R–X with M_t^n, the RATRP starts from the passivation reaction of the radical $I\cdot$ or $I-P\cdot$ with XM_t^{n+1}. In the initiation phase, once the free radical $I\cdot$ or $I-P\cdot$ is initiated, the halogen atom can be abstracted from the excessive metal halide XM_t^{n+1} in the oxidation state to form the reduced transition metal ion M_t^n and the dormant species I–X or I–P–X. Later, the transition metal ion M_t^n acts the same as in the conventional ATRP.

RATRP was also first reported by Dr. Matyjaszwski and Dr. Wang. In 1995, they successfully achieved RATRP of styrene using $AIBN/CuCl_2/bpy$. Because of the heterogeneous reaction, the amount of Cu(II) is high to control the polymerization well, and the reaction rate is very slow. This heterogeneous RATRP is difficult to control the polymerization of (meth)acrylate monomers. Later, Teyssie et al. developed it into the $AIBN/FeCl_3/pph_3$ system and successfully achieved the active controlled polymerization of methyl methacrylate. Zhu et al. used $BOP/CuCl_2/bpy$ for RATRP of styrene and methyl methacrylate, found that it also has active controllable polymerization characteristics, and proposed a reverse atom different from AIBN. The mechanism of transfer radical polymerization. First, the BPO molecule decomposes into two free radicals. Since the primary free radicals derived from BPO and its decomposition have strong oxidizing properties, it is impossible for Cu^{2+} to directly capture electrons from the BPO primary free radicals to form dormant species. After the addition of one or several monomer molecules, $CuCl_2$ can capture electrons on the free radicals, generate halogenation reaction, and produce dormant species and Cu^+. Since Cu^+ has reducibility, it can undergo redox reaction with BPO, which in turn generates a primary free radical, the reaction process is shown in the following formula.

$$\text{Ph-}\overset{O}{\overset{\|}{C}}\text{-O-O-}\overset{O}{\overset{\|}{C}}\text{-Ph} \longrightarrow 2\ \text{Ph-}\overset{O}{\overset{\|}{C}}\text{-O}^{\bullet}$$

$$\text{Ph-}\overset{O}{\overset{\|}{C}}\text{-O}^{\bullet} + n\text{M} \longrightarrow \underset{\text{H}}{\text{Ph-}\overset{O}{\overset{\|}{C}}\text{-O-M}_n^{\bullet}}$$

$$\text{Ph-}\overset{O}{\overset{\|}{C}}\text{-O-M}_n^{\bullet} + \text{CuCl}_2\cdot\text{bpy} \longrightarrow \text{Ph-}\overset{O}{\overset{\|}{C}}\text{-O-M}_n\text{-Cl} + \text{CuCl}\cdot\text{bpy}$$

$$\text{Ph-}\overset{O}{\overset{\|}{C}}\text{-O-O-}\overset{O}{\overset{\|}{C}}\text{-Ph} + \text{CuCl}\cdot\text{bpy} \longrightarrow \text{Ph-}\overset{O}{\overset{\|}{C}}\text{-O}^{\bullet} + \text{Ph-}\overset{O}{\overset{\|}{C}}\text{-O}^{-}\cdot\overset{2+}{\text{CuCl}}\cdot\text{bpy}$$

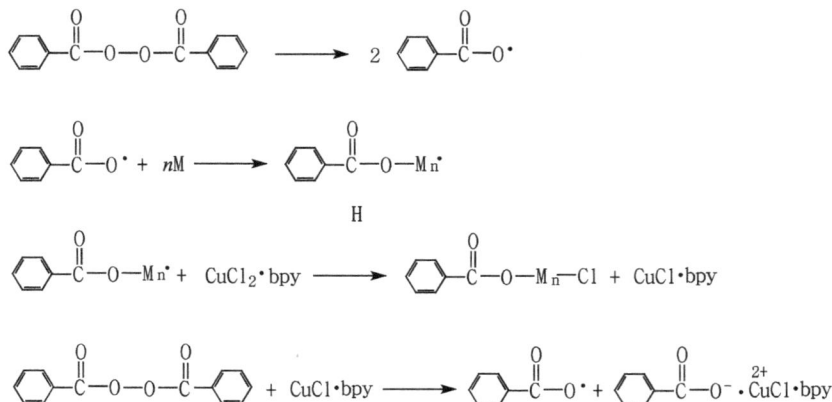

3.5 The focus of ATRP research

After the invention of ATRP, many scholars have carried out in-depth and extensive research, and its research focus can be summarized into the following four aspects.

3.5.1 Exploration on the essential problem of ATRP active species

Since the birth of the ATRP concept and technology, the nature of ATRP active species has been an open question. The Matyjaszewski research group has always believed that the nature of ATRP active species is the same as ordinary free radicals, based on: ① adding a large amount of free radical scavenger (such as TEMPO) can achieve ATRP inhibition, and adding a small amount of polar additives (such as water) does not affect the ATRP; ② the stereoregularity of the ATRP product of MMA is consistent with ordinary free radicals; ③ the classic free radical initiator azobisisobutyronitrile and copper halide successfully achieved reverse ATRP; and ④ a strong Cu^{2+} signal was observed in the electron paramagnetic resonance experiment.

However, these are only evidence of the free radical nature of the active species, and there is still no strong direct evidence that the active species of ATRP are free radicals. Haddleton et al. found that the commonly used phenolic free radical inhibitors have no inhibitory effect on ATRP, so they questioned the nature of the active species and tried to control the stereoselectivity of the reaction with a chiral catalyst, but the results were quite different from the original intention. The stereoregularity of the product is identical to that of free radical polymerization. In addition, they also proposed the possibility of carbocyclic bond heterocracking to produce carbocations. The Key Laboratory of Living Polymerization of East China

University of Science and Technology has conducted in-depth research on the catalytic activity center and reactive species of ATRP. It has been proved in ATRP by UV-visible spectrum analysis of CuX/bpy in organic solvent (EAC, acetonitrile, and acetone). The catalytically active Cu(I) complex structure during the reaction is [Cu(I)/bpy$_2$]X, and the catalytically active center is preferentially formed prior to the initiation step. At the same time, they also proposed the inference that there are many active species in the initial stage of polymerization. On this basis, they also believe that the polymerization reaction may occur in a coordination cage formed by the growth chain radicals and the catalyst. In short, there are many different opinions on the nature of ATRP active species, which is still inconclusive and needs further exploration.

3.5.2 Conversion of heterogeneous reaction system to homogeneous reaction

The early ATRP catalytic system was a heterogeneous system with low catalytic efficiency and low controllability. Therefore, many scholars have focused their research on finding homogeneous catalytic systems. At present, people mainly try to achieve homogeneous reaction from three aspects.
(1) Changing the copper halide ligand. In the ATRP reaction, the ligand functions to form a complex with the catalyst to stabilize the catalyst and increase the solubility of the catalyst in the polymerization system. Therefore, the ligand has a great influence on the entire polymerization process. In 1995, the ATRP system originally reported by Wang Jinshan and others was based on a halogenated alkane/bromide/bipyridine. However, it has been found that the complex formed by cuprous halide and bipyridine is slightly soluble in the reaction system, and the reaction is heterogeneous. Effective control is unfavorable. In order to improve the solubility of cuprous halide in the polymerization system, soluble side chains were introduced into the 4,4′ position of ligand bi-pyridine. After studying, they found that an alkyl chain with at least four carbons can meet this requirement. Matyjaszewski et al., 4,4′-di-tert-butyl-2,2′-bipyridine, 4,4′- in the ligand bipyridine, di-*n*-heptyl-2,2′-bipyridine and 4,4′-di-(5-fluorenyl)-2,2′-bipyridine in place of bipyridine with homogeneous ATRP, the relative molecular mass distribution of the obtained polystyrene and polyacrylate is significantly reduced, and some can be as low as 1.04–1.05. They also found that cheap and readily available polydentate linear alkylamines such as TMEDA, pentamethyldiethylenetriamine, hexamethyltriethylenetetramine and the like can also be used as a ligand for cuprous halide to form a complex with a cuprous halide to catalyze the ATRP reaction. Moreover, since the redox reaction site of the complex formed by the alkylamine and the cuprous halide is lower than the complex energy of the complex formed by the bipyridine and the cuprous halide, the reaction of the complex is lower than that of the bipyridine. The rate is significantly faster.

(2) Using transition metals other than copper, many scholars have attempted to use some soluble metal catalytic systems other than copper to achieve homogeneous ATRP, such as ruthenium, palladium, iron, and nickel. For example, with *p*-methoxybenzenesulfonyl chloride as the initiator, the complex reaction of ruthenium chloride, or a complex of palladium chloride and triphenylphosphine catalyzes the ATRP reaction of styrene to achieve a homogeneous system.

(3) Adding the appropriate solvent, some researchers have achieved a homogeneous reaction of ATRP by adding a solvent that is highly soluble in the catalyst. For example, Wang Xiaosong of East China University of Science and Technology chose several organic solvents with different solubility of cuprous halide/bipyridine, such as xylene, acetonitrile, and tetrahydrofuran, using ethyl 2-bromopropionate/CuCl/bpy system. The polymerization of methyl methacrylate was initiated, and the solubility of the catalytic system in acetonitrile was found to be large. The free radical polymerization of the solution with acetonitrile as solvent showed that the molecular mass distribution of the obtained polymer was significantly lower than that of other solvent systems. Similarly, in the Matyjaszewski patent, the ATRP reaction of *p*-chloromethylstyrene with benzene as solvent has a polymer relative molecular mass distribution which is much lower than that of the bulk polymerization without molecular weight distribution under the same conditions.

3.5.3 RATRP research

The organic halide initiator used in the conventional ATRP reaction is toxic, and the low-valent metal halide is not easily produced and is difficult to store and is easily oxidized. In order to solve these problems, many scholars have thought of a RATRP method using a high-valent metal halide as a catalyst. The principle has been discussed earlier.

The concentration of free radicals generated by the initiation of ordinary free radical initiators such as azobisisobutyronitrile cannot be controlled at a very low level, so the molecular weight distribution of the polymer obtained by the RATRP reaction is generally broad. In order to solve this problem, Qin Dongqi et al. proposed to add a thermally inducible thermal intermerator such as 2,3-dicyano-2,3-di in a RATRP initiation system. *Para*-Methyl phenylsuccinate was used to control the concentration of free radicals in the initiation reaction. The results of the study indicate that the addition of a thermally initiated transfer terminator is indeed beneficial to the control of the reaction. Thus, instead of using any other ligand, only DCDPST and tris(diethyldithiocarbamate) are used. Iron acts as an initiation system to initiate a RATRP reaction of styrene, resulting in a polymer relative molecular mass distribution as low as 1.09, which is already quite low in RATRP reactions.

3.5.4 Preparation of polymers of various structures by ATRP

After the monomer is subjected to ATRP reaction, the obtained polymer still contains a halogen atom at the end, and these polymers are used as a macroinitiator to continue the polymerization of the monomer to obtain a polymer having various structures, and the most common one is to prepare by ATRP. The content of block copolymer will be introduced later.

3.6 Application of ATRP technology

ATRP-controlled polymerization has been discovered for less than 5 years, and it has been synthesized to obtain a variety of polymers with specified structures. The process is simple and efficient, which is unbelievable. The application of the ATRP technology will be described below by way of specific examples.

3.6.1 Preparation of narrow relative molecular mass distribution polymers

As an active controllable polymerization, ATRP can obtain a polymer having a relatively narrow molecular mass distribution. For example, an organic halide/CuX (X is Cl, Br)/2, 2'-bpy initiation system is used to initiate polymerization of styrene to obtain a homopolymer having a relative molecular mass distribution index of 1. 1–1.2. However, such initiating systems are heterogeneous even at high temperatures (100–120 °C), so the relative molecular mass distribution of the polymer is unlikely to be close to the relative molecular mass distribution of typical anionic and cationic active polymers (the cationically active polymer has $M_w/M_n < 1.1$). However, if a certain oil-soluble substituent such as n-butyl group and tert-butyl group is carried on the 2,2'-bpy heterocyclic ring, the above initiating system can be changed to a homogeneous system. The relative molecular mass distribution of the polymer thus obtained can be as low as $M_w/M_n \sim 1.04$. This is the lowest relative molecular mass distribution obtained by free radical polymerization in human history.

3.6.2 Preparation of terminal functional group polymers

According to the principle of ATRP, when an organic halide RX (X is Cl or Br) is used as an initiator, the terminal of the product carries a halogen atom, and the halogen atom itself is a functional group. For example, polymerization of styrene is carried out using 1-phenylchloroethane or 1-phenylbromoethane/CuCl/2, 2'-bpy as an initiation system, and the product is a polystyrene having a terminal halogen atom. If 1-phenylhaloethane is changed to 1,4-dichloro(bromo)methylbenzene, the resulting product is a bifunctional polystyrene having a halogen atom at both ends of the molecular chain. The halogen atom at the end of the polymer molecular chain can be further evolved into other functional groups such as an amine group, a carboxyl group, an azide group, and an allyl group. For example, under the catalysis of $TiCl_4$, the halogen atom at the end of the polystyrene reacts with allyltrimethylsilane, and the terminal group can be converted into an allyl group. Of course, the terminal allyl polystyrene can also be obtained by ATRP of styrene directly using allyl chloride (bromine) as an initiator.

Similarly, in the presence of tetrabutylammonium fluoride, a halogen atom at the end of the polymer molecular chain can be reacted with azidodimethylsilane to give a terminal azide polystyrene.

If an organic halide having another functional group Z (e.g., -OH, -COOH, -CH = CH) is used as the initiator, 100% of the polymer ends carry the functional group Z. For example, 2-vinyl-vinyl acetate is used as an initiator to initiate polymerization of styrene, and the resulting polymer end is provided with a vinyl acetate unit, which is a macromonomer which can be used to prepare a graft copolymer. Another example is the use of hydroxyethyl 2-bromoisobutyrate as an initiator to carry out ATRP of methyl methacrylate under the catalysis of CuBr/2-pyridine acetal n-propylamine to prepare a hydroxyl-terminated polymethyl group, methyl acrylate. The literature reports that initiators with functional groups mainly include 4-cyanobenzyl bromide, 4-bromobenzyl bromide, chloromethyl naphthalene, allyl chloride (bromine), tert-butyl 2-bromopropionate, hydroxyethyl 2-bromopropionate, glycidyl 2-bromopropionate, 2-bromobutyrolactone, vinyl chloroacetate, and chlorinated allyl acetate, chloroacetamide.

If Z is a labeling group, various labeling polymers can be conveniently prepared for use in physicochemical studies.

3.6.3 Preparation of block copolymer

Block copolymers are the most widely studied and widely used polymers in molecular structure regular polymers. Up to now, only living copolymerization has been possible to synthesize block copolymers which are free of homopolymer, relative molecular mass, and composition controllable.

Diblock and triblock copolymers can be prepared directly by the ATRP process. There are two ways to use it in practice. The first one is to prepare a homopolymer of the first monomer by the ATRP method. After the first monomer is reacted, the second monomer is directly added to obtain a diblock copolymer. Since the product at the end of the ATRP reaction is a halogen with a terminal group, it is very stable. Therefore, after the first monomer reaction is completed, the reaction flask can be opened, the second monomer can be added, deoxidized, sealed, heated, and the copolymerization reaction proceeds, and finally a pure block copolymer is obtained. In the second method, a macromolecule containing a halogen atom is obtained by an ATRP method, and then the macromolecule is used as an initiator to initiate polymerization of the second monomer to obtain a diblock copolymer. The preparation of the block copolymer by this method is not possessed by the ionic living polymerization reaction. If the initiator is difunctional, the triblock copolymer can be obtained by both methods. For example, 1,4-dichloro(bromo)methylbenzene/CuBr/bpy is used as the initiation system, and then ATRP of BA is carried out, and then AN is added to continue the reaction to obtain PAN, a -b-PBA-b-PAN triblock copolymer.

Some monomers cannot undergo ATRP, but it is not very difficult to introduce the end of ATRP into the polymer chain. Therefore, a macroinitiator capable of initiating ATRP can be prepared by a certain method, and then ATRP is used. The method of synthesizing block polymers is the so-called semi-ATRP method. Professor Ying's group of East China University of Science and Technology has successfully introduced ATRP initiator groups into the ends of polyether, polyester, and polybutadiene to produce a series of macroinitiators, and with St, MA, and MMA. The body was subjected to ATRP block copolymerization to produce a series of special block copolymers. Polyvinyl acetate with a halogen atom can be prepared by radical polymerization using a carbon tetrachloride chain transfer agent. Then the polyvinyl acetate is used as a macroinitiator; block copolymers such as PVAc-b-PtBMA, PVAc-b-PS and PVAc-b-PVP can be prepared by ATRP method, for example, by reacting a terminal vinyl polydimethylsiloxane with p-dimethylsilylethylenebenzyl chloride, a polydimethylsiloxane having a benzyl chloride at both ends is used as a macromolecule. The initiator CuCl/bpy is used as a catalyst and is subjected to ATRP with monomers such as styrene and BA, respectively, to obtain an ABA-type block copolymer having polydimethylsiloxane as a middle block:

The other living polymerization and ATRP are mutually converted, and the monomers of different polymerization mechanisms can be combined on the same molecular chain to form a block copolymer. For example, Matyjaszewski et al. used 1-PECl/SnCl$_4$ as an initiator to carry out activated carbocation polymerization of styrene in methylene chloride at −15 °C in the presence of nBu$_4$NCl to obtain a polystyrene having $M_w/M_n = 1.17$. ^{1}H NMR analysis confirmed that the terminal group was -CH$_2$-CH(Ph)-Cl. This end group can actually act as an initiator for ATRP. Therefore, the polystyrene obtained earlier is purified and dissolved in toluene, and CuCl/bpy is used as a catalyst, and a new monomer is added to carry out ATRP to obtain a block copolymer.

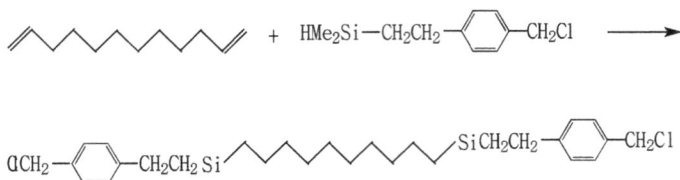

Matyjaszewski et al. also studied the preparation of block copolymers by transferring living ring-opening disproportionation polymerization to ATRP. First, using MoMe(CHCPhMe$_2$)(NAr)(Ot-Bu)$_2$ (Ar = 2,6-diisopropylbenzene) as an initiator, norbornene or dicyclopentadiene in toluene at room temperature, active ring-opening disproportionation polymerization occurs. The obtained polymer was treated with *p*-bromomethylbenzaldehyde to obtain a benzyl bromide-terminated polymer, which was then subjected to ATRP of styrene as a macroinitiator to obtain a block copolymer.

Similar studies have been reported, such as the reaction of a polystyrene active anion terminated with ethylene oxide with acetyl chloride to obtain a macroinitiator having an ATRP initiating group, followed by ATRP of styrene to obtain a block co-polymer. Another example is the use of 2-hydroxy-1,1,1-tribromoethane as an initiator, ring-opening anion polymerization of caprolactone initiated by hydroxyl group catalyzed by triethylaluminum, and then utilized under NiBr$_2$(PPh3)$_2$ catalysis. -CBr$_3$ initiates ATRP of MMA to give a block polymer.

3.6.4 Preparation of star polymers

The simplest preparation of star polymers by the ATRP method is the use of polyfunctional compounds as initiators. This method is called the "first nuclear hind arm" method, and the star polymer produced is a terminal polyfunctional polymer, which should have many applications. For example, the compound shown in Fig. 3.14 is used as a polyfunctional initiator to initiate styrene or methyl methacrylate to give a star-shaped PS or a star-shaped PMMA.

At present, the preparation of multiarm star polymers by ATRP "posterior arm nucleus" method is also being studied. That is, a homopolymer having a terminal group is first prepared by an ATRP method, and then reacted with a polyfunctional compound to obtain a multiarm star polymer. For example, the ATRP method is used to prepare polystyrene with a halogen atom at the end, and then divinylbenzene is added to continue the ATRP reaction, thereby obtaining a multiarmed star-shaped polyphenylene having a network cross-linked divinylbenzene as a core ethylene.

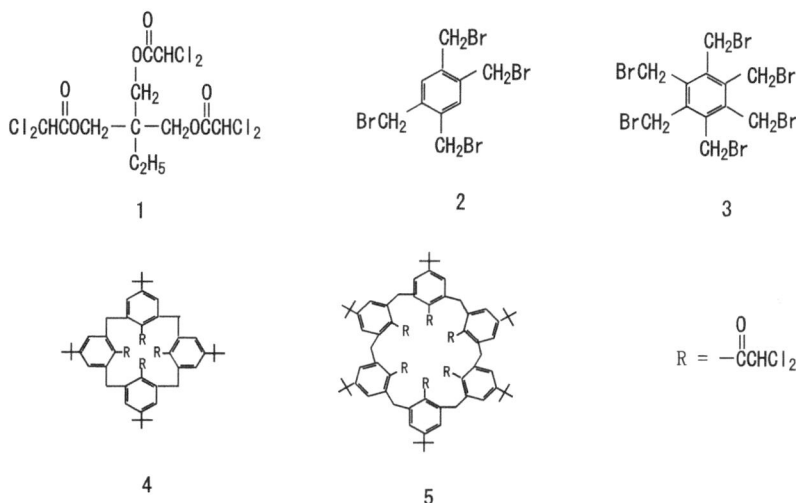

Fig. 3.14: Polyfunctional initiator for the preparation of star polymers.

3.6.5 Preparation of grafted and comb polymers

The graft copolymer can be synthesized by a variety of methods, but a comb polymer having a uniform side chain can usually be produced only by a macromonomer technique. ATRP technology provides two extremely convenient ways to efficiently synthesize comb polymers.

The first is macromolecular technology. It has been mentioned earlier that the use of ATRP produces a polystyrene macromonomer with vinyl acetate, which is free-radically polymerized to obtain the corresponding comb polymer. For example, 2-chlorovinyl acetate is used as an initiator and CuCl/bpy is used as a catalyst for ATRP of styrene. The obtained polymer is a macromonomer with a vinyl acetate unit end group. The comb copolymer can be prepared by simple radical polymerization.

The second is the macroinitiator technology. A homopolymer containing a plurality of ATRP-initiating pendant groups can be used as an ATRP initiator, and a comb polymer having a substantially uniform side group length can be obtained by normal ATRP. Moreover, such a comb polymer contains a plurality of terminal functional groups, and a special polymer can be further prepared. For example, free-radical polymerization using *p*-chloromethylbenzene as a monomer, the resulting product has a benzyl chloride group on each unit which is a very effective active site for ATRP. Using this polymer as a macroinitiator, a second monomer is added to the ATRP again to obtain a graft copolymer or a comb polymer.

ATRP of styrene or MMA using chloroprene rubber containing allylic chloride or brominated ethylene propylene diene monomer, SBS, natural rubber, and so on as a macroinitiator and CuCl/bpy as a catalyst, graft copolymers were obtained.

○○○○○○○○○○○○○○○○○○○○

○ Monomer A ○ Monomer B

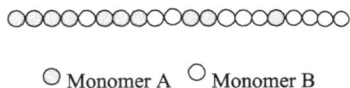

Fig. 3.15: Schematic diagram of the molecular structure of the gradient copolymer.

3.6.6 Preparation of gradient copolymer

The molecular structure of the gradient copolymer can be visualized as shown in Fig. 3.15. This is a more effective polymer blending compatibilizer than block co-polymers or random copolymers.

There are two methods for using ATRP to synthesize a gradient copolymer. One is to directly copolymerize two monomers with a large difference in reactivity ratio, and the other is to continuously feed one of the monomers. Matyjaszewski has studied the copolymerization of MMA with methyl acrylate, styrene and BA, styrene and AN in CuX/bpy catalysis. The relative molecular mass of the obtained gradient co-polymer is in agreement with the theoretical prediction, and the relative molecular mass distribution is also quite narrow ($M_w/M_n < 1.2$).

Sawamoto et al. used $RuCl_2/Al(O-i-Pr)_3$ as a catalyst and 1-PEBr as an initiator. First, a St/MMA monomer mixture with a composition ratio of 3/1 was added, and MMA was added twice during the reaction. An ABC-type triblock gradient copolymer having a St/MMA composition of 3:1–1:1–1:4 and a relatively narrow molecular weight distribution (M_w/M_n of about 1.5) was prepared.

3.6.7 Preparation of hyperbranched polymers

In recent years, the synthesis and characterization of hyperbranched polymers have been very active. Early hyperbranched polymers were mainly obtained by polycon-densation techniques and ionic polymerization techniques. The emergence of ATRP technology has opened up a new way for the synthesis of hyperbranched polymers. Self-condensing vinyl polymerization using ATRP technology allows for the convenient synthesis of hyperbranched polymers.

For example, with *p*-chloromethylstyrene as the initiating monomer and Cu/2,2'-bpy as the catalytic complex, a hyperbranched polymer having a chlorine atom at each end group can be obtained. During the polymerization, the polymer contains two different chlorine atoms, namely benzyl chloride and chain end chlorine. The latter C–Cl bond teaches easy breakage and generates free radicals, so the polymerization process cannot be "seek the fork", and the degree of branching cannot be high. The ATRP process for *p*-chloromethylstyrene is shown in the following reaction formula. The hyperbranched polymer is used as a macromolecular polyfunctional initiator to further initiate the polymerization of tert-butyl methacrylate to obtain star-shaped polymer with a poly-*p*-methylstyrene-based polymer as a core and polybutyl methacrylate as an arm. Based on this, an amphiphilic hyper-branched star-shaped polymer with poly-*p*-methylstyrene-based polymer as core and poly-methacrylic acid as arm was prepared by acidic hydrolysis.

3.7 Removal and recovery of ATRP system catalyst

ATRP catalysts are inefficiently initiated and are retained in the polymer to impart color and toxicity to the polymer. Therefore, the removal and recycling of catalysts is an important research topic for the further development of ATRP technology, and the only way to achieve industrialization. At present, there is no very effective method for the removal of catalysts in ATRP products. It is generally believed that the most promising methods are as follows.

3.7.1 Ion exchange resin technology

Before the ion exchange resin technology was applied to the ATRP, the alumina separation column was mainly used, and after removing the copper complex with NaOH or Na_2S, the polymer was precipitated with a poor solvent. Although the residual amount of catalyst in the polymer is very low after treatment by this method, the high viscosity polymer consumes a large amount of energy and solvent through the separation column, and is not suitable for large-scale industrial production (generally only suitable for the polymer to be treated). The amount is less than 10 g. Matyjaszewski et al. successfully achieved the above objectives by using crosslinked polystyrene ion exchange resin, which not only reduced energy consumption, but also saved a lot of solvent.

3.7.2 Stationary phase catalyst technology

The catalyst is adsorbed on a fixed carrier such as functionalized crosslinked polystyrene or silicone, and the ATRP reaction is carried out under heterogeneous conditions, and the amount of catalyst remaining in the obtained polymer is small, and no posttreatment is necessary. The supported catalyst can be reused to increase the catalytic efficiency. However, due to the steric hindrance caused by the undrainability of the carrier and the heterogeneity of the system, the reaction is more difficult to control than the homogeneous system, and the relative molecular mass distribution of the product tends to be broadened.

3.7.3 Removal of catalyst by difference in solubility

The end of the molecule is functionalized with respect to the molecular weight of the polyethylene, which is brought to the appropriate ATRP coordinating group to form a polyethylene ligand. At the temperature of the polymerization reaction, the ligand is soluble to form a homogeneous system. Once the reaction is completed, the system is cooled, and the polyethylene ligand is first settled out of the system to achieve the purpose of removing the catalyst. Because it is a homogeneous system, the reaction is more stable and easier to control than the stationary phase catalyst.

3.8 Prospects for ATRP academic research and industrial development

The emergence of ATRP technology has opened up a new field of living polymerization. ATRP technology combines the advantages of free radical polymerization and living polymerization. It can perform bulk, solution, suspension, and emulsion polymerization as in free radical polymerization, and can also synthesize polymers with specified structure, and the monomer is suitable for a wide range of applications. All monomers suitable for other living polymerization systems and some monomers that are currently incapable of living polymerization are included. In addition, there is a very useful feature that does not require a complicated synthetic route and therefore has a very broad application prospect.

Due to the great potential of ATRP in the synthesis of new materials, it has not only aroused the attention of the academic community but also caused great interest in the industry. With the advent of ATRP technology, 12 major companies such as Bayer, Green, and Rohm & Hass joined forces to fund the ATRP study at Carnegie Mellon University. At the same time, DuPont, Dow, BASF, ICI, 3M, IBM, Kodak, and other chemical products production and application companies are also concentrated to use this technology to develop new polymer materials of interest.

It should be noted that the industrialization of ATRP still has a long way to go. The main problem is that the catalytic system activity is not too high, the amount is large, and the metal salt as a catalyst is unfavorable to environmental protection. Therefore, the development of highly active catalysts or the development of economical and practical catalytically active system recovery technology is the key to the industrialization of ATRP. In addition, the industrialization of ATRP depends on whether this method can be used to synthesize products of commercial value. Therefore, research and development of new ATRP products will also be an important aspect of ATRP research in the future.

Even if industrialization takes a while, ATRP academic research will remain the most promising research topic in the field of polymer chemistry. Its important research areas may focus on the following aspects: ① develop a highly active catalytic system. The monomer can be polymerized using a very small amount of catalyst at a lower temperature (40–80 °C); ② broaden the monomer surface to simple olefins such as vinyl acetate, vinyl chloride, and ethylene; ③ develop an ATRP catalytic system free of metals; and ④ ATRP controls the stereoregularity of polymers.

It can be expected that the academic ATTRP research fever will not only lead to the discovery of a series of new catalytic systems but also lead to the discovery of new polymerization reactions. Once a breakthrough is made, it will produce incalculable social and economic benefits.

References

[1] Szwarc, M, Levy, M, Milkovich, R. Polymerization initiated by electron transfer to monomer. a new method of formation of block polymerization. J Am Chem Soc. 1956;78(1):2656–2657.
[2] Szwarc, M. Carbanion, living polymers, and electron transfer processes. New York: Wiley-Interscience, 1968.
[3] Miyamoto, M, Sawamoto, M, Higashimura, T. Living polymerization of isobutyl vinyl ether with hydrogen iodide/iodine initiating system. Macromolecules. 1984;17(3):265–268.
[4] Faust, R, Kennedy, JP. Living carbocationic polymerization. Polymer Bulletin. 1986;15(4): 317–323.
[5] Aida, T, Inoue, S. Living polymerization of epoxides with metalloporphyrin and synthesis of block copolymers with controlled chain lengths. Macromolecules. 1981;14(5):1162–1166.
[6] Wang, JS, Matyjaszewski, K. Controlled/"living" radical polymerization. atom transfer radical polymerization in the presence of transition-metal complexes. J Am Chem Soc. 1995;117(20): 5614–5615.
[7] Wang, JS, Matyjaszewski, K. Transition metal catalyzed atom transfer radical polymerization (ATRP): principle and mechanism. Polym Mater Sci Eng. 1995;73:414–415.
[8] Wang, JS, Matyjaszewski, K. Controlled/"Living" Radical Polymerization. Halogen Atom Transfer Radical Polymerization Promoted by a Cu(I)/Cu(II) Redox Process. Macromolecules. 1995;28(23):7901–7910.
[9] Odian, G. Principles of Polymerization, 2nd. New York: John Wiley & Sons Inc, 1981.
[10] Otsu, T, Yoshida, M. Role of initiator-transfer agent-terminator (iniferter) in radical polymerizations: Polymer design by organic disulfides as iniferters. Macromol Chem Rapid Commun. 1982;3(2):127–132.
[11] Georges, MK, Veregin, RPN, Kazmaier, PM Narrow molecular weight resins by a free-radical polymerization process. Macromolecules. 1993;26(11):2987–2988.
[12] Otsu, T, Matsunaga, T, Doi, T, et al. Features of living radical polymerization of vinyl monomers in homogeneous system using N, N-diethyldithiocarbamate derivatives as photoiniferters. Eur Polym J. 1995;31(1):67–78.
[13] Otsu, T. Iniferter concept and living radical polymerization. J Polym Sci, Part A Polym Chem. 2000;38(12):2121–2136.
[14] Otsu, T, Tazaki, T. Living radical polymerization in homogeneous system with phenylazotriphenylmethane as a thermal iniferter. Polymer Bulletin. 1986;16(4):277–284.
[15] Otsu, T, Mstsunaga, T, Kuriyama, A, et al. Living radical polymerization through the use of iniferters: Controlled synthesis of polymers. Eur Polym J. 1989;25(7–8):643–650.
[16] Otsu, T, Matsumoto, A, Tazaki, T. Radical polymerization of methyl methacrylate with some 1,2-disubstituted tetraphenylethanes as thermal iniferters. Polymer Bulletin. 1987;17(4):323–330.
[17] DeLeon, ME, Gnanou, Y, Guerrero, R. Control of free-radical polymerization of methyl methacrylate by a diphenylethane-type initer. Polym Prepr. 1997;38(1):667–668.
[18] Bledzki, A, Braun, D, Titzschkau, K. Polymerisationsauslösung mit substituierten ethanen, 6†. Polymerisation von methylmethacrylat mit verschiedenen tetraphenylethanen. Makromol Chem. 1983;184(4):745–754.

[19] Bledzki, A, Braun, D, Titzschkau, K. Polymerisationsauslösung mit substituierten Ethanen, 1. Polymerisation von Methylmethacrylat mit 1,1,2,2-Tetraphenyl-1,2-diphenoxyethan. Makromol Chem. 1981;182(4):1047–1056.

[20] Bledzki, A, Balard, H, Braun, D. Initiation of polymerization with substituted ethanes, 11† Chromatographic separation of methyl methacrylate oligomers and NMR analysis of the diastereoisomers. Macromol Chem Phys. 1988;189(12):2807–2822.

[21] Otsu, T, Mstsunaga, T, Kuriyama, A, et al. Living radical polymerization through the use of iniferters: Controlled synthesis of polymers. Eur Polym J. 1989;25(7–8):643–650.

[22] Otsu, T, Ogawa, T, Yamamoto, T. Solid-phase block copolymer synthesis by the iniferter technique. Macromolecules. 1986;19(7):2087–2089.

[23] Otsu, T, Yoshioka, M, Tanaka, T. Synthesis of telechelic polymers through radical polymerization with a two-component iniferter system. Eur Polym J. 1992;28(11):1325–1329.

[24] Kroeze, E, Boer, B, Brinke, G, et al. Synthesis of SAN-PB-SAN triblock copolymers via a "living" copolymerization with macro-photoiniferters. Macromolecules. 1996;29(27):8599–8605.

[25] Yang, XM, Qiu, KY. Radical polymerization of styrene initiated with alkyl N,N-diethyldithiocarbamylacetate photoiniferters. J Macromol Sci, Part A. 1997;34(2):315–325.

[26] Yang, XM, Qiu, KY. Block copolymerization of vinyl monomers with macrophotoiniferter. J Macromol Sci, Part A. 1997;34(3):543–549.

[27] Qin, SH, Qiu, KY. Radical polymerization of styrene initiate with a new multifunctional iniferter. Polymer Bull. 2000;44(2):123–128.

[28] Qin, SH, Qiu, KY. A facile method for synthesis of ABA triblock copolymers with macro-iniferter technique. Polymer. 2001;42(1):3033–3042.

[29] Qin, SH, Qiu, KY. Synthesis of triblock copolymers containing polyisoprene segments using iniferier technique. Acta Polym Sin. 2001;4:549–552.

[30] Qin, SH, Qiu, KY. Living radical polymerization and copolymerization of vinyl monemers initiated with novel iniferters. Acta Polym Sin. 2002;2:127–136.

[31] Qin, SH, Qin, DQ, Qiu, KY, et al. Bulk Polymerization of styrene with diethyl 2,3-dicyano-2,3-di (p-tolyl)-succinate as a thermal iniferter. J Macromol Sci Part A Chem. 2001;38(1):57–65.

[32] Clouet, G, Kayser, D. Macrothiuram disulfide for the free radical synthesis of PDMS-vinyl triblock copolymers. I. syntheses and polymerization kinetics. J Polym Sci, Part A: Polym Chem. 1993;31(13):3387–3396.

[33] Mahesh, GN, Sivaraman, A, Tharanikkarasu, K, et al. Synthesis and characterization of polyurethane–polyvinylbenzyl chloride multiblock copolymers and their cationomers using a polyurethane macroiniferter. J Polym Sci, Part A: Polym Chem. 1997;35(7):1237–1244.

[34] Otsu, T, Yamashita, K, Tsuda, K. Synthesis, reactivity, and role of 4-vinylbenzyl N,N-diethyldithiocarbamate as a monomer-iniferter in radical polymerization. Macromolecules. 1986;19(2):287–290.

[35] Qin, SH, Qiu, KY. Synthesis of poly (methyl methacrylate) macromer by radical polymerization. Acta Polym Sin. 1999;4:509–512.

[36] Qin, SH, Qiu, KY. Synthesis of well-defined functionalized polystyrene by a new monomer iniferter. Acta Polym Sin. 2000;3:375–378.

[37] Qin, SH, Qiu, KY. Syntheses of well-defined macro-monomers by living radical polymerization. Acta Chimica Sinica. 2001;59(10):1776–1782.

[38] Qin, SH, Qiu, KY. Synthesis of graft copolymers with a polymerizable photoiniferter. Acta Polym Sin. 2002;1:102–107.

[39] Qin, SH, Qiu, KY. Synthesis of macromonomer from radical polymerization of styrene with a polymerizable photoiniferter. J Appl Polym Sci. 2000;75(11):1350–1356.

[40] Qin, SH, Qiu, KY. A new polymerizable photoiniferter for preparing poly(methyl methacrylate) macromonomer. Eur Polym J. 2001;37(4):711–717.

[41] Georges, MK, Veregin, RPN, Kazmaier, P M, et al. Narrow molecular weight resins by a free-radical polymerization process. Macromolecules. 1993;26(11):2987–2988.

[42] Georges, MK, Veregin, RPN, Kazmaier, PM, et al. ACS polymer division. Polymer Preprints. 1994;35(2):870–871.

[43] Odell, PG, Veregin, RPN, Michalak, LM, et al. Rate enhancement of living free-radical polymerizations by an organic acid salt. Macromolecules. 1995;28(24):8453–8455.

[44] Li, I, Howell, BA, Matyjaszewski, K, et al. Kinetics of decomposition of 2,2,6,6-tetramethyl-1-(1-phenylethoxy)piperidine and its implications on nitroxyl-mediated styrene polymerization. Macromolecules. 1995;28(19):6692–6693.

[45] Percec, V, Barboju, B. "Living" radical polymerization of styrene initiated by arenesulfonyl chlorides and CuI(bpy)nCl. Macromolecules. 1995;28(23):7970–7972.

[46] Yoshida, E, Ishizone, T, Hirao, A, et al. Synthesis of polystyrene having an aminoxy terminal by the reactions of living polystyrene with an oxoaminium salt and with the corresponding nitroxyl radical. Macromolecules. 1994;279(12):3119–3124.

[47] Meijs, GF, Rizzardo, E. Living free-radical polymerization by reversible addition –fragmentation chain transfer: the RAFT process. Macromolecules. 1988;31(16):5559–5562.

[48] Chiefari, J, Chong, YK, Ercole, F, et al. Living free-radical polymerization by reversible addition –fragmentation chain transfer: the RAFT process. Macromolecules. 1998;31(16):5559–5562.

[49] Greszta, D, Mardare, D, Matyjaszewski, K. "Living" radical polymerization. 1. Possibilities and limitations. Macromolecules. 1994;27(3):638–644.

[50] Percec, V, Barboju, B. "Living" radical polymerization of styrene initiated by arenesulfonyl chlorides and CuI(bpy)nCl. Macromolecules. 1995;28(23):7970–7972.

[51] Ando, T, Kato, M, Kamigaito, M, Sawamota, M. Living radical polymerization of methyl methacrylate with ruthenium complex: formation of polymers with controlled molecular weights and very narrow distributions. Macromolecules. 1996;29(3):1070–1072.

[52] Granel, C, Dubois, P, Jerome, R, Teyssie, P. Controlled radical polymerization of methacrylic monomers in the presence of a bis(ortho-chelated) arylnickel(II) complex and different activated alkyl halides. Macromolecules. 1996;29(27):8576–8582.

[53] Wei, ML, Xia, JH, MeDermott, NE. Atom transfer radical polymerization of styrene in the presence iron complexes. Polym Prep. 1997;38(2):231–232.

[54] Patten, TE, Xia, J, Abemathy, T, Matyjaszewski, K. Polymers with very low polydispersities from atom transfer radical polymerization. Science. 1996;272(5263):866–868.

[55] Haddleton, DM, Jasieczek, CB, Hannon, MJ, et al. Atom transfer radical polymerization of methyl methacrylate initiated by alkyl bromide and 2-pyridinecarbaldehyde imine copper(I) complexes. Macromolecules. 1997;30(7):2190–2193.

[56] Cheng, GL, CP, Hu, Ying, SK. A novel initiator system of atom transfer radical polymerization for styrene. China Synth Rubber Ind. 1997;20(2):116.

[57] Nakagawa, Y, Gaynor, SG, Matyjaszewski, K. The synthesis of end functional polymer by "living" radical polymerization. Polym Prep. 1996;37(1):577–578.

[58] Beers, KL, Gaynor, SG, Matyjaszewski, K. The use of "living" radical polymerization to synthesize graft copolymers. Polym Prep. 1996;37(1):571–572.

[59] SM, Jo, Gaynor, SG, Matyjaszewski, K. Homo- and ABA block polymerization of acrylonitrile, n-butyl acrylate, and 2-ethylhexyl acrylate using ATRP. Polym Prep. 1996;37(2):272–273.

[60] Kotani, Y, Kato, M, Kamigaito, M, Sawamoto, M. Living radical polymerization of alkyl methacrylates with ruthenium complex and synthesis of their block copolymers. Macromolecules. 1996;29(22):6979–6982.

[61] Zhang, YM, Luo, N, Ying, SK. Synthetic study on the block copolymers of vinyl acetate and vinyl monomers. China Synth Rubber Ind. 1997;20(5):314.

[62] Coca, S, Matyjaszewski, K. Block copolymers by transformation of "living" carbocationic into "living" radical polymerization. Macromolecules. 1997;30(9):2808–2810.

[63] Chen, XY, Ivan, B, Kops, J. Polystyrene-b-polyisobutylene-b-polystyrene block copolymers by combining living cationic and living free radical polymerization. Polym Prep. 1997;38(1):715–716.

[64] Wang, XS, Luo, N, Ying, SK. Synthesis of block copolymers through atom transfer radical polymerization. China Synth Rubber Ind. 1997;20(2):115.

[65] Liu, F, Liu, B, Luo, N, Ying, SK. Block copolymers by the transformation of living anionic polymerization into atom transfer radical polymerization. China Synth Rubber Ind. 1998;21 (5):303.

[66] Liu, B, Liu, F, Luo, N, Ying, SK. Synthesis of chloroacetyl oligomer as macroinitiator for ATRP. China Synth Rubber Ind. 1998;21(5):304.

[67] Liu, YF, Wang, XL, Ying, SK. Poly(styrene-b-2-methyl oxazoline) block polymer synthesized by the transformation of "living" ATRP to living cationic ring opening polymerization. China Synth Rubber Ind. 1998;21(5):305.

[68] Liu, YF, Wang, XL, Ying, SK. Conversion of halogen terminal group in polymer from ATRP via THF cationic ring opening reaction. China Synth Rubber Ind. 1998;21(5):306.

[69] Hawker, CJ, Karklay, RB., Grubbs, R B. Architectural and structural control of styrene polymerizations by novel "living" free radical procedures. Polym Prep. 1996;37(2):515–516.

[70] Gaynor, SG, Edelman, SZ, Matyjaszewski, K. From hyperbranched to crosslinked polymers by atom transfer radical polymerization. Polym Mater Sci Eng. 1996;74:236–237.

[71] Gaynor, SG, Edelman, SZ, Matyjaszewski, K. Branched and hyperbranched macromolecules by atom transfer radical polymerization. Polym Prep. 1996;37(2):413–414.

[72] Gaynor, SG, Edelman, SZ, Matyjaszewski, K. Synthesis of branched and hyperbranched polystyrenes. Macromolecules. 1996;29(3):1079–1081.

[73] Fréchet, JMJ, Ledue, MR, Weimer, M, et al. Living free radical polymerization and dendritic polymers. J Polym Prep. 1997;38(1):756–757.

[74] Wang, GJ. Preparation of functional multicomponent polymers by active/controllable polymerization. School of Chemistry and Chemical Engineering. Shanghai Jiaotong University [Doctoral dissertation]. 2000.

[75] Zhu, SM. Study on a new catalytic system for atomic transfer radical polymerization and its application in the preparation of new materials. School of Chemistry and Chemical Engineering, Shanghai Jiaotong University [Doctoral dissertation]; 2001.

[76] Yang, XM, Xu, SJ, Qiu, KY. The development of living radical polymerization. Chin Polym Bull. 1996;3:166–173.

[77] Zheng, SY, Lin, J, Li, WL. Synthesis of poly (methyl methacrylate) by stable free radical polymerization. J Xiamen Univ (Natural Science). 2002;41(4):468–471.

[78] WL, Li, Zhuang, RC, Zheng, SY. Living free radical microemulsion polymerization of methyl methacrylate. J Xiamen Univ (Natural Science). 2002;41(6):768–772.

Exercises

1. **What is living controlled polymerization of free radical and its four characteristics?**
   The chain polymerization method can control the structure and size of the polymer molecule, and has a narrow relative molecular mass distribution. There are four characteristics that distinguish the living polymerization from the traditional polymerization reaction: ① the initiation reaction rate is much larger than the growth reaction rate, and there is no chain termination and chain transfer reaction. All the polymer chains grow at the same time, and the number of growth chains is always the same. Molecular mass distribution is very narrow ($\bar{M}_w/\bar{M}_n \leq 1.1$); ② the relative molecular mass of the polymer is proportional to the ratio of the concentration of the

consumed monomer to the initial concentration of the initiator; ③ the relative molecular mass of the polymer increases linearly with the conversion rate, and the degree of polymerization of the obtained polymer can be controlled by controlling the amount of monomer and initiator charged; ④ when the conversion rate of the first monomer reaches 100%, another monomer is added to synthesize a block copolymer having a predetermined structure.

2. **How to implement living controlled polymerization of free radical?**
    Three ways to achieve the above goals have been found.
    Reactive growth of free radicals and stable free radicals to form dormant covalent compounds; growth of free radicals and nonradical materials reversibly form dormant and persistent free radicals; reversible deuteration transfer between growth free radicals and chain transfer agents.

3. **What is TEMPO initiation system?**
    During the polymerization process, TEMPO is a stable free radical, which only undergoes a coupling reaction with a growing free radical to form a covalent bond, and this covalent bond can be decomposed to generate free radicals at high temperatures. Therefore, after TEMPO captures the growth of free radicals, it is not the true death of the active chain, but only temporarily inactivated and becomes a dormant species, as shown in the following formula.

4. **What is the mechanism of RAFT radical polymerization**
    The mechanism of RAFT radical polymerization can be expressed by the following reaction formula:

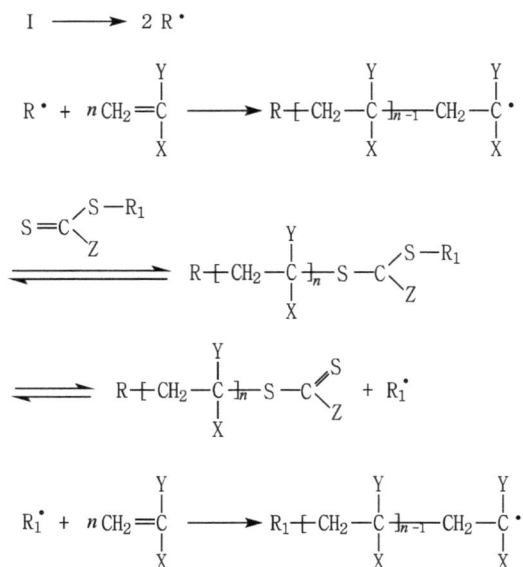

$$I \longrightarrow 2\,R^{\bullet}$$

$$R^{\bullet} + n\,CH_2{=}\underset{X}{\overset{Y}{C}} \longrightarrow R{-}\!\left(CH_2{-}\underset{X}{\overset{Y}{C}}\right)_{\!n-1}\!\!CH_2{-}\underset{X}{\overset{Y}{C}}{}^{\bullet}$$

$$S{=}C\!\!\overset{S-R_1}{\underset{Z}{\diagdown}} \underset{\longleftarrow}{\overrightarrow{\qquad}} R{-}\!\left(CH_2{-}\underset{X}{\overset{Y}{C}}\right)_{\!n}\!\!S{-}C\!\!\overset{S-R_1}{\underset{Z}{\diagdown}}$$

$$\underset{\longleftarrow}{\overrightarrow{\qquad}} R{-}\!\left(CH_2{-}\underset{X}{\overset{Y}{C}}\right)_{\!n}\!\!S{-}C\!\!\overset{S}{\underset{Z}{\diagdown}} + R_1^{\bullet}$$

$$R_1^{\bullet} + n\,CH_2{=}\underset{X}{\overset{Y}{C}} \longrightarrow R_1{-}\!\left(CH_2{-}\underset{X}{\overset{Y}{C}}\right)_{\!n-1}\!\!CH_2{-}\underset{X}{\overset{Y}{C}}{}^{\bullet}$$

The correctness of this mechanism can be confirmed by the analysis of nuclear magnetic resonance and UV-visible spectroscopy, and the presence of fragments of chain transfer agent molecules at the polymer chain ends.

It is not difficult to see from the above reaction mechanism that the two groups Z and R in the chain transfer agent play a crucial role. Wherein, Z should be a group capable of activating

C=S addition to a radical, usually an aryl group or an alkyl group; and R should be an active radical leaving group, and the radical R· generated after breaking the bond should be effective. In reinitiation polymerization, the commonly used are cumene, cyanoisopropyl, and the like.

5. **What is the application of ATRP?**
   Preparation of narrow relative molecular mass distribution polymers; preparation of terminal functional group polymers; preparation of block copolymer; preparation of star polymers; preparation of grafted and comb polymers; preparation of gradient copolymer; preparation of hyperbranched polymers

Chapter 4
Synthesis of dendrimers and hyperbranched polymers

4.1 Introduction

The study of highly branched macromolecules is one of the hottest topics in the field of polymer chemistry for more than a decade. Due to their novel structures, unique properties, and potential application, such polymers have received widespread attention from the scientific community and industry since their inception.

Distinguishing from their structure features, highly branched macromolecules can be divided into two categories: one is a dendrimer and the other is a hyperbranched polymer. Dendrimer molecules have regular and controllable branching structure. Usually they must be prepared by continuous synthesis in multiple steps. After each step, they are separated and purified. The process is very cumbersome. Hyperbranched polymers are often obtained by direct polymerization of AB_x monomers in one step, which is simple and easy to obtain. However, the branching structure of hyperbranched polymers is incomplete and difficult to control. There are differences in structure and properties between these two polymers, but they also have many similar chemical and physical properties. For example, they have high functionality on the surface of the molecular structure; they have a large solubility in organic solvents; their melts and solutions have lower viscosities than the corresponding linear molecules, and their molecular structure does not affect their glass transition temperature.

Since hyperbranched polymers can be obtained directly by one-step polymerization, it is obviously more likely to achieve large-scale industrial production than dendrimers and has more potential application. Therefore, people have shown greater interest in them in recent years. According to their characteristics and performance, it is expected to find use in coatings, adhesives, rheology aids, linear polymer modifiers, crystal nucleating agents, structural control agents for organic–inorganic dopants. To this end, many scientists have made great efforts and achieved fruitful results.

The proposal of hyperbranched polymer can be traced back to more than 50 years ago. In 1952, Flory mentioned such polymers in his paper, discussed their structure, gave their relative molecular mass distributions, and wrote about it in his famous textbook *Polymer Chemistry*, which he published in the second year. However, due to the lack of excellent mechanical properties and the long-standing misunderstanding of branched polymers, hyperbranched polymers had been useless for a long time. Until 1990, people discovered the relationship between their unique structure and performance and reraised them, and carried out a lot of research work and eventually establishes a proper position for hyperbranched polymers in the field of polymer

https://doi.org/10.1515/9783110597097-004

science. Among these studies, Y. H. Kim, O. W. Webster, C. J. Hawker, J. M. J. Frecher et al. made outstanding contributions.

4.2 Polymer structure and branched polymer

In classical textbooks, polymers have long been divided into three categories: linear, branched, and bulk. In addition to a small number of branched polymers with special structure (such as star, comb, and graft copolymers), it is generally believed that the properties of the branched polymers are not as good as the corresponding linear polymers. Therefore, the attitude toward branched polymer is basically repulsive and negative. Although Flory proposed the concept of hyperbranching 45 years ago, theoretically studied it and got the relative molecular mass distribution of such polymers, hyperbranched polymers had been in existence for many years.

In recent years, with the advancement of polymer science, the relationship between polymer structure and properties has been deepened, which has led to a new understanding of branched polymers. Branching has a huge impact on the properties of polymers. The properties of the branched chain, the length of the branch, the distribution of the branch, the degree of branching, and other physical parameters is closely related to its chemical, physical, mechanical and rheological properties. Whether these effects are positive or negative is determined entirely by the purpose of the study of the application.

Another reason why hyperbranched polymers have not been valued for a long time is the lack of characterization. Evidence shows that hyperbranched polymers appeared as byproduct in many accidental cases, long before Kim and Webster used 3,5-dibromophenylboronic acid as monomer to synthesis the first hyperbranched polymer. For example, in 1920, Hunter and Woollett used the silver salt reaction between ethyl iodide and trihalophenol to obtain a product with high molecular weight and random structure. Unfortunately, due to the lack of means of analysis skills at that time, it was impossible to characterize the structure of this product. Today, we can speculate that the product is probably a hyperbranched polymer based on the ingredient they used.

4.3 The synthesis of dendrimers

The dendrimers were developed in 1985 by Dr. Tomilia of the Michigan Institute of Chemistry and Professor Newkome of the University of South Florida at the same time. They are highly ordered in three dimensions and we can independently control the size, the shape, the structure and the functional groups of the molecule in molecular level. Their highly branched structure and unique monodispersity give them special properties and functions. This type of polymer has many molecular ends on its chain skeleton and its molecular structure is like a tree. The name "dendrimer" comes from the Greek word δένδρον (dendron) which translates to "tree." Compared with linear polymers, the structure of dendrimers is very symmetrical, the periphery of the molecules is relatively tight, and the number of terminal groups increases geometrically with algebra, so they have some special physicochemical properties. For example, the functional density of the molecular surface is extremely high; the shape of the molecule is usually quite unique and numerous cavities spread inside the molecular. Because of these special properties, they have broad application prospects.

In the past 30 years, the synthesis, properties, and applications of dendrimers have been fully studied. Compared to linear polymers, dendrimers have a nearly spherical structure with less intermolecular intertwining and a lower viscosity in solution. Due to its stable structure and narrow molecular mass distribution, dendrimers are likely to serve as a size standard in gel permeation chromatography (GPC). Dendrimers have a large size, and their size and the number of functional groups can be controlled which makes it an ideal building block for larger nanomaterials. Block copolymer with hydrophilic polymer and lipophilic dendrimer can be used as a surfactant. Some dendrimers with special structures are electrically conductive, while others are useful as catalysts, as molecular recognition markers, and so on.

Since the 116th volume in 1992, the *American Chemical Abstracts* had created a new title called dendritic polymers in the general subject index, indicating the scientists' high interest in such polymers. In the past 30 years, many dendrimers have been synthesized, and the research on their properties has been deepened, and become one of the hot spots in the field of polymers at home and abroad. At present, research hotspots for dendrimers have shifted from synthesis to application fields. Besides, research paper and patents on application have occupied the majority in this field. Since the mid-1990s, Peking University and the Beijing Institute of Chemistry of the Chinese Academy of Sciences have also carried out basic research work in this emerging field. Many high-level academic journals have published review articles on dendrimers.

4.3.1 Basic method for dendrimer synthesis

The synthesis of dendrimers mainly includes two basic methods: inside-out and outside-in. In 1985, Tomalia and Newkome first developed the inside-out synthesis route, and in 1990 Frecher et al developed the outside-in synthesis route.

The inside-out synthesis route is carried out by expanding outward from the center point of the dendrimer. First, the central core molecule is reacted with 2 mol or more reagent containing two or more protected branch active sites. Then the protecting group is removed, and the activated group is further reacted, and thus repeated until the desired dendrimer is synthesized (Fig. 4.1). The disadvantage of this route is that the larger the number of reaction growth stages, the more functional groups are required to react, the less likely the growth reaction is to proceed completely, and the more easily the molecules are defective. To ensure complete reaction, excess reagents and harsh reaction conditions are often required, which makes the separation and purification of the product difficult.

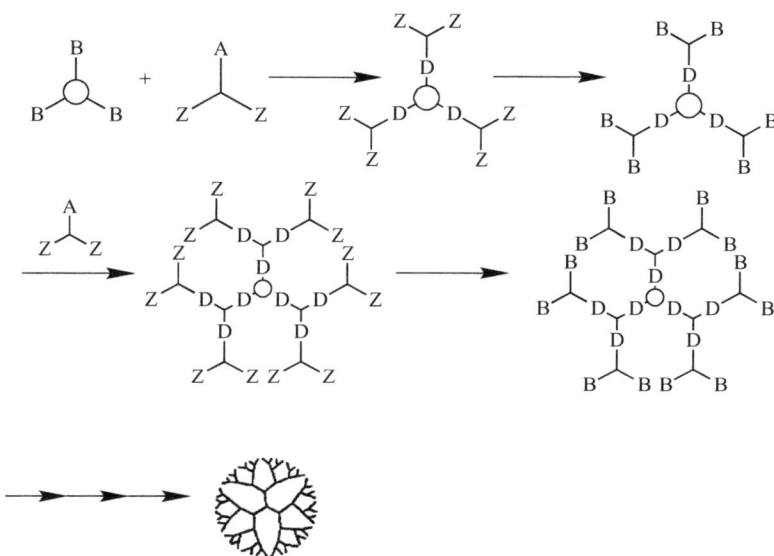

Fig. 4.1: The inside-out synthesis route.
A, B: reactive group; Z: the protected form of B; D: functional group formed by reaction of A and B

The outside-in synthesis route starts from the edge portion of the desired dendrimer molecule and proceeds stepwise inward. A portion of the dendrimer is first synthesized to form a wedge, which is then linked to the central molecule to form a dendrimer (Fig. 4.2). The outside-in synthesis route is superior to the inside-out synthesis route in synthesizing monodisperse dendrimers, purification, and characterization, and involves less reactive functional groups in each growth process. However, as the growth order increases, the steric hindrance of the reactive group at the central point increases, which hinders further reaction, and that makes the polymer generally not as large as that synthesized by the first method.

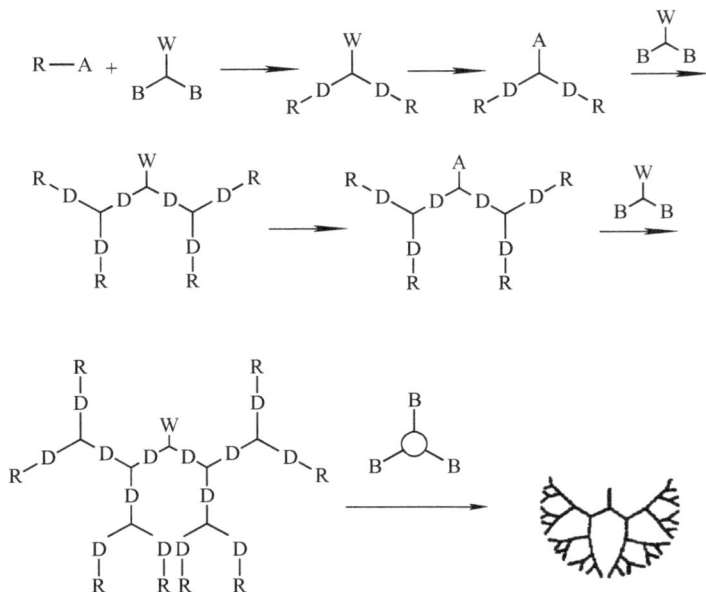

Fig. 4.2: The outside-in synthesis route.
A, B: reactive group; W: the protected form of B; D: functional group formed by reaction of A and B

The following specific examples describe some synthesis routes of dendrimers.

4.3.2 Dendritic hydrocarbon polymer

The synthesis of the dendrimer is a plurality of repeated steps, and the yield of each reaction should be high in order to ensure a high overall yield of the synthesis. Miller combined inside-out and outside-in methods: the reaction that converts aryltrimethylsilane to arylboronic acid and the coupling reaction of arylboronic acid and aryl bromide, and successfully synthesized aromatic dendritic hydrocarbon polymer containing multiple aromatic rings. The reaction process is displayed in equation 1.

The aromatic dendrimer hydrocarbon polymer obtained has an amorphous structure and is soluble in many organic solvents such as tetrahydrofuran, toluene, dichloromethane, and chloroform. It also has good heat resistance, and will not decompose until it reaches 500 °C or higher. Dendritic polyphenylene can be formed by self-phase coupling of 3,5-dibromophenylboronic acid under the catalysis of Pd(PPH$_3$)$_4$. The dendritic polyphenylene is soluble in tetrahydrofuran and o-chlorobenzene, and is insoluble in dichloromethane and ether solvents. Since the edge of polyphenylene is rich in bromine atoms, it can be easily converted into a carboxyl or carboxyl salt.

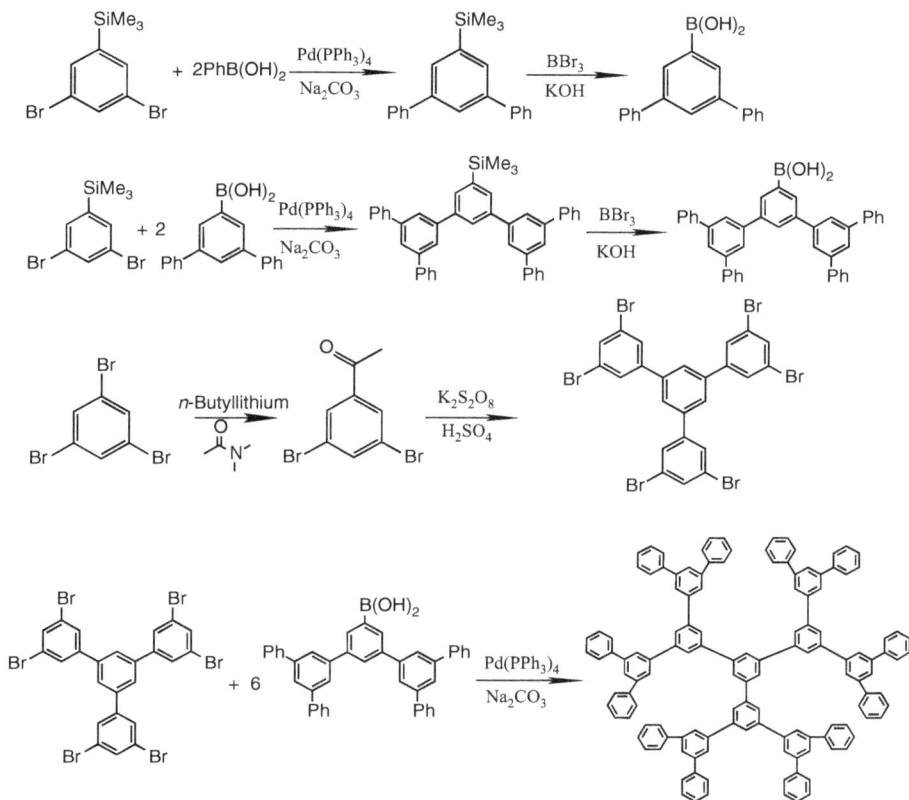

Chemical reaction equation 1

4.3.3 Dendritic polyether

Dendritic polyethers are synthesized by polyol as monomers, for example, 4,4-dihydroxybenzene-1-pentanol and 3,5-dihydroxybenzyl alcohol are common monomers. Taking 3,5-dihydroxybenzyl alcohol as an example, the reaction formula for synthesizing dendritic polyethers with high molecular weight are displayed in equation 2.

It can be seen from the earlier reaction formula that the polymerization process firstly protects the phenolic hydroxyl group of the polyhydroxy compound, then converts the benzylic hydroxyl group into a bromide, and then reacts with the polyhydroxy compound. The earlier two-step reaction process is repeated many times to obtain a dendritic polyether compound having a relatively high molecular weight. Ten generations of dendritic polyether have been synthesized.

Sometimes we couple high growth grade dendrimers with polyfunctional central molecules to obtain highly branched macromolecules. In general, those soft polyfunctional center molecules are more likely to set off the coupling reaction

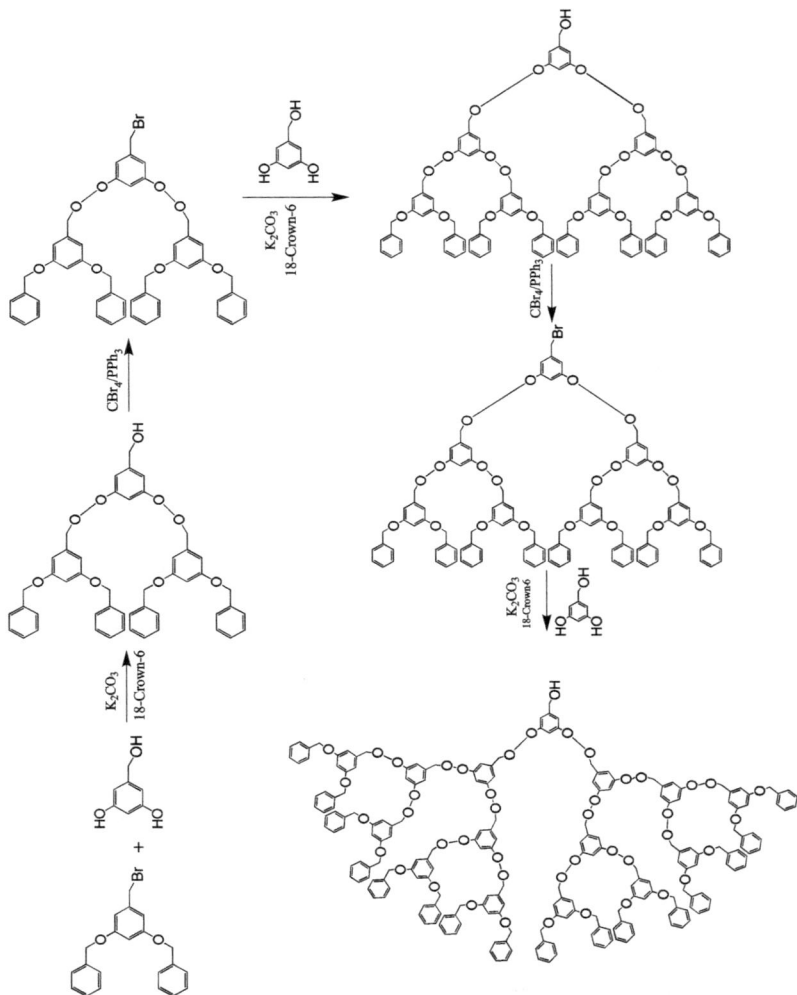

Chemical reaction equation 2

than those rigid, compact polyfunctional molecules. The use of higher growth order benzyl bromide to couple with higher growth order polyhydroxy compounds is a much faster synthesis of dendrimers.

Further, an optically active dendrimer can be synthesized by a polyphenol and a D-tartaric acid derivative or an L-tartaric acid derivative, and its optical activity is proportional to the number of optically active units contained therein. The combination of a porphyrin ring in the interior of the dendrimer has special photophysical and electrochemical redox properties. There are many examples of modifying dendritic polyethers in a similar manner.

4.3.4 Phosphorus-containing dendrimer

The dendritic polymer containing phosphorus on the branches is generally synthesized by an inside-out method by repeating a two-step or three-step reaction process. The central molecule is $SPCl_3$ or a six-membered ring $N_3P_3Cl_3$. The synthetic route is as follows:

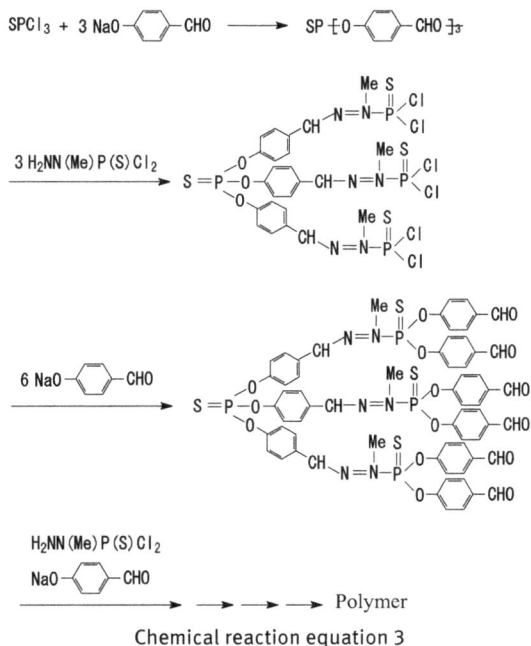

Chemical reaction equation 3

The surface of the dendrimer synthesized by this route contains a plurality of aldehyde groups or $-PCl_2$ groups. The aldehyde group or the $-PCl_2$ group has a large reactivity, and can be used to introduce other reactive groups on the surface of the dendrimer, such as mercapto, decyl, hydroxy, crown ether, and conjugated unsaturated groups. Usually the larger the group introduced, the greater the steric hindrance and the slower the reaction.

4.3.5 Dendritic polyamide-amine

Dendritic polyamide-amine (PAMAM) is the world's first synthetic dendrimer and is the most extensive studied dendrimer. The synthesis of PAMAM is carried out through Mickeal addition and amidation condensation reaction with ethylenediamine as the core, and its synthesis route represents a typical inside-out synthesis method for preparing dendrimers. As shown in the following equation, the PAMAM prepared by inside-out synthesis method is open and relatively loose at low generation (below 3.0 G, G stands for generation) and is close-packed on the surface at high generation (4.0 G or more). Nowadays 10 G PAMAM has been synthesized.

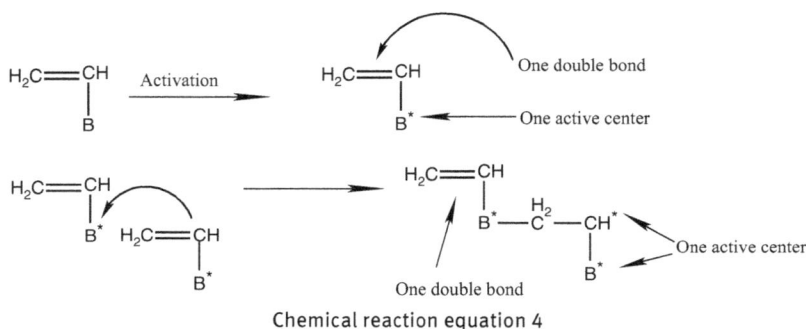

Chemical reaction equation 4

A plurality of amino groups is distributed on the surface of the PAMAM, and they can be further modified to obtain a functional polymer material. For example, using a 1–5 G PAMAM and 3,5-dihydroxybenzoic acid (3,5-DAC reaction), a series of peripheral modifications by small phosphor molecule 3,5-DAC can be obtained. The PAMAM is highly fluorescent and can be used as a fluorescent functional material. In addition, PAMAM have important applications in biomedical applications, as described in more detail in the following sections.

4.4 Synthesis of hyperbranched polymers

4.4.1 Basic points of the Flory theory

In his 1952 paper and 1953 work, Flory theoretically proved that a monomer containing an A functional group and n B functional groups ($n \geq 2$) can form highly branched polymer through intermolecular reaction without gelation. Each step of the reaction between A and B will regenerate $n-1$ B functional groups. The basic requirements for this monomer are as follows:

(1) The functional groups A and B can be activated in some way, for example, by catalyst or by removal of protecting groups.
(2) The activated functional groups A and B can react with each other, but they do not react with each other.
(3) The reactivity of functional groups A and B does not change as the reaction proceeds.
(4) The reactivity of the functional groups A and B should be high and specific enough to synthesize product with high relative molecular mass and to inhibit the production of by-products.
(5) No cyclization reaction occurs in the molecule

On this basis, Flory derived the number average degree of polymerization, the weight average degree of polymerization, and the degree of polymerization distribution index formula.

4.4.2 Growing process of AB$_x$-type monomer

The growing process of AB$_x$-type monomers is a typical polycondensation reaction process. Taking the polymerization of 3,5-dibromophenylboronic acid monomer proposed by Kim and Webster as an example, the first step of the reaction can be written as follows:

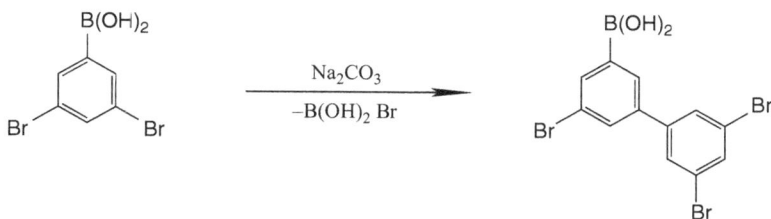

Chemical reaction equation 5

As can be seen from the earlier formula, the first step is the condensation of two monomers to form a dimer. In this reaction, as the molecule $B(OH)_2Br$ is released, a C–C bond is formed between the two benzene rings participating in the reaction. This dimer therefore contains one $B(OH)_2$ group and three bromine atoms. If expressed by the general formula, that is, one group A and three groups B, one more group than the monomer.

There are multiple possibilities for further growth. As shown in Fig. 4.3, the trimer can only be obtained by reacting a dimer II with a monomer I, resulting in the formation of two isomers: III a and III b. So far, no branching structure has appeared.

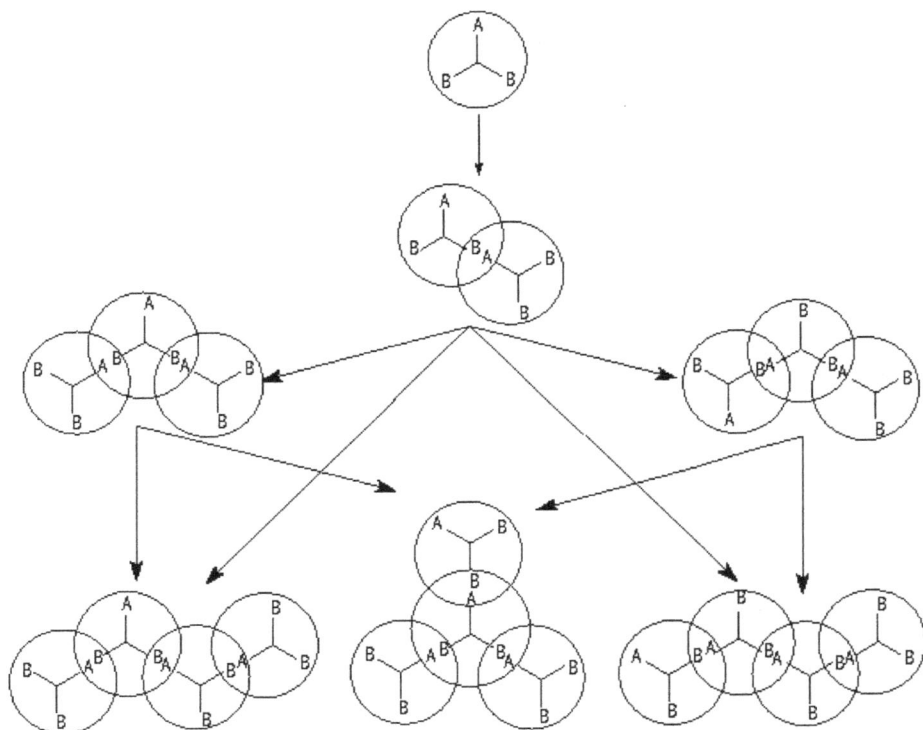

Fig. 4.3: Reaction possibilities for the formation of tetramers by AB_2 monomers.

The formation of a tetramer can be achieved by a combination of two dimers or a reaction between a trimer (III a or III b) and a monomer I. Tetramers have three isomers (IV a –IV b), of which only one is branched. The pentamer has 20 isomers, at least 6 of which are branched structures.

Once the branched structure is formed, the subsequent reaction is statistically increased in all directions. That is to say, the molecular structure of the product obtained is incompletely branched and not completely symmetrical. Here comes the concept of degree of branching. The degree of branching of the fully branched dendrimer is 1, the degree of branching of the linear polymer (completely unbranched) is 0, and the degree of branching of the hyperbranched polymer is between 0 and 1.

It can be seen from the earlier reaction rule that each growth step produces $x-1$ functional groups. Then, n AB_x monomers are reacted in the $n-1$ step to finally obtain a hyperbranched polymer having a polymerization degree of $P = n$, which has $(x + 1)n - 2n + 1$ B functional groups and one A functional group on the molecule.

As can be seen from Fig. 4.3, functional group A has a special significance for the growth process. Regardless of the degree of polymerization of the product, each molecule has one (and only one) reactive functional group A and remains until the end of the reaction.

Hawker et al. conducted a detailed study on the problem of intramolecular cyclization in the synthesis of hyperbranched polyesters. It is believed that the properties of ester groups and catalysts have a great influence on the relative molecular mass of hyperbranched polyesters. But for the degree of branching, they have almost no effect. In other words, the possibility of intramolecular cyclization is small. When the hyperbranched polyester prepared with 4,4-(4'-phenol)pentanoic acid, the functional group A which undergoes intramolecular cyclization is less than 5%.

When the degree of polymerization is high, the hyperbranched polymer molecule formed a spheroid. It is not difficult to understand that functional groups on the surface (or edge) of the molecule react more readily than those inside the molecule. Therefore, the degree of branching of the finally formed hyperbranched polymer is unlikely to reach 1, and the degree of polymerization is unlikely to reach the theoretical value.

4.4.3 Research on synthesis of hyperbranched polymers by AB_x monomers

Shortly after Kim and Webster used 3,5-dibromophenylboronic acid in 1990 to prepare the first hyperbranched polyphenyl end group with Br, Frechet and his colleagues used a similar method in 1991 with 3,5-bis(trimethylsiloxy)benzoyl chloride and obtained a hyperbranched polyester whose terminal group is hydroxyl group.

Between 1990 and 2000, researches on the synthesis of hyperbranched polymers with AB_x monomers have become a hot topic. The typical methods are as follows.

(1) In 1992, Kim used an aromatic compound containing a chloroamino group and an acid chloride group as a monomer to obtain a hyperbranched polyamide whose terminal group is amino group or carboxyl group. Since the molecule has a rigid benzene ring and a highly polar amide group, it fits the basic conditions for becoming a polymer liquid crystal. Studies have shown that the solution containing 40% of this polymer exhibits the properties of a nematic liquid

crystal at room temperature. At about the same time, Percec reported a hyperbranched polyether with thermotropic liquid crystal properties and studied the viscoelasticity of its solution. The viscosity of the solution was found to be much lower than that of the same type of linear polymer, and the degree of dissimilarity was also much higher.

(2) In 1993, Frechet synthesized a hyperbranched polyurethane. Since the polyurethane is obtained by the reaction between alcohol and isocyanate, the AB_x-type monomer is required to both have a hydroxyl group and an isocyanate group in the same molecule. The isocyanate group has a large chemical activity. The self-reaction of it can form a dimer, and if it reacts with a hydroxyl group, a carbamate will form, with water, it will liberate CO_2. Therefore, the monomer needs to protect the isocyanate during storage. Frechet used 3,5-dibenzamide phenol as raw material which will decompose at high temperature and produce an isocyanate group to react with the hydroxyl group in phenol and successfully synthesized hyperbranched polyurethane. In the same year, Kumar and Ramakrishnan used 3,5-diphenol azidobenzoyl as raw materials and successfully prepared hyperbranched polyurethanes.

(3) In 1994, Turner et al. converted 5-hydroxyisophthalic acid, which was easily decomposed at high temperature and could not be directly melt-condensed, into 5-acetoxyisophthalic acid and 5-hydroxyethoxyisophthalic acid. It then undergoes melt polycondensation to give a wholly aromatic and alkane-aryl polyester hyperbranched polymer having an acetoxyphenyl group and a phenol group. When this hyperbranched polyester is blended with linear polycarbonate, polyester, and polyamide, both tensile strength and compressive modulus increase, but percentage of breaking elongation and toughness decrease.

(4) In 1995, Malmstroem et al. used a melt polycondensation of 2,2-dimethylolpropionic acid to form a peralkyl polyester hyperbranched polymer. Since the application of hyperbranched polymers with similar structures in coatings has been reported, such hyperbranched polymers are promising as binders for coatings with good film formation and workability.

(5) In 1996, Hawker et al. synthesized polyethylene glycol hyperbranched polymers by AB_x macromonomers with linear oligomer units. These polymers are polymer electrolytes with high ion conductivity. Kakimoto et al. successfully produced hyperbranched polyphenylene sulfide by polysulfide cation method. The T_g of this polyphenylene sulfide is 20 – 40 °C higher than linear polyphenylene sulfide.

(6) In 1997, Wooley et al. protected a phenolic group on 1,1,1-tris(4-phenolyl)ethane with a silane group, converted the other two to imidazolium ester, and then used $CeF/AgNO_3$ as catalyst to prepare a hyperbranch. Hay et al. used Cs_2CO_3 or $Mg(OH)_2$ as a condensing agent to prepare hyperbranched poly(phenylene ether sulfone) with fluorine end groups and phenolic end groups, with a T_g of 277 °C and T_d of 377 °C. The mechanical relaxation behavior of hyperbranched polyesters with different end groups was studied by means of dielectric

spectrum, Differential Scanning Calorimeter (DSC), and dynamic mechanics. It was found that the hyperbranched polyester with terminal groups of benzoyl had only one secondary relaxation (β) below the glass transition temperature. It originates from the rotation of the ester group. The hyperbranched polyesters with oxime hydroxyl and acetyl end group have a lower secondary relaxation (γ) in addition to the secondary relaxation β. The secondary slack strength of hyperbranched polymers is much lower compared to their linear analogs.

(7) Moore et al. attached a starting unit to a carrier for so-called self-limited growth polymerization, which effectively controlled the relative molecular mass and relative molecular mass distribution. Before this, if we use one-pot method to prepare hyperbranched polymers with AB_x monomer, it was impossible to control the relative mass fraction relative to the molecular mass distribution.

(8) Zhang et al. introduced organic chromophoric groups, such as aniline group and carbachol group, into hyperbranched polymers to prepare a batch of hyperbranched polymers with nonlinear optical properties and photoconductivity. Fomina et al. synthesized a novel monomeric β, β'-dibromo-4-(10-undecynyloxy) styrene, which homopolymerized or copolymerized with β, β'-dibromo-4-ethynylethylene gives a hyperbranched polymer having photoluminescent function and discrete conjugated units. Maciejewski et al. converted phenolic resin into a hyperbranched structure, which is beneficial to broaden the application prospect of phenolic resin.

It can be said that various types of functional groups and monomers capable of undergoing polycondensation have been used to prepare hyperbranched polymers, and thus various types of hyperbranched polymers have been synthesized.

4.4.4 New developments in hyperbranched polymer synthesis – self-condensing vinyl polymerization

In 1995, Frechet et al. reported a new process for preparing hyperbranched polymers, namely self-condensing vinyl polymerization (SCVP). The monomer used in this method is completely different from the AB_x monomer described earlier.

The basic idea of this method is that the monomer $A = B-C^*$ contains one vinyl group with the ability for polymerization and one active center group which can be activated to initiate polymerization. In the process of forming a dimer, the active center C^* attacks the vinyl group on the second monomer and undergoes an addition reaction. During the reaction, as a C^* and a double bond are consumed, a new active center is generated on the second carbon atom of the double bond and a C^* is introduced. This new active center also can react with double bonds on another monomer. Therefore, the dimer contains a vinyl group and two active centers (Fig. 4.4). Two active centers can be further reacted. The trimer formed will have one vinyl group and three active centers, the tetramer contains one vinyl group and four active centers. The result of repeated reactions is the formation of hyperbranched polymers.

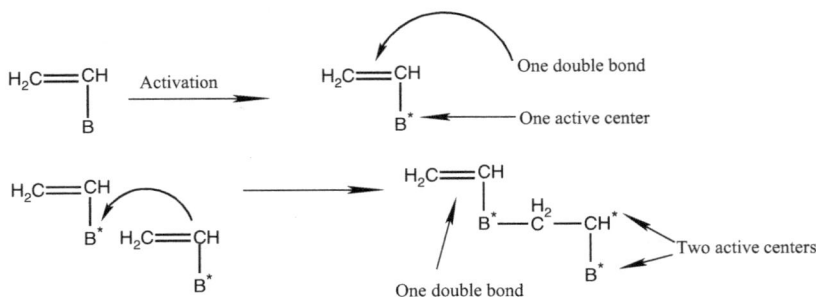

Fig. 4.4: Schematic diagram of self-condensation vinyl polymerization to form dimer.

The following formula is the reaction equation for preparing a hyperbranched polymer by cationic polymerization using 3-(1-chloroethyl) styrene as a monomer.

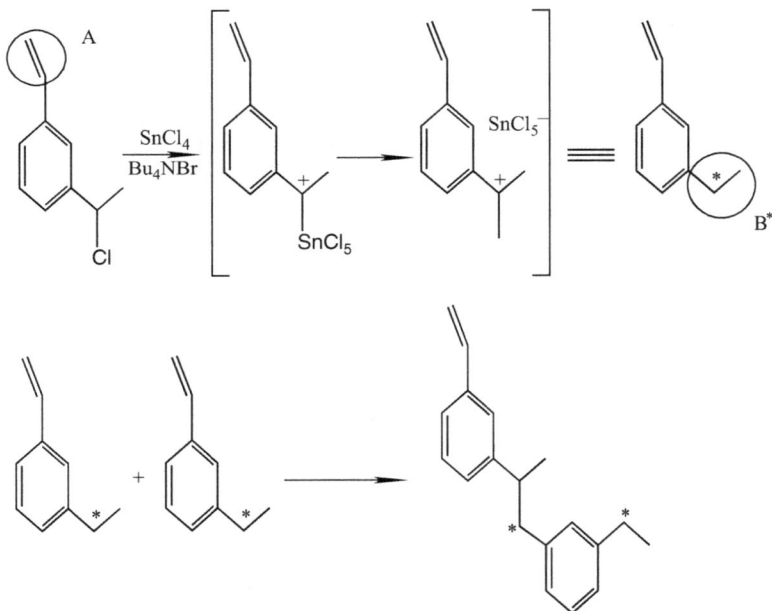

Chemical reaction equation 6

In principle, the concept of SCVP can be applied to various types of vinyl polymerization mechanisms (e.g., cationic polymerization, anionic polymerization, free radical polymerization, atom transfer polymerization, group transfer polymerization, and ring-opening polymerization). The structure of products depends on the nature of the B group and the type of activation. In fact, these mechanisms have been used to prepare hyperbranched polymers.

In 1995, Frechet reported another hyperbranched polymer prepared by (SCVP) and patented it. The monomer of this hyperbranched polymer is a vinyl ether containing a double bond and an acetoxy group which can be activated by a Lewis acid to form a cation. In addition, Frechet has successfully produced hyperbranched polymers from vinyl initiated by living radicals. Weber et al. used ruthenium to catalyze the polymerization of 4-acetylstyrene cations. Matyjaszewski et al. and Mueller et al. also obtained hyperbranched polymers by atom transfer polymerization and group transfer polymerization, respectively.

Since self-condensation vinyl polymerization is almost suitable for various polymerization mechanisms of vinyl monomers, some scholars have calculated that many new and practical materials can be prepared by such polymers with good industrial applications, such as hyperbranched polyfluoropolymers, hyperbranched liquid crystal polymers, thermoplastic elastomers, and polyolefins with novel structures.

4.4.5 The degree of branching of hyperbranched polymers

Since the branch structure of the hyperbranched polymer is imperfect, here comes the concept of the degree of branching. The degree of branching of the fully branched dendrimer is 1, and the degree of branching of the completely unbranched linear polymer is 0. The degree of branching of the hyperbranched polymer defined by Frechet and Hawker is the fraction of the number of monomer units and terminal units that are branched in the system.

$$\text{Degree of branching}(\text{DB}) = \frac{\text{Terminal units} + \text{Branched units}}{\text{Terminal units} + \text{Branched units} + \text{Linear units}} \quad (4.1)$$

However, this definition does not consider that the degree of branching of the linear polymer should be equal to 0. For example, according to this definition, the branching degree of the linear dimer is 1, and the degree of branching of the linear trimer is 1/2 which obviously does not match the concept of degree of branching. Therefore, Yan and Mueller corrected eq. (4.1). They believe that since the linear polymer contains a starting group and a terminal group, the error of eq. (4.1) can be avoided by subtracting 1 from the numerator and the denominator (if the functionality of the starting unit is f, then subtract f). Therefore, the degree of branching is defined as:

$$\text{Degree of branching}(\text{DB}) = \frac{\text{Terminal units} + \text{Branched units} - 1}{\text{Terminal units} + \text{Branched units} + \text{Linear units} - 1} \frac{n!}{r!(n-r)!}$$

$$(4.2)$$

Hoeter et al. also made similar considerations for the definition of the degree of branching of hyperbranched polymers and obtained the same results.

According to the earlier definition, Yan and Mueller derived the relationship between the average degree of branching and the conversion rate of functional group A from the polymerization dynamics differential equation. Specifically, for the AB_x-type polycondensation reaction,

$$\overline{DB} = \frac{x^3/2}{1-(1-x)(1+x-x^2/4)} \tag{4.3}$$

When conversion rate $\bar{x}=1$, $\overline{DB}_\infty = 0.5$. For self-condensing vinyl polymerization, the relationship between average degree of branching and double bond conversion rate is as follows:

$$\overline{DB} = \frac{2(1-e^{-x})(x-1+e^{-x})}{1-(1-x)(2-e^{-x})} \tag{4.4}$$

Assuming that the reactivity of the two active centers in the system is the same, then the conversion rate $\bar{x}=1$, $\overline{DB}_\infty = 0.465$. Yan and Mueller's research indicates that the average degree of branching of the self-condensing vinyl polymer product is related to the reactivity of the two active centers in the system. That is, the average degree of branching is the ratio of the reaction rates of the two active centers, $r = k_A/k_B$. When the double bond conversion rate $\bar{x}=1$, $\overline{DB}$ reached its maximum at $r \approx 0.259$. Currently, $\overline{DB}_\infty = 0.5$ which is like the case of the polycondensation reaction of the AB_x-type monomer.

4.4.6 Characterization of hyperbranched polymers

Characterization of hyperbranched polymers primarily focuses on the determination of degree of branching and relative molecular mass. So far, the determination of these parameters has largely relied on the characterization of polymeric compounds. For example, GPC is mainly used for the determination of the relative molecular mass, and the degree of branching is mainly determined by nuclear magnetic resonance (NMR) or the like.

Frechet, in his paper, details the method for determining the degree of branching of hyperbranched polymers by ^{1}H and ^{13}C NMR. He believes that in his synthetic hyperbranched polyester structure, there are four monomeric structural units:

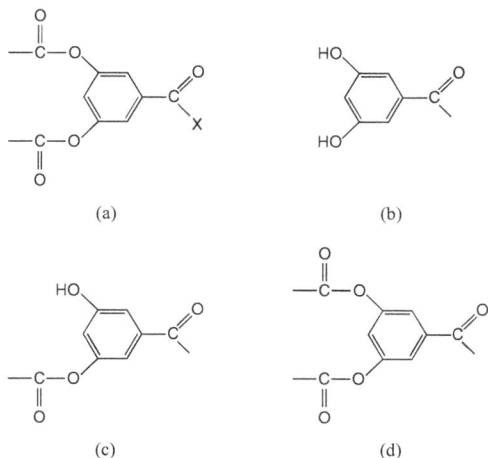

Chemical reaction equation 7

(a) in the earlier formula is the starting monomer unit of the hyperbranched polymer, and there is only one in each polymer molecule, so that it is negligible when
the molecular weight is large. (b) is an end group monomer unit having two phenol
groups. (c) is a linear monomer unit having a phenol group and an ester group. (d)
is a branched monomer unit having no phenolic group and only two ester groups.
These structural units have a special response to NMR.

In order to confirm these structures of the hyperbranched polymer, Frechet synthesized several model compounds (c) –(g) similar in structure to the earlier monomer
units, as shown in the following structural formula. It is apparent that these compounds should have a similar response to NMR as the earlier monomer units.

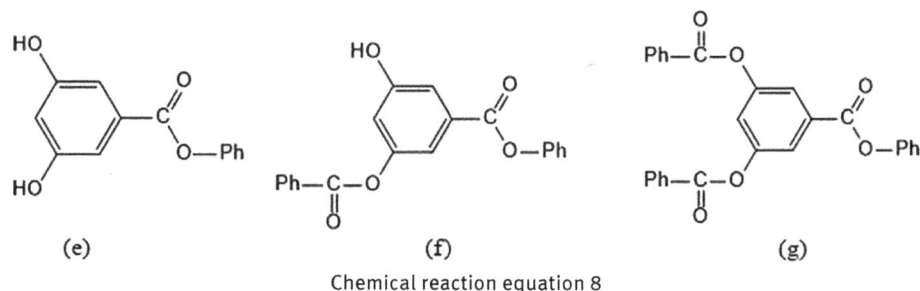

Chemical reaction equation 8

Experiments have shown that the ^{1}H NMR spectrum of the Frechet-synthesized
polyester hyperbranched polymer is in good agreement with the spectrum of the
earlier chemical. Frechet used this method to determine the degree of branching of
the polyester hyperbranched polymer: 0.55–0.60.

Hawker et al. also successfully determined the degree of branching of the fluo-
rine-terminated hyperbranched polyether ketone by 19F NMR analysis.

However, the earlier methods for determining the degree of branching of a polymer are limited and are effective only for certain specific types of hyperbranched polymers whose structures have a clear response to NMR. However, for some polymers, their NMR spectra are difficult to discern. Therefore, the degree of branching of many hyperbranched polymers has not been effectively measured to date.

Therefore, Hawker et al. used another method to determine the degree of branching. They modified the terminally hydroxylated hyperbranched polyester synthesized from 4,4-bis(4'-phenolic) valerate with methyl iodide to give a polymer with a terminal group an ether bond. Hydrolysis is carried out to obtain three kinds of low molecular hydrolysis products, and the structural formulas (h)–(j) are shown in the following figure. These three products correspond directly to the terminal units, branching units, and linear units in the polyester hyperbranched polymer. The relative contents of these three products (h)–(j) were determined by GPC and HPLC to be 24%, 51%, and 25%. Therefore, the degree of branching of this hyperbranched polyester is 49%.

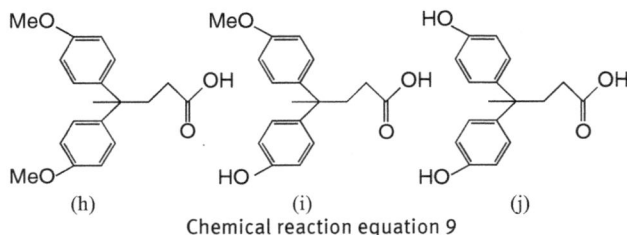

Chemical reaction equation 9

Wooley et al. used the same method to measure the degree of branching of hyperbranched polycarbonate to 53%.

It is not difficult to see that this method is limited, because not all polymers can be modified and hydrolyzed. Especially in the hyperbranched polymer molecules obtained by self-condensation vinyl polymerization, there are often no characteristic groups available for hydrolysis or degradation, so the method proposed by Hawker cannot be used to determine the degree of branching of the polymer.

In the study of hyperbranched polymers, the determination of relative molecular mass and relative molecular mass distribution is also a complex problem. Mueller and Yan derived the calculation formula of the average relative molecular mass and relative molecular mass distribution of hyperbranched polymers from the dynamic equations. It is pointed out that the hyperbranched polymer obtained by self-condensation vinyl polymerization has a molecular weight distribution which is much wider than that of its linear analog and is about twice that of the polycondensation reaction product of AB_x-type monomer. For the self-condensing vinyl polymerization, its average relative molecular mass and relative molecular mass expression is as follows ($k_A = k_B$, x is double bond conversion):

$$\overline{P_n} = \frac{1}{1-x} \tag{4.5}$$

$$\overline{P_w} = \frac{1}{(1-x^2)} \tag{4.6}$$

$$\frac{\overline{P_w}}{\overline{P_n}} = \frac{1}{1-x} = \overline{P_n} \tag{4.7}$$

That is, the relative molecular mass distribution index is equal to the number average degree of polymerization. For the AB_x-type polycondensation reaction, the average relative molecular mass and relative molecular mass distribution are expressed as (x is the conversion rate of the functional group A)

$$\overline{P_n} = \frac{1}{1-x} \tag{4.8}$$

$$\overline{P_w} = \frac{1-x^2/2}{(1-x)^2} \tag{4.9}$$

$$\frac{\overline{P_w}}{\overline{P_n}} = \frac{1-x^2/2}{1-x} = \frac{\overline{P_n}^2 + 2\overline{P_n} - 1}{2\overline{P_n}} \approx \overline{P_n}/2 \tag{4.10}$$

The relative molecular mass distribution index is approximately equal to half the number average polymerization degree. These findings help to judge the polymerization mechanism and determine whether a hyperbranched polymer is obtained.

The determination of the relative molecular mass of hyperbranched polymers generally still employs GPC with narrow distribution linear polystyrene (PS) as a standard. However, a hyperbranched polymer is a spherical molecule and the relationship between its relative molecular mass and the hydrodynamic radius R_h or radius of rotation is quite different from a linear molecule. Moreover, the R_h of the hyperbranched polymer is generally lower than that of the corresponding linear polymer with the same molecular mass. Therefore, the relative molecular mass of the hyperbranched polymer measured with the PS standard is much lower than the actual value. For example, Frechet et al. found that the weight average relative molecular weight $\overline{M}_w$ of a polyether hyperbranched polymer as determined by low angle laser light scattering is 3–5 times higher than that measured by the GPC method. In addition, since the surface of the hyperbranched polymer spheroidal structure tends to have many functional groups, the degree of solvation when they act with a solvent also greatly affects the R_h of the polymer. Newkome et al. observed that the R_h value of the carboxyl end group dendrimer was affected by the change in solvent pH as high as 50%. Kim measured the relative molecular mass of the polyamide hyperbranched polymer with a mixed solvent of N,N-dimethylacetamide/LiBr/H_3PO_4/tetrahydrofuran is 24,000 –46,000, while with pure N,N-dimethylformamide (DMF), the relative molecular mass measured is 700,000–1,000,000. This indicates that the aggregation state of the hyperbranched polymer molecules is completely different in different solvents. However, Turner et al. believe that the hydrodynamic radius R_h of the hyperbranched polymer can be estimated by comparison with linear PS standards. In the study of the wholly aromatic polyester hyperbranched polymer, Turner's research concluded that the R_h of the hyperbranched polymer with a weight average molecular weight $\overline{M}_w$ of 20,000 is 2.5 nm, and when the $\overline{M}_w$ is 100,000, the R_h is 9.0 nm.

4.5 Application of dendrimers and hyperbranched polymers

4.5.1 Application of dendrimers in gene transfection

In the early 1970s, the birth of genetic engineering enabled people to reorganize genetic material at the molecular level, making gene therapy possible. In order to achieve the purpose of preventing and treating diseases by using recombinant DNA, scientists have been hoping to find effective introduction of genetic material (antisense oligonucleotide, antisense expression plasmid, DNA vaccine, and therapeutic nucleic acid) into cells and promote the way it is expressed for more than 30 years. Exposed DNA has low infectivity for eukaryotic cells in in vitro and in vivo. Even if the DNA that has entered the cell, the retention and expression time is short due to factors such as nuclease in the cell, and it is difficult to meet the needs of the treatment. Therefore, the establishment of a safe, effective, and excellent DNA delivery vector system is a necessary means to develop gene therapy and transfection methods.

In order to obtain the pharmacological effects of an in gene or an antisense nucleic acid (antisense oligonucleotide or antisense DHA expression plasmid), the intravenously injected DNA reagent must pass through a series of biological barriers before it finally reaches the cytoplasm or nucleus of the target cell. These biological barriers include capillary beds, macrophages of the reticuloendothelial system, blockade of the capillary endothelium at the molecular level, inhibition of fusion by the extracellular matrix, and the cell membrane itself. Based on the earlier considerations, the ideal reagent for transporting genetic material in or in vivo should have the following characteristics: (1) stable binding to nucleic acid materials, (2) small enough to be present in the blood, and (3) promote the penetration of cells.

In the past 10 years, many centuries have been used in in vitro cell culture to enhance the transfection efficiency of genetic material such as plasmid DNA and antisense nucleic acids. These agents include cationic lipid complexes, polypeptides, surfactants, liposomes, and other agents. They are superior to viral vectors in many ways, such as being nonimmunogenic, without the potential for infection and is an easy way to produce. However, the transfection efficiency of these reagents is usually not as good as that of recombinant vectors such as adenovirus which significantly limits their use.

The new cationic polymer, PAMAM, which has emerged in recent years, brings hope to solve the earlier problems.

PAMAM is a new type of polymers with a rigid spherical structure which is radially symmetric. Many amino groups are distributed on the surface of the spherical macromolecule. PAMAM from first to tenth generation has been prepared, and its molecular diameter is 1–13 nm, which belongs to the nanometer scale range.

Under physiological conditions, the amino group at the end of the dendrimer PAMAM can be fully protonated into a positively charged $-NH_{3+}$. Therefore, the surface of the molecule has a high positive charge density and can electrostatically interact with the negatively charged bioactive phosphoric acid group on the main chain of the DNA molecule existing in the natural state to form a complex. The complex can achieve high levels of DNA transport efficiency in in vivo or in vitro, efficiently transfected into cells of a variety of cell lines, including meta-generational cells. At the same time, PAMAM also has a series of excellent features. First of all, the PAMAM/DNA complex has high stability and solubility. It can be stably present in aqueous solution for several weeks and is stable in a large pH range and under different buffer conditions without dissociation. It can be widely used in various biological and medical fields as an antibody, nucleic acid substance, contrast agent, and radiopharmaceutical binding substrate.

Currently, in more than six generations, degraded PAMAM products are commonly used in cell gene transfection experiments. This degraded PAMAM is obtained by heating and treating the intact dendrimer. The heat treated PAMAM still retains many terminal amino groups because the original positive charge is retained; at the same time, the degraded structure is highly branched, and the residual branches are no longer affected by the steric hindrance of the adjacent branches. The time of heat degradation varies from 1 to 25 h, so the transfection efficiency of the resulting degraded PAMAM is also different. In addition, it is also related to the solvent used in the heat degradation. When degraded in solvents like water and butanol, the transfection efficiency after degradation is related to the dissociation of these solvents. In some nondissociating solvents (such as DMF), heat degradation treatment does not improve transfection efficiency.

4.5.2 Application of dendrimers as carriers for drug and drug delivery

The dendrimer can be used as a potential drug delivery carrier in at least two aspects: ① the drug molecule can be physically captured inside the dendritic structure, and ② the drug molecule is covalently bound to the surface of the dendrimer or other group to form a dendrimer–drug conjugate.

(1) Physical capture

The dendrimer is appropriately designed to form a large cavity inside to capture the drug, making it possible to carry out a subsequent controlled release process. For example, the surface of a polypropylene imine dendrimer contains 64 functional groups that can permanently capture certain types of small molecule dyes, and under other specific conditions, dye molecules can also be released.

These properties suggest that it is possible to design an extremely sensitive selective drug delivery carrier that will be released when a certain signal or other stimulus near the target tissue is received. The drawback is that the release of its capture molecules is a "full" or "none" reaction; a slow, controlled release process has not yet been observed. Although this concept is very attractive, the dendrimer can only contain a limited amount of drug molecules, and it is difficult to remove the tightly surrounding polymer backbone to release the drug, thereby losing practical value in drug delivery. If the drug molecules are noncovalently bound to the dendrimer monomolecular particles by hydrogen bonding or by hydrophobic interaction, it is more promising due to the simple design.

(2) Dendrimer-drug conjugate for multivalent intermolecular adsorption

Multivalent adsorption is often used to explain the high affinity and selectivity phenomena encountered in the biomolecular recognition process of carbohydrate mediators, and for single sugar-receptor complexes, this adhesion is strong and nonselective. Many things in the body, such as cell adhesion, endocytosis of receptor mediators, pathogenesis of viruses and bacteria, and cell-to-cell recognition processes, play a key role in sugar–protein interactions. This inspires people to study the potential of polymers with sugar side chains as drugs. In this regard, the dendrimer appears to be unique because it can contain a plurality of end groups with similar properties and the corresponding ligands can form a higher concentration locally.

For example, a polylysine dendrimer with a saccharide group on the surface inhibits the binding of influenza viruses to glycoproteins on the surface of red blood cells. This is due to the binding of glycodendrimers to wheat germ agglutinin, which mediates the interaction between red blood cells, thereby significantly inhibiting red blood cells from viral infection. With similar dendrimers, the possible interactions between these compound and other glycoproteins have also been investigated. It has been found that the saccharide dendrimer contains about 12 saccharide end groups which are capable of binding to plant lectins to inhibit binding of the lectin to human glycoproteins. And this degree of inhibition is directly related to the number of sugar residues on the surface of the dendrimer, further supporting the concept of multivalent intermolecular adsorption.

Bioactive macromolecules such as polysaccharides, antibodies, and drug molecules can be bound to the terminal group of dendrimers to form dendrimer-drug conjugates. Due to the diverse functional groups of dendrimers, a dendrimer can carry a variety of drug molecules. By using dendrimers with different generation or changing the coupling conditions, the number of drug molecules per conjugate can vary. For example, PAMAM-platinate conjugates exhibit antitumor activity in all tested tumor models, including tumor models resistant to platinum.

5-Fluorouracil was bound to the fourth- and fifth-generation PAMAM, and then the conjugate was hydrolyzed in phosphate buffer, resulting in 5-fluorouracil release. Experiments on boron neutron capture therapy were also performed on dendrimers. The boron complex and the antibody are adsorbed on the surface of the dendrimer. When the antibody targets the cancer cell, the boron complex can capture neutrons of external origin and release the radiant energy to kill the cancer cells.

4.5.3 Application of dendrimers in medical diagnosis

Dendrimers are used in medical imaging fields such as magnetic resonance imaging (MRI), mainly because of their multiplicity of functional groups at the end, which have multivalent molecular adsorption effects. Many contrast agents can be combined on a single molecule in a precontrolled manner for site or tissue specific reactions and enhanced responses, while enhancing imaging sensitivity. The dendrimer acts as a carrier to provide an integrating group for the MRI reagent. For example, the chelating agent diethylenetriaminepentaacetic acid (DTPA) is covalently bonded to the sixth-generation PAMAM via a thiourea bond to form a PAMAM-DTPA complex, which can bind an average of 170 cesium ions (Gd^{3+}), which greatly exceeds the number of other macromolecular carriers. These dendrimer contrast media greatly enhances the imaging capabilities of the heart, blood vessels, and many other organs. Another example is PAMAM with folate on the surface, which is targeted to tumor cells expressing folate receptors. The strontium element was combined on the fourth-generation folate–PAMAM to detect the imaging system, and the reaction rate was found to be enhanced by 109%.

4.5.4 Application of hyperbranched polymers in the coatings industry

In recent years, domestic and foreign companies have been committed to the development of low-energy, high value-added, low-pollution, or nonpolluting coatings. The use of hyperbranched polymers in coatings has attracted the attention of the coatings industry. Due to the unique molecular structure of the hyperbranched polymer, the molecules are highly branched and the atoms in the molecule are closely packed, giving them many unique properties.

(1) Low viscosity. Compared to linear macromolecules of the same chemical composition and the same molecular weight, the hyperbranched polymer has a much lower viscosity at the same concentration. Therefore, it is very suitable for the preparation of high solids coatings. The addition of hyperbranched

polymer to the linear polymer system can greatly reduce the viscosity of the system and improve the fluidity of the coating and is used as a rheology modifier.

(2) High solubility. Hyperbranched polymers tend to be much more soluble than linear polymers of similar chemical composition due to the distribution of many functional groups around their molecules. In practical applications, the amount of solvent can be reduced, the cost can be reduced, and emissions can be reduced.

(3) Excellent film formation. The hyperbranched polymer has a dendritic macromolecular structure, which determines the noncrystallinity and nonentanglement of the polymer, so that the polymer has excellent film-forming properties. At the same time, many polar groups in the molecule give the paint film excellent adhesion.

(4) Easy to modify. The hyperbranched polymer molecule contains many terminal functional groups, which can be used for further chemical modification, grafting, block modification, and so on and is applied to prepare coatings with various properties.

Therefore, hyperbranched polymers are undoubtedly the ideal coating resin with great potential. The following is a review of some of the applications of hyperbranched polymers in coatings.

1) Polyester

The polycondensation of polyhydroxymonocarboxylic acid monomers and their use in coatings is one of the early patents. In recent years, many patents have reported hyperbranched polymers prepared from polyols and AB_2 monomers and their use in coatings.

At present, most of the reported aliphatic hyperbranched polyesters are synthesized from 2,2-dimethylolpropionic acid. The polymer end group is modified with acryloyl chloride to obtain a coating for ultraviolet (UV) light curing, but the process is complicated but the reaction time is long. A photocurable hyperbranched polymer synthesized with trimethylolpropane, 2,2-dimethylolpropionic acid, p-toluenesulfonic acid, and excess maleic anhydride has recently been reported. This new UV-curable coating does not require any solvent during the coating process, so it does not cause environmental pollution, and the cured product has excellent comprehensive properties, providing a possible way for the application of hyperbranched polymers.

A photocurable powder coating can be prepared by blocking an aliphatic polyester containing 16 terminal hydroxyl groups with methacrylic anhydride. The hydroxyl terminated hyperbranched polyester synthesized from pentaerythritol and dimethylolpropionic acid is modified with tall oil fatty acid to obtain a hyperbranched alkyd resin for use in alkyd paint formulations with similar molecular mass. Compared with the traditional alkyd resin with acid value and hydroxyl value, it has low viscosity, fast drying, and so on and has strong outdoor weather resistance. In addition, hyperbranched polyesters synthesized from 3,5-bis(trimethylsiloxy)benzoyl chloride have broad prospects in the application of specialty coatings.

2) Polyether

The hyperbranched polyether was synthesized by self-condensation of 5-bromomethyl-1, 3benzenediol in a crown ether solution of potassium carbonate. The polymer with different end groups of the same skeleton had a glass transition temperature between 38 °C and 70 °C. Another dendritic ether/amide polymer can self-assemble to form micelles in aqueous solution and can be applied to heat-curing coating systems due to its low melt viscosity, good compatibility, and more modified end groups.

3) Polyamine/ester

A hyperbranched polymer of a polyamide/ester structure synthesized from a series of cyclic carboxylic anhydrides and diisopropanolamine can be used as a curing agent for powder coatings. If the terminal hydroxyl group is further modified, such as modification with a fatty acid, a coating resin like an alkyd resin of different oiliness can be obtained. Coating resins modified with silicones, isocyanates, and acrylates have also been reported. Therefore, this is a kind of coating resin with great application prospects.

The polyacrylonitrile hyperbranched polymer is used together with benzophenone or isopropyl thioxanthone as a photoinitiation system, and the phenoxyethyl acrylate can be rapidly cured under nitrogen protection to obtain a higher degree of reaction.

4) Polyacrylate

At present, a variety of (meth) acrylate hyperbranched polymers which can be rapidly cured under UV light irradiation have been synthesized. Because of its low viscosity, the coating system can be diluted or solvent-free, and the cured film has excellent mechanical properties. It is a kind of environmentally friendly green paint.

4.5.5 Application of dendrimers as surfactants

As the generation of dendritic polymer increases, the molecular structure gradually approaches spherical shape. Although different from the molecular structure of conventional surfactants, these dendrimers tend to have the basic properties of surfactants because the dendrimers in the molecule may also contain lipophilic groups and hydrophilic groups. In addition, compared with the conventional polymer, the dendrimer has a well-defined structure, small crystallinity, low viscosity, good solubility, and many reactive or functional groups can be introduced at the end. Therefore, it has broad application prospects as a novel surfactant.

4.5.5.1 Surface activity of dendrimers

Many dendrimers are water-soluble polymers whose aqueous solutions have the characteristics of a typical aqueous surfactant solution. For example, a typical PAMAM, because its hydrophilic group is a carbonyl group and an amine group on the surface of the molecule, and the lipophilic group is an internal hydrocarbon chain and an external methyl group. Therefore, the products of different branched generations have certain surface activity, but their surface activity is related to the configuration of the molecule itself, in addition to the branching generation. It has been found that dendrimers with lysine as the core and tert-butoxycarbonyl as the terminal group not only have a certain surface activity, but also have good foaming ability.

4.5.5.2 Solubilization of dendrimers

The PAMAM has both a hydrophilic surface group and a hydrophobic inner layer, which is a single macromolecule linked by covalent bonds. As the number of branching generations increase, the functional groups on the surface of the molecule will become denser and denser, but there are still many cavities inside the molecule, which can accommodate lipophilic small molecules, thus macroscopically increasing the solubility of small molecules, that is, solubilization. The number of small molecules solubilized by each molecule varies with the molecular structure. The solubilization of dendrimers has important value as a drug delivery carrier and a catalyst carrier.

4.5.5.3 Demulsification of dendrimers

Since the end of the dendrimer contains many active groups, the oil–water interface can be strongly adsorbed to replace the original protective layer, and the strength of the new interface film is greatly reduced, and the protective effect is weakened, which is advantageous for demulsification. In addition, due to the relatively large molecular mass of the dendrimer, it can be dispersed in the emulsion, so that the fine droplets flocculate into loose micelles, which are then aggregated into large droplets to separate the oil and water, thereby achieving the goal of emulsion breaking.

For example, the third-generation PAMAM prepared by the divergence method has high demulsification properties for the O/W type crude oil emulsion and can quickly remove the oil phase in the emulsion. At 50 °C, the amount added was 100 mg/L, and the dehydration rate exceeded 90%.

4.5.5.4 Application of dendrimer surfactants
(1) Biomedicine

The earliest research of dendrimers is in the field of biomedicine. Tomalia has used a lot of text to discuss the significance of such synthetic polymers in the simulation of life sciences. He refers to the molecular size, topology, shape, surface chemistry, and flexibility of the polymer as "critical molecular design parameters," and believes that accurately controlling these parameters is the primary function of any entity in a variety of life processes. The evolution of natural biological entities involves a more highly ordered structure from atoms to point molecules and finally to biological networks. The assembly of small dot molecules into more "smart" biopolymers can be achieved by controlling "critical molecular design parameters". Therefore, dendrimers are of great significance in simulating life sciences.

In addition, dendrimers can be used in pharmaceutical carriers because certain drugs have been weakened or difficult to directly act on the affected part before entering the affected part due to poor solubility of the drug. The specific structure and properties of the dendrimers are very effective in controlling the delivery and release of the drug. A comb polymer such as a highly branched generation can be used as a carrier for a drug. After modification, the end group of low-branched (1.5 G) PAMAM, the dendritic polymer obtained has excellent solubility in benzoic acid, o-hydroxybenzoic acid, and so on. The solution also has extremely high stability and is an ideal drug delivery system. At present, foreign research in this field has made a breakthrough.

(2) Material modification

Since the molecular chain and terminal groups of dendrimers can be changed, these functional groups mainly include amine groups, ester groups, hydroxyl groups, amide groups, carboxyl groups, metal chelates, and hydrocarbons, which can be used as the structure and surface modifiers. For example, a dendrimer obtained by linking tin to a C_1-C_{10} alkyl group as a part of a dendrimer molecule and then connected to other hetero atoms can be used as an X-ray developing material.

A perfluorospherical polymer can be obtained by treating a dendrimer with fluorine. Due to its modification of the perfluorochain and end groups, it has a good lubricating effect in a complex system, and a special surface film can be formed on the surface of the magnetic recording material. In addition, dendrimer surfactants can also be used in the coatings industry. After partial fluorination and anionization of the terminal groups of the dendrimer, the resulting polymer is suitable for use in highly cross-linked and nonadhesive coatings to improve the surface properties of the coating.

(3) Industrial catalysis

The immobilization of liquid phase catalysts has always been a problem that chemists are trying to solve. Since the dendrimer has a broad cavity inside the molecule and a large amount of reactive functional groups outside the molecule, the active center of the catalyst can be introduced inside the dendrimer to complete the entire catalytic process inside the cavity. It is also possible to couple the active center of the catalyst to the outside of the molecule by utilizing the activity of the molecular surface end groups. In addition, the special molecular structure of dendrimers can function as a catalyst carrier in many cases. For example, a PAMAM having a branching degree of 4.0 G can be used as a template to disperse the transition metal Cu, Pt, and Pd on the surface to function as a carrier, and the catalyst can be used for hydrogenation of an olefin. This will provide a new class of support for precious metal catalysts.

The PAMAM is phosphated with diphenylphosphine methanol and complexed to form a Pa-bismethyl-tetramethylenediamine complex. This complex catalyzes the reaction of aryl bromides with acrylates and styrene and can be recycled.

(4) Petroleum industry

Oil extraction requires a large number of surfactants, many of which are polymeric compounds. Since the dendrimer contains a large amount of reactive functional groups and has a high relative molecular mass, it can be used as an emulsifier for a highly efficient O/W emulsion. For example, a 3.0 G PAMAM is used as a crude oil emulsifier, and when added in an amount of 1 mg/kg, the O/W emulsion containing 5% of crude oil can be completely broken. When the amount is 100 mg/kg at 50 °C, the dehydration rate exceeds 90%. If the end group is modified, the amount of emulsifier added can be reduced to 0.5 mg/kg, which is a new type of high-efficiency crude oil emulsifier. At present, most oil fields in China use new oil recovery technology, which cause a significant increase in water content in the production fluid, and the proportion of O/W emulsion is increasing. The polyether emulsifiers that were commonly used cannot solve the problem of demulsification of complex emulsions now. Using emulsifiers with new chemical structures to solve the problems caused by the adoption of new oil recovery technologies will be the main research direction of emulsifier development.

4.5.6 Application of dendrimers and hyperbranched polymers in other applications

4.5.6.1 Preparation of special structural polymers

The block copolymer of the hyperbranched polymer and the linear macromolecule has a good comprehensive property, thereby adding a new content to the block copolymer. By the polycondensation reaction of 4-fluoro-2'-methyl-4'-(trimethylsilane) benzophenone, a telechelic oligomeric (ether-ketone) which contains two trimethylsilyl end groups and one methyl group per repeat unit can be prepared. 1H NMR end group analysis showed two α, ω-bis(acetyloxy) oligo(ether-ketones) having a degree of polymerization of 14 and 28, respectively. These oligomeric (ether-ketone) and 70- or 140-times molar mass of silylated 3,5-bis(acetoxy)benzoic acid are bulk polycondensed at 270 °C, and then the product is dissolved in tetrahydrofuran and separated to obtain ABA triblock copolymer. Due to the long period of oligomerization (ether-ketone), low crystallization can be observed after annealing.

If the polycondensation is carried out at 290 °C, a completely soluble amorphous ABA triblock copolymer is obtained.

A two-stage polymer was synthesized by copolymerization of a hydrophilic linear molecular polyethylene oxide with a hydrophobic hyperbranched polyamide-amine. The glass transition temperature is determined by the terminal functional group of the hyperbranched polymer, and the added linear segments (polyethylene oxide having a molecular weight of 2,000 and 5,000, respectively) change the properties of the block copolymer. The viscosity behavior of short-chain polyethylene oxide copolymer molecules is like that of linear molecules, while long-chain polyethylene oxide copolymer molecules can form single-molecule micelles. The micelles are regularly arranged at the water–air interface. The hydrophobic hyperbranched portion is directed toward the air, and the polyethylene oxide segment is dissolved in the water. Such monolayers can be prepared as ultrathin nanoporous membranes on solid substrates.

By the reaction between a hyperbranched poly(aromatic ether phenylquinoxaline) or a poly(aryl ester) reactive functional group and a substituted oligomeric trioxirane disiloxane, an organic–inorganic heterochain polymer with high modulus, excellent crack resistance, and nanometer phase-separated transparent can be prepared. This organic–inorganic heterochain polymer remained stable at 400 °C and showed good toughness after introduction of the hyperbranched polymer. In addition, the degree of phase separation of such organic–inorganic heterochain polymers is controlled by the nature of the chain-end functional groups of the hyperbranched macromolecules. If a reactive group of triethoxysilane is used as a terminal group, nanosized phase separation can be obtained.

4.5.6.2 Polymer blending

The structure of the hyperbranched polymer is between the linear form and the spherical shape, and when blended with other polymers, the rheological properties of the blend can be greatly changed. The molecular model of the hyperbranched polymer indicates that there are a large number of voids between the chain. The molecular complexes present in the voids have been confirmed by the NMR study of the interaction between *p*-toluidine and hyperbranched polyphenylene. Such an effect can provide the free energy required for polymer compatibility. In the glassy state, the complexation is the same as the physical reversible cross-linking of the polymer chain. In the molten state, this static interaction is no longer maintained and the spherical hyperbranched polymer will affect the rheology of the matrix polymer. By adding a small amount of highly branched polyester to the PS, the rheology and thermal stability of the PS will change significantly. When the content of the hyperbranched polymer is 5%, the melt viscosity of the blend at 120 °C is 80% of the original and becomes 50% at 180 °C. The higher the temperature, the greater the shear rate and the more pronounced the effect. The addition of a hyperbranched polymer can also alter the thermal stability of the PS. Hyperbranched polymers also have a significant effect on the mechanical properties of PS blends. The flexural modulus of the 2% blend did not change, but the initial modulus increased by 24% and the maximum stress decreased by 15%. The reason for the high initial modulus is the formation of weak cross-links between the PS and the hyperbranched polymer by aryl–aryl interactions.

Hyperbranched polymers can also act as molecular enhancers for blends. The blend of highly branched polyester and polycarbonate can greatly increase the modulus of the polycarbonate, but its elongation is reduced. Adding 22.5% of the highly branched polyester to the PC, the tensile modulus of the film increased by 32%, the compressive modulus increased by 12%, and the elongation at break decreased from 82% to less than 3%. This reinforcement of the hyperbranched polyester is due to the large amount of aryl structure.

The thermoplastic polymer alloy with excellent compatibility and bonding properties can be prepared by blending hyperbranched polyester synthesized from polypropylene, maleic anhydride polypropylene, ethoxylated pentaerythritol, and 2,2-dimethylpropionic acid with nylon 6.

4.5.6.3 Hyperbranched polymer film

Hyperbranched polymers have great selectivity in forming Langmuir–Blodgett (LB) films due to their large amount of end groups that can be modified. By introducing different functional groups, LB films of different characteristics can be obtained.

Some amphiphilic derivatives of highly branched polyphenyls can be assembled at the water/air interface to form a unique Langmuir monolayer. For example, a terminal carboxyl derivative is assembled at the water/air interface to form a monolayer, and forms a surface of a fatty LB film as formed by inorganic crystallization nucleation at the water and air interface.

The structure of these amphiphiles is completely different from the amphiphilic molecules that form the traditional Langmuir membrane. For traditional amphiphiles, it must now be integrated into both hydrophilic and hydrophobic parts at the water/air interface. Correspondingly, the structure of the hyperbranched polymer is completely different from that of the conventional amphoteric substance, but to form a stable monolayer film, it still needs to be hydrophilic and lipophilic.

Crooks et al. reported the preparation of surface grafted hyperbranched polyacrylic acid films on self-assembled monolayers. This film produced by layer-by-layer deposition has a highly branched structure and unique physicochemical properties, and its thickness can exceed 0.1 μm. Due to the branched structure of the polymer, the thickness of the film does not increase linearly with an increase in the number of layers. In addition, this polymer contains a high density of carboxylic acid groups which can selectively bind metal ions or serve as a reaction point for subsequent modification. For example, molecular modification of a highly branched polyacrylic acid film results in a fluorescent, electroactive, hydrophobic, and molecular recognition property. Therefore, highly branched polymer films will provide a new approach to chemical sensor applications and tailoring polymer surface properties for a wide range of engineering applications.

4.5.6.4 Hyperbranched polymer liquid crystal

After in situ alkylation of a chain terminal phenol group of a homopolymer of 10-bromo-1-(4-hydroxy-4′-biphenyl)-2-(4-hydroxyphenyl) decane, the hyperbranched polymer obtained exhibits the characteristics of a thermally enantiomerically nematic liquid crystal. Meanwhile, 3,5-diacid chloride sulfinamide and its derivatives can produce a lyotropic hyperbranched aromatic polyamide liquid crystal in a low temperature amide solution. Studies have shown that a polyamide solution containing more than 40% of the earlier polymer exhibits a nematic liquid crystal structure at room temperature. The solution containing 60% of the earlier polymer is a hard gel under static conditions, but becomes a flowable liquid under shear. These solutions did not lose birefringence properties until 150 °C. The methyl ester of this polymer exhibits birefringence properties at room temperature, but becomes isotropy at around 70 °C. After being kept at room temperature for nearly one day, it again obtained birefringence properties. This behavior may be related to the tendency of the polymer to form aggregates in the amide solution. However, these polymers did not exhibit birefringence properties in an amide solution containing only 0.1% $CaCl_2$. There is currently no clear explanation for this phenomenon of hyperbranching.

However, the phase transition temperatures of the hyperbranched polymer liquid crystals reported in the literature are generally limited, and some even overlap with the glass transition temperature. This has affected the application of hyperbranched polymer liquid crystals. Recently, Percec designed a "willow"-like hyperbranched polymer based on the idea that rod-shaped molecules are arranged in parallel with each other to form a nematic liquid crystal. The so-called willow-like hyperbranched polymer refers to a molecule whose branching point is conformationally flexible, as shown in Fig. 4.5. Due to the flexibility of this molecule, its branches change conformation and are aligned parallel to each other to reduce free volume. This hyperbranched polymer liquid crystal also exhibits the behavior of nematic liquid crystals over a wide temperature range (50 °C–132 °C).

A cholesteric hyperbranched polymer liquid crystal having a chiral end group has been developed. Preparation of cholesteric hyperbranched polymers utilizes two different block units: ① AB_2 type monomer containing one mixed anhydride and two biphenyl acetate groups and ② 3,4-disubstituted benzoyl chloride end group unit. Both block units were synthesized from methyl 3,4-dihydroxybenzoate. After the diacetate monomer is treated with p-toluenesulfonic acid, a bulk polycondensation reaction is carried out, and then an acid catalyst and a chiral blocking agent are added to prepare a cholesteric hyperbranched polymer. This cholesteric hyperbranched compound exhibits cholesteric liquid crystal behavior in the range between 95 °C to 151 °C.

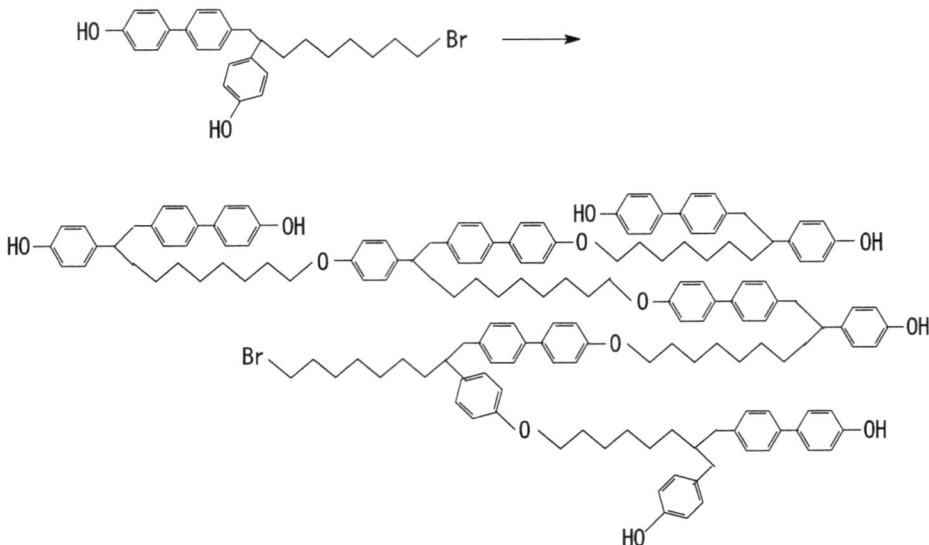

Fig. 4.5: Crystal structure of a willow-like hyperbranched polymer liquid crystal. Reprinted with permission from [57]. Copyright ACS Publications, 1994.

4.5.6.5 Degradable hyperbranched polymer

Since the hyperbranched polymer molecule has a spherical structure like a dendrimer, and its surface is also determined by the topological arrangement of the end groups, the degradable hyperbranched polymer can be used as a template for generating nanosized cavities. If the structural end groups are still attached to the substrate, a precisely controlled cavity with a chemical gene can be obtained. This is very important for simulating enzyme catalysts and molecular separation.

Alkoxysilanes are readily hydrolyzed under acidic conditions and such polymers are generally degradable. For example, the degradable hyperbranched poly (didodecyloxy) methylsilane can be readily degraded to methyldialkoxysilicic acid in methanol and HCl solutions. Its "shell" contains a large amount of vinyl which can be further chemically modified, which makes sense for the preparation of functionalized cavities.

4.5.6.6 Thermosetting resin

Highly branched polymers as thermosetting resins have several advantages: ① The same highly branched polymer backbone can be modified by end groups to give different resin structures. This structural change can cause a huge change in resin properties. These easily changeable properties are mainly due to resin viscosity, T_g, and wettability. ② The high solubility of the resin gives it good compatibility with other resins, and the amount of volatile solvent is greatly reduced in coating applications. ③ Since the functional group is generally a terminal group, polymerization easily occurs. Therefore, this resin can be cured quickly. In addition, the cured film of the highly branched polyester also has a low internal stress. This is because the highly branched polyester has a high initial relative molecular mass, and therefore, the amount of shrinkage during curing is relatively small, and thus the internal stress is also low.

4.5.6.7 Rheological properties modifier

The processing property of polymers is an important factor in determining whether a polymer has application value. Hyperbranched polymers have a lower melt viscosity. It has a good application prospect in the modification of the rheological behavior of traditional polymers.

It has been found that various hyperbranched polymers have different effects on the melt viscosity of PMMA, but have little effect on PS. When transmission electron microscope (TEM) was used to study the rheological modification mechanism of hyperbranched polymers, it was found that after the addition of hyperbranched polymers, large-scale phase separation occurred between the two components, while small components (hyperbranched polymers) migrated toward the surface. This indicates that the decrease in the melt viscosity of the polymer is not caused by the melt effect, and the hyperbranched polymer acts only as a surfactant which causes a decrease in apparent viscosity during migration to the surface of the polymer.

The content of the different hyperbranched polymers does not have a large effect on the melt viscosity of the polymer. This indicates that a very small amount of hyperbranched polymer can greatly change the melt viscosity of the linear polymer. In terms of use value, this has not only been greatly reduced, but also greatly reduced the degree of deterioration of the mechanical properties of the polymer due to phase separation.

By using a hyperbranched polymer as a processing aid for linear low-density polyethylene, it was found that the viscosity of the mixture was lowered, and the flow was accelerated at a fixed extrusion speed, so that the energy required for extrusion was remarkably lowered. At the same time, it also eliminates the shark skin phenomenon that often occur in the previous processing. Grafting the hyperbranched polymer onto the polypropylene also greatly reduces the interfacial tension between the nylon 6 and the polypropylene, making the two more compatible.

4.5.6.8 Drug sustained release agent

Hyperbranched polymers have been studied as drug carriers and are expected to be used in agriculture, cosmetics, and medical industries. A hyperbranched star-shaped polymer with a 1,1,1-trihydroxyphenylethane hyperbranched polymer as a core and a long chain of polyethylene glycol as an arm is synthesized, and the core of it is highly hydrophobic and can carry hydrophobic drug. The hydrophilicity of the arm molecular chain around the molecule is better, which increases the solubility of the hydrophobic drug in the polar medium. Controlling the size and shape of the hyperbranched polymer molecules can effectively control the distribution of the slow release drug in the body. If a hyperbranched polymer that is physically cross-linkable (e.g., hydrogen bonding) to a slow release drug is designed, it can produce a biocompatible small molecule drug after hydrolysis.

4.5.6.9 Conductive polymers and optical polymers

The PAMAM is modified with a cationic naphthalene diimide and then reduced to an anion with sodium thiosulfate in an aqueous solution or in formamide. Thereafter, the water is volatilized under the protection of argon to form a conductive hyperbranched polymer powder. The modified 3rd-generation PAMAM was prepared into a film by using formamide as a solvent. The fully reduced membrane has a conductivity of 10^{-3} s/cm in a neutral environment and a conductivity of 10^{-2} s/cm in a semireduced membrane. The conductivity of the polymer increases with increasing humidity, and when the relative humidity reaches 90%, the conductivity is 18 s/cm. After the PAMAM is methylated and quaternized, the PAMAM quaternary ammonium salt obtained has an electrical conductivity of 10^{-6}–10^{-5} s/cm.

A double-layered light-emitting diode was prepared by using hyperbranched polyimidazole as an electron-conducting layer and poly 9-tetradecyl-3,6(dibutadienyl)carbazole as a hole-conducting layer. The ionization potential of the hyperbranched polyimidazole was determined to be 5.9 eV, and the ionization potential of poly 9-tetradecyl-3,6(dibutadienyl)carbazole was 5.4 eV, and the electron affinities were 3.4 and 2.4, respectively. Both polymers are capable of functioning as an electron conducting layer and a hole conducting layer. High quality films can be prepared by spin coating, and with these novel electron emitter materials, new light emitting diodes can be successfully fabricated.

The crossl-inkable polymethacrylate and hyperbranched polyester are mixed for use as a stable nonlinear optical material.

4.5.6.10 Preparation of metal nanomaterials

Metal nanomaterials are a high-tech new material developed in the 1980s and can be used as a catalyst for petrochemicals. Because when the metal becomes a nanoparticle, the surface area is increased, the surface activity is increased, and thus the catalytic activity and the adsorption capacity are also improved. It can also be used as an excellent additive for ultrafine metal conductive adhesives, ultralow temperature heat exchangers, and composite materials. The melting point of nanometals is lower than that of ordinary metals, and the hardness is 3 to 5 times harder than that of conventional coarse-grained materials. Adding metal nanoparticles to chemical fibers can act as an antistatic agent.

In the PAMAM dendrimer with NH_2 and COOH end groups, the silver nanoparticles can be photochemically synthesized, and nanosilver with an average diameter of about 7 nm can be observed by spectrophotometry and transmission electron microscopy. This nanosilver can be used for the friction-reducing layer material, added to the chemical fiber, and has the function of sterilization and deodorization.

The charged PAMAM dendrimer is used to form an organic–inorganic hybrid colloid, which can be used as a template for the host–guest nanoscale synthesis. PAMAM dendrimers between 6.0 and 9.0 G can give a gold colloid template to prepare gold nanomaterials. A 10.0 G PAMAM dendrimer can be used to prepare smaller gold particles.

The PAMAM is used as a nanometer-scale "nanoreaction cell", for example, Cu^{2+} is reduced to Cu particles by a cavity inside the dendrimer. That is, Cu^{2+} inside the dendrimer molecule is chemically reduced to a cluster having a particle diameter between 4 to 64 nm. By changing the structure and size of the dendrimer molecules, it is possible to control the generation of nanoparticles of different sizes, which is expected to be used to prepare transition metal nanomaterials.

4.5.6.11 Nanocomposites

"Nanocomposite" was proposed by Roy et al. in the early 1980s. Different from single phase nanocrystalline materials and nano phase materials, it is a composite material composed of two or more solid phases, and one of them is composed of nanometer size (1–100 nm) in at least one direction. There have been many reports on nanocomposites in the field of PAMAM dendrimers in recent years.

A stable gold dendrimer was prepared by reducing PAMAM-tetrachloroaurate with hydrazine using a PAMAM dendrimer as a template for polymerization. Poly (4-sulfonate sodium styrene) (PSS) was used as the oppositely charged polyelectrolyte to assemble into a uniform multilayer gold-dendritic polymer nanocomposite by electrostatic layer by layer. The surface modified PAMAM dendrimer can be used to prepare dendrimer-metal nanocomposites such as Cu-PAMAM, Ag-PAMAM, Au-PAMAM.

As a semiconductor, a single CdS has always been favored for its optical properties. According to research, the CdS-dendritic PAMAM nanocomposite formed by combining CdS with dendrimer has good stability and can emit blue light. This composite is added to the SiO_2 to produce a stable, blue-emitting glass.

4.5.6.12 Adsorption resin

The complexation of 4.0 G PAMAM dendrimer and its derivatives with Cu^{2+} was studied by spectrophotometry. It was found that the terminal amine-based PAMAM dendrimer mainly has two coordination modes: $Cu\text{-}N_4$ and $Cu\text{-}N_2$, and the terminal hydroxyl group PAMAM dendrimer mainly has the coordination mode of $Cu\text{-}N_2$. As algebra increases, the number of Cu^{2+} that PAMAM dendrimers can complex increases. According to this feature, PAMAM dendrimer can be used as a complexing agent for toxic metal ions to treat water pollution caused by toxic metal ions.

References

[1]   Flory, PJ. J Am Chem Soc. 1952;74:2718.
[2]   Flory, PJ. Cornel University Press, 1953;361.
[3]   Wang, GJ, Yan, DY. Chin Polym Bull. 1999;2:1.
[4]   Fréchet, JM, Ledue, MR, Weimer, M, et al. J Polym Prep. 1997;38(1):756.
[5]   Fréchet JMJ Science. 1994;263:1710.
[6]   Johansson, M, Malmstroem, E, Hult, A. J Polym Sci, Polym Chem Ed. 1993;31:619.
[7]   Mekelburger, HB, Jaworek, W, Vogtle, F. Angew Chem. 1992;104(12):1609.
[8]   Gauthier, M, Moeller, M. Macromolecules. 1991;24:4548.
[9]   Kricheldorf, HR, Stobers, O, Lubbers, D. Macromol Chem Phys. 1995;196:3549.
[10] Gitsov, I, Wooley, KL, Fréchet, JMJ. Angew Chem. 1992;104(9):1282.
[11] Tomalia, DA, Hall, M, Hedstrand, DM. J Am Chem Soc. 1987;109:1601.
[12] Uchida, H, Kabe, Y, Yoshino, K. J Am Chem Soc. 1990;12:7077.
[13] Wooley, KL, Hawker, CJ, Fréchet, JMJ. J Chem Soc, Perkin Trans 1. 1991:1059.
[14] Moore, JS, Xu, Z. Macromolecules. 1991;24:5893.
[15] Hawker, CJ, Fréchet, JMJ. Polymer. 1992;33(70):1507.
[16] Weng, LL, Zheng, H. Acta Chim Sinica. 1997;55:595.
[17] Nagasak, T, Ukon, M, Arimori, S, Shinkai, S. Chem Soc Chem Commun. 1992;608.
[18] Jansen, JF, Brabander, EM, Meijer, EW. Science. 1994;266:1226.
[19] Fréchet, JMJ. Pure Appl Chem. 1996;A33(10):1399.
[20] Lu, P, Paulasaari, J, Weber, WP. Polym Prepr. 1996;37(2):342.
[21] Aoshima, S, Fréchet, JMJ, Grubbs, RB. Polym Prepr. 1995;1:531.
[22] Hawker, CJ, Farrington, P, Mackay, M, et al. J Am Chem Soc. 1995;117:4409.
[23] Malmstroem, E, Johansson, M, Hult, A. Macromolecules. 1995;28:1698.
[24] Hawker, CJ, Fréchet, JMJ, Grubbs, RB, Dao, J. J Am Chem Soc. 1995;117:10763.
[25] DeSimone, JM. Science. 1995;269(25):1060.
[26] Gaynor, SG, Edelman, SZ, Kulfan, A, Matyjaszewski, K. Polym Prepr. 1996;37(2):413.
[27] Simon, PFW, Radke, W, Müller, AHE. Macromol Rapid Commun. 1997;18:865.
[28] Maciejewski, M, Keolzierski, M, Bednarek, E. Polymer Bull. 1997;38:613.
[29] Johansson, M, Hult, A. J Coat Technol. 1995;67(849):35.
[30] Massa, DJ, Shriner, KA, Tnrner, SR. Macromolecules. 1995;28:3214.
[31] Hedrick, JL, Hawker, CJ, Miller, RD. Macromolecules. 1997;30:7607.
[32] Campagna, S, Giannetto, A, Serroni, S. J Am Chem Soc. 1995;117:1754.
[33] Fréchet, JMJ, Aoshima, S. US 5587441, 1996.
[34] Fréchet, JMJ, Aoshima, S. US 5587446, 1996.
[35] Fréchet, JMJ, Aoshima, S. US 5663260, 1997.
[36] Hawker, CJ, Fréchet, JMJ. J Am Chem Soc. 1990;112:7638.
[37] Wooley, KL, Hawker, CJ, Lee, R. Polym J. 1994;26:187.
[38] Chu, F, Hawker, CJ, Pomery, PJ. J Polym Sci A: Polym Chem. 1997;35:1627.
[39] Yan, D, Müller, AHE. Macromolecules. 1997;30:7024.
[40] Fréchet, JMJ, Henmi, M, Gitsov, I. Science. 1995;269:1080.
[41] Chang, HT, Fréchet, JMJ. J Am Chem Soc. 1999;121:2313.
[42] Magnusson, H, Malmstrom, E, Hult, A. Macromol Rapid Commun. 1999;20:453.

[43] Bednarek, M, Biedron, T, Helinski, J. Macromol Rapid Commun. 1999;20:369.
[44] Sunder, A, Hanselmann, R, Frey, H. Macromolecules. 1999;32:4240.
[45] Sunder, A, Hanselmann, R, Frey, H, et al. Macromolecules. 2000;33:309.
[46] Suzuki, M, Ii, A, Saegusa, T. Macromolecules. 1992;25:7071.
[47] Tsukahara, Y, Inoue, YO, Kohjiya, S. Polymer. 1994;35:5785.
[48] Wei, HY, Shi, WF. Chem J Chin Univ. 2001;22(2):338.
[49] Ning, M, Huang, PC. Polym Mater Sci Eng. 2002;18(6):11.
[50] Zhu, MG, Zhang, QZ, Hou, ZS. Chin Polym Bull. 2002;(4):32.
[51] QR, Han, Lei, ZQ, Yang, ZW. J Funct Polym. 2002;15(4):418.
[52] Wang, J, Yang, JZ, Chen, HX. Chin J Synth Chem. 2019(1):62.
[53] Xiao, J, HZ, Lu, Zou, P. China Med Eng. 2002;22(4):6.
[54] Xu, L, Yi, M, Xu, HY. Chin J Cancer Biother. 2003;10(10):1.
[55] Wang, J, Yang, JZ. China Surfactant Deterg Cosmet. 2002;32(1):35.
[56] Wang, ZG, Tong, SY. China Coatings. 2003;2:27.
[57] Virgil, P, Peihwei, C, Masaya, K. Toward "Willowlike" Thermotropic Dendrimers
      Macromolecules. 1994;27(16):4441–4453.

Exercises

Q1: What are the common synthetic methods for dendrimers? What's their advantages and disadvantages?

A1: The inside-out method and the outside-in method

The disadvantage of inside-out method is that the larger the number of reaction growth stages, the more functional groups are required to react, the less likely the growth reaction is to proceed completely, and the more easily the molecules are defective. To ensure complete reaction, excess reagents and harsh reaction conditions are often required, which makes the separation and purification of the product difficult.

The outside-in synthesis route is superior to the inside-out synthesis route in synthesizing monodisperse dendrimers, purification and characterization, and involves less reactive functional groups in each growth process. However, as the growth order increases, the steric hindrance of the reactive group at the central point increases, which hinders further reaction, and that makes the polymer generally not as large as that synthesized by the first method.

Q2: For AB_x monomers, what conditions do the structures need to meet so that it can form hyperbranched polymers?

A2: The basic requirements for AB_x monomer are

(1) The functional groups A and B can be activated in some way, for example, by catalyst or by removal of protecting groups.

(2) The activated functional groups A and B can react with each other, but they do not react with each other.

(3) The reactivity of functional groups A and B does not change as the reaction proceeds.

(4) The reactivity of the functional groups A and B should be high and specific enough to synthesize product with high relative molecular mass and to inhibit the production of by-products.

(5) No cyclization reaction occurs in the molecule.

Q3: **What is the most commonly used method for measuring the relative molecular mass of hyperbranched polymers? What is its flaw? What causes this flaw?**

A3: The determination of the relative molecular mass of hyperbranched polymers generally usually employs GPC with narrow distribution linear PS as a standard.

The problem is that the relative molecular mass of the hyperbranched polymer measured with the PS standard is much lower than the actual value.

Reasons: 1. A hyperbranched polymer is a spherical molecule and the relationship between its relative molecular mass and the hydrodynamic radius R_h or radius of rotation is quite lower than a linear molecule. 2. Since the surface of the hyperbranched polymer spheroidal structure tends to have a large number of functional groups, the degree of solvation when they act with a solvent also greatly affects the R_h of the polymer.

Q4: **For 21 AB$_3$ molecular, after 20 steps of reactions, a hyperbranched polymer with a polymerization degree of P = 21 is obtained. How much A and B functional group does it has?**

A4: Three functional groups B and one functional group A.

Q5: **Please list some practical applications of hyperbranched polymers.**

A5: In gene transfection, as carriers for drug and drug delivery, in medical diagnosis, in the coatings industry, as surfactants, and so on.

Chapter 5
Open-loop disproportionation polymerization

5.1 Introduction

Cyclic olefins are an important component of olefinic compounds, including mono-cyclic and polycyclic olefin compounds. The former is represented by cyclopentene, and the latter is represented by norbornene.

Ring-opening polymerization of cyclic olefins has been a topic of interest in the field of polymer chemistry. Especially in the past 10 years, the research on the ring-opening deuteration polymerization of cycloolefins has received more and more attention, and it has become a research hotspot in the field of polymer synthesis.

The study of ring-opening metathesis polymerization (ROMP) of cyclic olefins began in the 1950s. In 1955, Anderson and Merckling and others used $MgBr_2/TiCl_4$ and $LiAlH_4/TiCl_4$ as catalysts to ring-opening polymerization of norbornene. In the 1960s, Eleuterio used molybdenum oxide and aluminum oxide as catalysts, and carried out ring-opening polymerization of norbornene, and proved the special structure of the polymer by ozone oxidation. Later, Truett et al. demonstrated that the use of Zeigler–Natta catalysts can also polymerize norbornene into polymers of identical structure and propose a mechanism for cleavage and ring opening of the sigma bond (next to the double bond of the ring). The reaction formula of ring-opening polymerization of norbornene involved in the earlier work is shown in Fig. 5.1.

Fig. 5.1: Ring-opening disproportionation polymerization of norbornene.

It can be seen from the earlier reaction formula that the ring-opening polymerization of norbornene is substantially different from the mutual addition polymerization of the double bond of the olefin and is also different from the ring-opening polymerization of the heterocyclic ring of the lactam, the cyclic ether and the lactone, but the double bond is constantly translocated and the molecular chain is gradually enlarged. It is therefore also known as open-loop metathesis polymerization. After polymerization, the double bond, single bond, and ring structure in the monomer remain unchanged in the repeating unit of the obtained polymer, but the connection mode changes, so it is a new polymerization method completely different from the conventional polymerization. The resulting polymer has very specific properties.

https://doi.org/10.1515/9783110597097-005

In 1967, Calderon obtained 2-butene and 3-hexene from 2-pentene using WCl6-EtOH/Et2AlCl catalytic system, and successfully achieved homogeneous catalytic disproportionation of olefins. He named the reaction "the olefin metathesis reaction." Next, he carried out the ring-opening polymerization of the cyclic olefin with the same catalyst and proposed that the polymerization reaction is carried out by the mechanism of double-bond cleavage in the cyclic olefin molecule. Calderon basically explains the essence of open-loop disproportionation polymerization, and first relates the catalytic translocation of olefins to the ring-opening translocation of cyclic olefins, hence the name ROMP. In 1974, Dolgoplosk proposed a chain carbene mechanism for ring-opening disproportionation polymerization of cyclic olefins, which is currently accepted as a mechanism.

With the deepening of the theory and application of open-loop disproportionation polymerization and the development of the petrochemical industry, the open-loop disproportionation polynorbornene (PNBE) first realized industrial production in 1976, and then opened the disproportionated polycyclooctene and polybicyclo ring in the 1980s. Pentadiene has been industrialized one after another. The large-scale industrialization of the product proves the broad application prospect of the open-loop disproportionation polymerization reaction, and at the same time stimulates the research work on the theory of open-loop disproportionation polymerization.

In 1986, Gilliom et al. published a ring-opening deuteration polymerization of norbornene using a Ti-heterocyclic ring as a catalyst, and the relative molecular mass of the obtained product was narrow. They tracked the polymerization with deuterated norbornene and found that the relative molecular mass of the product increased with the addition of new monomers; there was almost no chain transfer reaction and chain termination reaction in the reaction. This indicates that the open-loop deuteration polymerization under this condition causes active polymerization, and this discovery has stimulated research enthusiasm. Since Szwarc proposed the concept of living polymerization in 1996, there is no other type, and the true active polymerization was discovered. In the following 10 years, polymer chemists have successfully synthesized many polymers and novel functional polymer materials with regular structure, controlled composition, monodispersity, and excellent performance by active ring-opening disproportionation polymerization.

Many studies have found that, in addition to obtaining a homopolymer of a specific structure, a ring-shaped deuteration polymerization can also obtain a strictly alternating copolymer, and the reaction speed is fast.

For example, a ring-opening deuteration polymerization of 5-acetoxycyclooctene is carried out using a ruthenium carbene complex $[(PCy_3)_2Cl_2Ru(=CH-CH=CPh_2)]$ (Cy is a cyclohexyl group) as a catalyst, and the obtained product is a ternary element containing both polyethylene ($-CH_2-CH_2-$), polyvinyl acetate $[-CH_2 CH(OCOCH_3)-]$, and poly 1,4 butadiene ($-CH_2-CH=CH-CH_2-$) structural unit alternating copolymers (as shown in Fig. 5.2), which is not possible with any other method to achieve. Because we know that if the three monomers of ethylene, vinyl acetate, and 1,3-butadiene

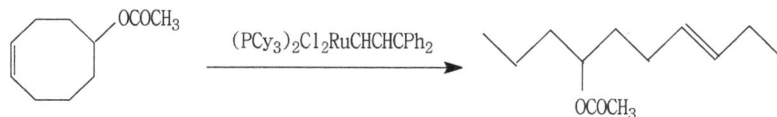

Fig. 5.2: Preparation of triblock copolymer by ring-opening deuteration polymerization of 5-acetoxycyclooctene. Reprinted with permission [12]. Copyright, 1994.

are copolymerized by free radicals, and there are nine chain-growth reactions, involving six different values of monomer aggregation, it is impossible to obtain a strictly alternating terpolymer completely. This is the charm of open-loop deuteration polymerization to attract the attention of scientists.

In the past decade, the use of ring-opening deuteration polymerization of cyclic olefins has developed many new polymer materials with excellent properties, such as reaction injection molding polydicyclopentadiene (new high impact plastic), PNBE, polycyclooctene (new thermoplastic elastomer), and so on and to achieve large-scale industrial production. In addition, the use of ring-opening deuteration polymerization, a method with strong molecular design capabilities, has led to the development of many new varieties of polymers that cannot be achieved by other methods, such as the preparation of the strictly alternating terpolymers described earlier. Therefore, the emergence of ring-opening deuteration polymerization provides a useful means for the structural design of polymer materials, especially functional polymers. Open-loop disproportionation polymerization has become a powerful tool for structural design, tailoring, and molecular assembly of new polymer materials.

5.2 Principle of open-loop disproportionation polymerization

5.2.1 Catalyst for ring-opening disproportionation polymerization

The process in which a cyclic olefin is cleaved in a molecule in the presence of a catalyst and joined to a macromolecule in a head-to-tail manner is called a ring-opening deuteration polymerization.

The focus of open-loop disproportionation polymerization research is the research and development of catalysts or catalytic systems. Over the past few decades, through unremitting efforts, hundreds of catalysts have been found to be available for ring-opening deuteration polymerization of cyclic olefins.

Generally, the catalyst for ring-opening disproportionation polymerization is a composite catalyst comprising a transition metal inorganic compound as a main catalyst, a main group metal organic compound as a cocatalyst, and sometimes a third component compound is required as an activator. At present, the most important open-loop deuteration polymerization catalysts have three major categories: The first is a traditional catalyst, which was developed by Anderson et al. since the 1950s, and later developed by Laverty et al. In such a catalyst, the main catalyst is MX_n (wherein M = W, Mo, Re, Os, etc. transition metals, X = Cl, Br, etc. halogen atoms). The second type is a water-soluble catalyst, which can be used for the ring-opening deuteration polymerization of an aqueous solution system, and mainly represents $K_2RuCl_3 \cdot H_2O$. The third type is a carbene-type or alkylene-type catalyst developed in recent years. The catalyst has clear initiation mechanism, high catalytic activity, and strong stereoselectivity of the product. Therefore, it is the most important open-loop enthalpy polymerization catalyst of research in development.

5.2.1.1 Traditional catalyst

The traditional catalyst for ring-opening disproportionation polymerization is the most thorough part of the research to date. Anderson et al. began to study the use of open-loop disproportionation polymerization. In addition to good catalytic activity for common cyclic olefins, such catalysts also have good catalytic activity for polar monomers such as conjugated dicarbon-based compounds. The stereoregularity of the polymer obtained by such a catalyst is controlled by the stereo configuration of the catalyst and can be adjusted within a certain range, and thus has good molecular structure design. Such catalysts are currently mainly used for the polymerization of cyclopentene, cyclopentadiene, norbornene, and derivatives thereof.

5.2.1.2 Water-soluble catalyst

The water-soluble catalyst which is currently an important ring-opening deuteration polymerization is $K_2RuCl_3 \cdot H_2O$. It is mainly used for the ring-opening deuteration polymerization of 2,3-difunctional substituted norbornene and 7-oxidized norbornene. The polymerization conditions for adjusting low temperature and high pressure can effectively control the stereo structure of the polymer.

5.2.1.3 Carbene- or alkylene-type catalyst

Carbene-type or alkylene-type catalysts are the new catalysts for the fastest-opening ring-opening deuteration polymerization of cyclic olefins in recent years. There are already six types available so far, as shown in Fig. 5.3.

Among the carbene-type or alkylene-type catalysts, the most important ones are the c type (Schrock type) and the e type (Grubbs Ru type) in Fig. 5.3. The two types of catalysts are mainly described.

The cocatalysts of such catalysts are mainly $CpTiCl_2/RMgX$, (wherein R = CH_3, X = Cl, Br, I, etc.), titanocene compound/MeLi, and the like.

Invented by Casey in 1974

a

Invented by Osborne in 1986

b

Invented by Schrock in 1986

c

Invented by Grubbs in 1986

d

Invented by Grubbs in 1992

e

Invented by Herrman in 1991

f

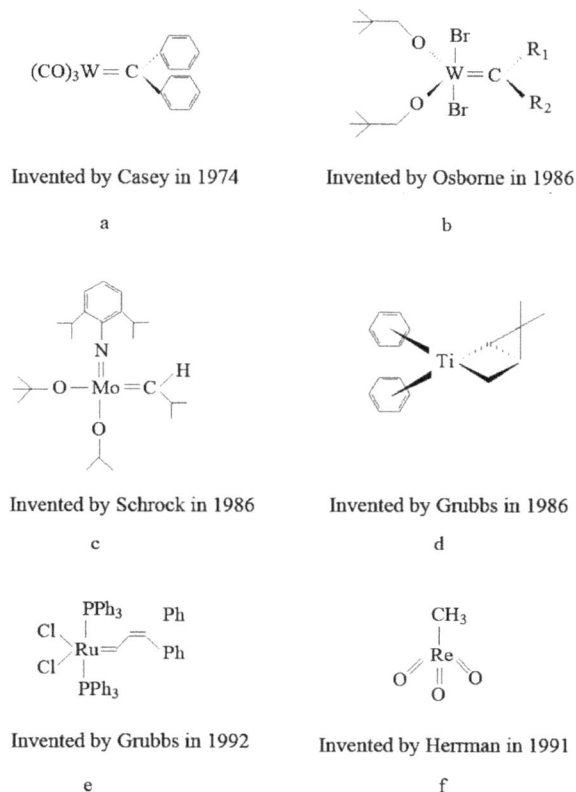

Fig. 5.3: Carbene or metal alkylene catalyst.

(1) Schrock-type catalyst

The main advantage of the Schrock-type catalyst is its clear initiation mechanism. Its catalytic activity, stereoselectivity of the product, and compatibility with heteroatoms depend on the nature of the alkoxy substituent. There are many kinds of ligands. There are fewer side reactions than the same type of W carbene catalyst.

Chiral ligands have recently been introduced into Schrock-type catalysts. A typical example is the disubstituted-linked (2,2'-dihydroxybinaphthalene) alkylene molybdenum. The ring-opening deuteration polymerization initiated by this catalyst produces an all-cis isotactic product. For example, for the ring-opening deuteration polymerization of 2,3-R_2-norbornene (R = CF_3 or CO_2Me), the tacticity of the product is > 95%.

The rate of polymerization of the monomers initiated by different chiral ligands is different, and the stereoregularity of the polymers is also different. The proportion of the cis-reverse structure in the polymerization product also varies depending on the ligand.

Studies have shown that Schrock-type catalysts with methacrylates cross-linked with methylaluminoxane as carriers have higher activity than unsupported catalysts and can control the tacticity, relative molecular mass, and distribution of polymers. Another advantage of this catalyst is that it can be reused multiple times without reducing its activity.

(2) Grubbs Ru-type catalyst

Ru salt has been used as a catalyst for ring-opening deuteration polymerization for a long time. In the late 1980s, Grubbs found that aqueous Ru salt compounds used as ring-opening deuteration polymerization catalysts not only are resistant to trace amounts of water, but also have high catalytic activity for ring-opening deuteration polymerization of aqueous systems. In order to clarify the catalytic mechanism of Ru salt compounds, Grubbs synthesized a carbene-type Ru catalyst with tricyclohexyl as a ligand in 1992 and confirmed its catalytic activity for ring-opening deuteration polymerization. He soon discovered that the use of triphenylphosphine instead of tricyclohexyl as a ligand greatly improved the catalytic activity, but the rate of initiation was slow. However, if a fluorinated phenyl group is introduced at the carbene position, not only the catalytic activity is improved, but also the chain initiation speed and the chain growth rate are greatly improved. The structure of the fluorocarbon-containing carbene-type Ru catalyst is shown in Fig. 5.4.

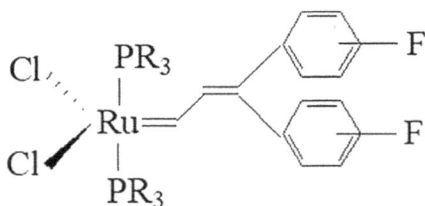

R is a cyclohexane group and F is in the meta or para position.

Fig. 5.4: 1-Ru-4,4-dichlorophenyl-1,3-butadiene used as a catalyst for ring-opening deuteration polymerization.

Although the early Ru salt catalyst has good catalytic activity for ring-opening deuteration polymerization, the catalytic mechanism is not clear, and it is impossible to systematically study and improve the catalytic activity. Therefore, a good catalytic system cannot be formed, and in recent years, there is a possibility of being gradually eliminated. The carbene-type Ru catalyst makes up for the deficiency of the following Ru salt catalyst and is stable in HCl solution of aldehyde, ketone, water, alcohol, and diethyl ether and has great catalytic activity for ring-opening deuteration polymerization of cyclic olefin. Such catalysts have been used in the ring-opening deuteration polymerization of various functional group-containing monomers, and various useful functional polymer materials have been prepared. For example, the structure of a novel carbene-type Ru catalyst is shown in Fig. 5.5.

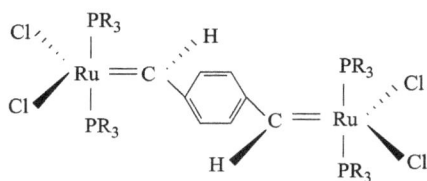

Fig. 5.5: New carbene-type Ru catalyst.

This catalyst contains a bis-carbene structure which initiates the reaction starting from both ends of the catalyst, so that the resulting polymer has two active ends which can be further used to prepare triblock copolymers.

The use of a carbene-type Ru catalyst is highly inert to many functional groups and can be used for the synthesis of polymers having pendant groups which are cross-linkable. These pendant cross-linkable polymers can be cross-linked under the action of ultraviolet light. Therefore, it can be used as a photosensitive polymer.

5.2.2 The active center of ring-opening deuteration polymerization and its initiation mechanism

There have been many discussions on the mechanism of ring-opening deuteration polymerization of cyclic olefins, and the currently accepted view is the mechanism of initiation and growth of metal carbene complexes.

Metal carbene complexes which initiate ring-opening deuteration polymerization of cyclic olefins can be obtained by several routes:
(1) Prepared in advance, as described earlier for the Schrock complex.
(2) By in situ reaction of a main catalyst (such as WF_6, [$Mo(CO)_5Cl$].) with a cocatalyst (such as Et_2AlCl, Me_4Sn), for example:

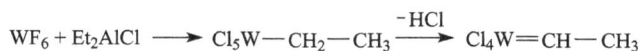

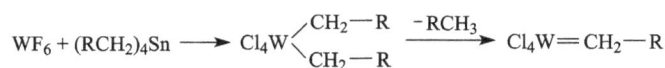

$$WF_6 + Et_2AlCl \longrightarrow Cl_5W-CH_2-CH_3 \xrightarrow{-HCl} Cl_4W=CH-CH_3$$

$$WF_6 + (RCH_2)_4Sn \longrightarrow Cl_4W{\Big\langle}{{CH_2-R}\atop{CH_2-R}} \xrightarrow{-RCH_3} Cl_4W=CH_2-R$$

(3) The monomer is complexed with the active center metal atom of the catalyst to undergo a hydrogen transfer reaction on the double bond to directly form a metal carbene complex.

The initiation reaction begins with the formation of a metal-heterocyclobutane transition state after the double bond in the cyclic olefin complexes with the metal carbene. After complexation, both the double bond and the metal carbene bond in the cyclic olefin are activated and then broken to form a new growth metal carbene complex.

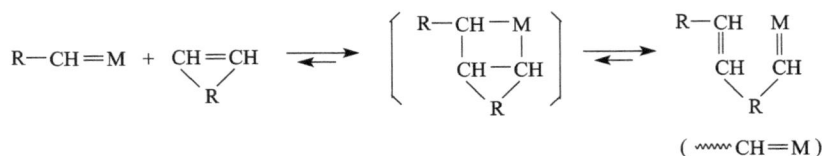

$$R-CH=M \; + \; \underset{R}{CH=CH} \; \rightleftharpoons \; \left[\begin{array}{c} R-CH-M \\ | \qquad | \\ CH-CH \\ \diagdown_{R}\diagup \end{array} \right] \; \rightleftharpoons \; \begin{array}{cc} R-CH & M \\ \| & \| \\ CH & CH \\ \diagdown_{R}\diagup \end{array}$$

$$(\text{\textasciitilde\textasciitilde} CH=M)$$

The newly formed metal carbene complex can continue to form a metal heterocyclic butane transition state with a double bond on the cycloolefin monomer, which upon cleavage forms a new growth metal carbene complex. This was repeated to obtain a polymer.

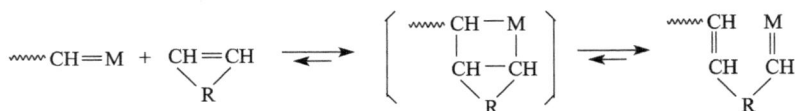

$$\text{\textasciitilde\textasciitilde} CH=M \; + \; \underset{R}{CH=CH} \; \rightleftharpoons \; \left[\begin{array}{c} \text{\textasciitilde\textasciitilde} CH-M \\ | \qquad | \\ CH-CH \\ \diagdown_{R}\diagup \end{array} \right] \; \rightleftharpoons \; \begin{array}{cc} \text{\textasciitilde\textasciitilde} CH & M \\ \| & \| \\ CH & CH \\ \diagdown_{R}\diagup \end{array}$$

The active center of the ring-opening deuteration polymerization is a metal heterocyclobutane transition state formed by a metal carbene complex and a double bond on the cycloolefin monomer.

In the polymerization process, in addition to the earlier polymerization reaction, the following chain transfer reaction may occur.

(1) Self-biting transfer. The growing macromolecular metal carbene complex chain end reverses the double bond on the polymer itself chain, resulting in the formation of a cyclic polymer.

The difficulty of self-back-biting transfer reaction depends on the rigidity of the metal carbine growth chain. The product of ring-opening polymerization of monocyclic monoolefins is prone to self-back-biting transfer reaction due to the flexibility of the metal carbene growth chain. However, the product of ring-opening deuteration polymerization of norbornene is less likely to undergo its own reverse bite transfer reaction due to the rigid structure of the molecular chain.

(2) The macromolecular metal carbene complex undergoes a cross-chemical reaction with other macromolecules in the system.

$$P_1 \text{\textsf{wwww}} CH-M + P_2 \text{\textsf{wwww}} CH=CH \text{\textsf{wwww}} P_3 \longrightarrow P_1 \text{\textsf{wwww}} CH=CH \text{\textsf{wwww}} P_3 + P_2 \text{\textsf{wwww}} CH=M$$

(3) The macromolecular carbene complex undergoes a cross-chemical reaction with the added acyclic olefin in the system.

$$P \text{\textsf{wwww}} CH=M + R_1-CH=CH-R_2 \longrightarrow P \text{\textsf{wwww}} CH=CH-R_1 + R_2-CH=M$$

Here, the addition of the acyclic olefin becomes a molecular weight regulator.

During the polymerization process, chain termination reactions that may occur are as follows:

(1) The α-hydrogen transfer reaction occurs at the active chain end of the growing macromolecular carbene complex to form a macromolecule with a terminal double bond and to reduce the metal atom.

$$P \text{\textsf{wwww}} CH_2-CH=M \longrightarrow P \text{\textsf{wwww}} CH_2=CH_2 + M$$

(2) The growing macromolecular carbene complex reacts with the terminator to form a metal carbene complex without initiation activity. Commonly used terminators are trimethylvinylsilane, stilbene, vinyl ether, and the like. For example:

$$P \text{\textsf{wwww}} CH_2-CH=M + Me_3Si-CH=CH_2 \longrightarrow P \text{\textsf{wwww}} CH_2=CH_2 + Me_3Si-CH=M$$

(3) The growing macromolecular carbene complex terminates with a cyclopropanation reaction with other molecules containing double bonds.

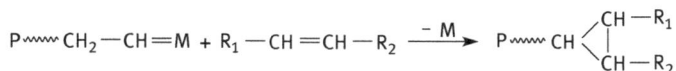

$$P \text{\textsf{wwww}} CH_2-CH=M + R_1-CH=CH-R_2 \xrightarrow{-M} P \text{\textsf{wwww}} CH \underset{CH-R_2}{\overset{CH-R_1}{<}}$$

5.3 Ring-opening disproportionation polymerization of norbornene and its derivatives

5.3.1 Ring-opening disproportionation polymerization of norbornene

The study of the ring-opening deuteration polymerization of norbornene can be traced back to the 1950s in the last century. Anderson et al. pioneered this important polymerization. Later, many scientists joined the research and began to look for various catalysts and made great progress.

French scientist Larroche et al. used a Group VII A metal-organic complex as a catalyst to carry out ring-opening polymerization of norbornene, and the content of the double-bonded cis structure was determined by 1 H NMR and 13 C NMR. The results show that the W catalyst has good stereoselectivity for the aggregation of norbornene. When the same catalyst is used to polymerize linear olefins, it exhibits a weak stereoselectivity. There are different opinions on the interpretation of this phenomenon. One view is that the stereoselectivity in the ring-opening deuteration polymerization process may be related to the certain directionality of the cyclic olefin double bond to the metal coordination and the ligand to cause a certain steric hindrance of the central metal. Another view is that stereoselectivity depends on the energy of both the coordination olefin and the metal cyclobutane structure. If the latter is larger than the former, the system will lose stereoselectivity. The Lewis acid in the catalyst can coordinate to the metal or their complex, thereby reducing the energy of the metal cyclobutane and increasing the stereoselectivity of the system.

Gilliom et al. published a ring-opening deuteration polymerization of norbornene with Ti-heterocyclobutane as a catalyst. This discovery is undoubtedly of great significance. Nguyen et al. conducted further in-depth research on this. They used a triphenylphosphine carbene-type Ru complex as a catalyst to polymerize norbornene. The polymerization was carried out in a polar solvent, and the polymerization process was followed by deuterated norbornene, and it was found that the polymerization was carried out according to the principle shown in Fig. 5.6.

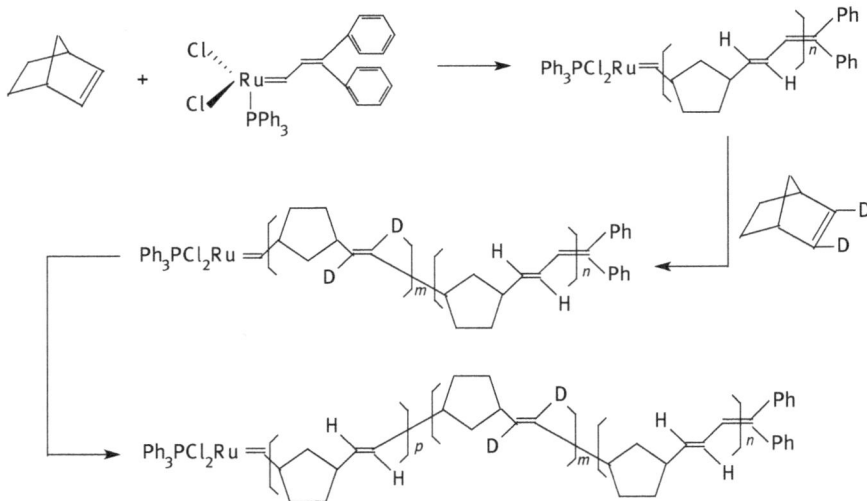

Fig. 5.6: Schematic diagram of norbornene living polymerization.

It has been found that the chain termination reaction and the chain transfer reaction are much slower than the chain growth rate during the polymerization, and the relative molecular mass distribution of the product is narrow, so it is enough to prove that it is a living polymerization.

The usual catalyst initiates the polymerization of norbornene and the product is a cis-double bond structure as shown in Fig. 5.1. However, when RuCl$_3$ is used as the catalyst, the product of the ring-opening deuteration polymerization of norbornene is a trans-double bond structure, as shown in Fig. 5.7.

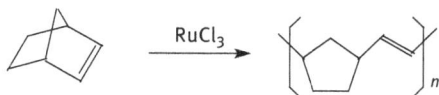

Fig. 5.7: Ring-opening deuteration polymerization of norbornene with RuCl₃ as catalyst.

5.3.2 Ring-opening deuteration polymerization of norbornene derivatives

The norbornene derivative is usually prepared by a Diels–Alder reaction of a substituted cyclopentadiene with an olefin. The ring-opening deuteration polymerization of these monomers provides many functional polymers and special polymers having various properties.

The key issue in achieving ring-opening deuteration polymerization of norbornene derivatives is to find a suitable catalyst. Generally, the activity of the catalyst is susceptible to the polar groups in the monomer molecule. However, since the ring tension of norbornene is large, the tension released during the ring-opening polymerization can make the reaction free energy negative, so it has been found that most of the catalysts for the ring-opening deuteration polymerization of norbornene are effective for the norbornene. The ring-opening deuteration polymerization of the ene derivative is also very effective, and thus there is basically no obstacle to achieving a ring-opening deuteration polymerization of the norbornene derivative.

The dicyclopentadiene (DCPD) molecule also carries a norbornene ring structure. It has two isomers, the internal (bridged) isomer (endo-DCPD) and the external (looped) isomer (exo-DCPD). Both isomers are capable of undergoing ring-opening polymerization. Polymers of different structures can be obtained with the same catalyst.

For example, $RuCl_3 \cdot 3H_2O$ is used as a catalyst for the ring-opening deuteration polymerization of endo-DCPD, and a full cis-double bond structure polymer is obtained. For open-loop deuteration polymerization of exo-DCPD, a polymer based on a trans-double bond structure is obtained. Figure 5.8 is a schematic diagram of the ring-opening deuteration polymerization of two biscyclopentadiene isomers.

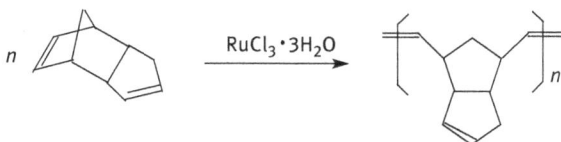

Fig. 5.8: Schematic diagram of ring-opening disproportionation polymerization of two dicyclopentadiene isomers.

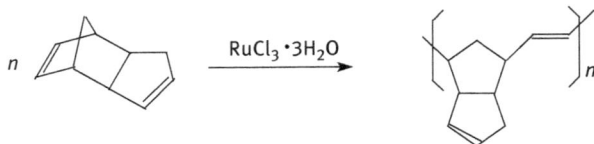

Fig. 5.8 (continued)

Interestingly, if ReCl₅ is used as a catalyst, a highly stereoregular syndiotactic polymer having an all-cis double bond structure can be obtained regardless of whether it is an internal formula or an external DCPD, as shown in Fig. 5.9.

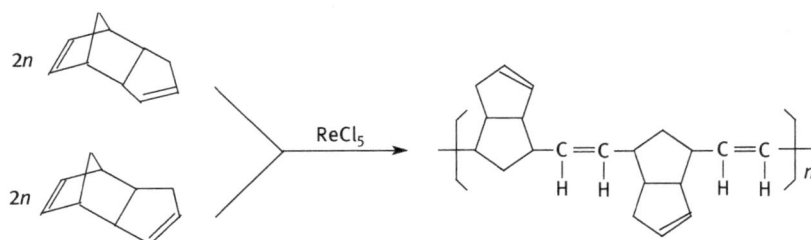

Fig. 5.9: Bicyclopentadiene ring-opening polymerization with ReCl₅ as catalyst.

Since the 1990s, with the development of optical fiber communication, research on the ring-opening deuteration polymerization of fluorine-containing norbornene has been carried out. For example, a ring-shaped deuteration polymerization of 2,3-bis(trifluoromethyl)-bicyclo(2,2,1)heptane-2,5 diene with a Schrock-type Mo catalyst gives a stereoregular polymer. When the earlier mentioned monomers are polymerized by using WCl₆/Me₄Sn, RuCl₃/Me₄Sn, MoCl₅/Me₄Sn, or Mo(CH-t-Bu) as a catalyst, a crystalline polymer is obtained, and the stereoregular ratio is greatly reduced.

The polyfluorinated norbornene derivative is subjected to ring-opening deuteration polymerization using WCl₆/Ph₄Sn as a catalyst, and the obtained product is an amorphous polymer having high T_g, low refractive index, excellent oxidation resistance, and heat stability. The reaction formula is shown in Fig. 5.10.

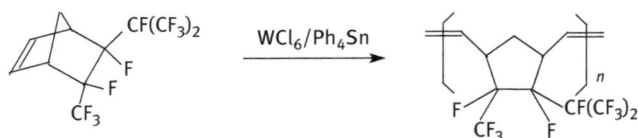

Fig. 5.10: Ring-opening deuteration polymerization of polyfluorinated norbornene derivatives.

In addition, extensive studies have been conducted on norbornene derivatives containing substituents such as boron, silicon, chlorine, nitrile, silica, siloxane, silonitrile, and amide to synthesize many useful polymers. For example, a boron-containing norbornene derivative (norbornene-8-bicycloindole borane) is subjected to ring-opening deuteration polymerization and hydrolyzed under strong alkali conditions to obtain a hydroxyl group-containing norbornene derivative. The reaction formula is shown in Fig. 5.11.

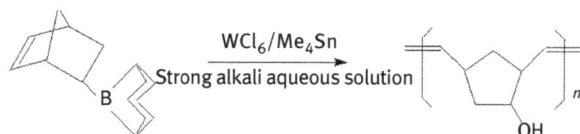

Fig. 5.11: Ring-opening deuteration polymerization of boron-containing norbornene derivatives [11].

The acetoxynorbornene can be subjected to ring-opening polymerization using a W-based catalyst, and the obtained polymer contains an acetoxy group. This laid the foundation for the synthesis of norbornene graft copolymers and functional polymers containing norbornene structural units, as shown in Fig. 5.12.

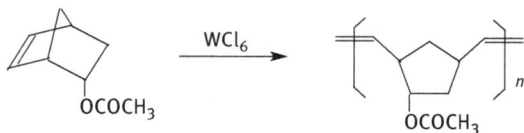

Fig. 5.12: Ring-opening deuteration polymerization of acetoxynorbornene.

5.3.3 Synthesis of copolymers of norbornene and its derivatives

The synthesis of norbornene and its derived copolymers includes three basic methods. One is to directly copolymerize two norbornene monomers by ring-opening deuteration polymerization to obtain a norbornene random copolymer. The other is that the ring-opening deuteration polymerization of the norbornene monomer is carried out by ring-opening deuteration copolymerization, and then the norbornene derivative active polymerization chain is used to polymerize another norbornene monomer. And a block copolymer is obtained finally. The third type is to carry out ring-opening deuteration polymerization of the norbornene monomer by ring-opening deuteration copolymerization, and then carry out other types by using the special structure of the norbornene derivative living polymer chain or norbornene derivative unit. Polymerization: the result is a graft copolymer or a block copolymer. These three types of polymers have important uses and are introduced separately in the following sections.

5.3.3.1 Direct ring-opening disproportionation copolymerization of norbornene and its derivatives

As mentioned earlier, the W-based catalyst can be used to effect ring-opening polymerization of norbornene or acetoxynorbornene. In fact, it is also very easy to copolymerize these two monomers under the catalysis of a W-based catalyst, the product being a thermoplastic random copolymer. DCPD can also be copolymerized with acetoxynorbornene in the same manner. Figure 5.13 is a schematic diagram of the synthesis of these two copolymers.

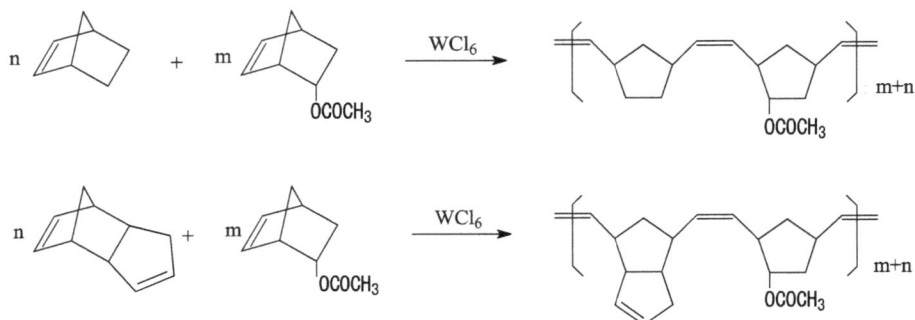

Fig. 5.13: Schematic diagram of the synthesis of norbornene random copolymer.

5.3.3.2 Preparation of block copolymers by ring-opening disproportionation of norbornene and its derivatives

The ring-opening disproportionation polymerization in the presence of specific polymerization conditions and catalysts is a living polymerization which has been previously discussed. The preparation of block copolymers is the most effective function of living polymerization, as is the active ring-opening deuteration polymerization. For example, a cyclopentene derivative of ferrocene is used as a catalyst to carry out active ring-opening polymerization of norbornene, followed by sequential addition of DCPD and norbornene, and an ABA-type triblock copolymer is successfully prepared. The reaction process is shown in Fig. 5.14.

Star polymers can also be prepared by this method. For example, Scheock complex is used as a catalyst to carry out active ring-opening deuteration polymerization of norbornene and then to initiate polymerization of dimerized norbornene to obtain a star-shaped polymer. Figure 5.15 shows the star-shaped polymer synthesis process.

The catalytic activity of WCl_6 is relatively low, and the rate of initiation of ring-opening deuteration polymerization is slow, so that the product of polymerization of DCPD is a linear polymer by using it as a catalyst. If $RxAlCl_3$-x is added as a cocatalyst to this system, the polymerization rate is greatly accelerated and the product is cross-

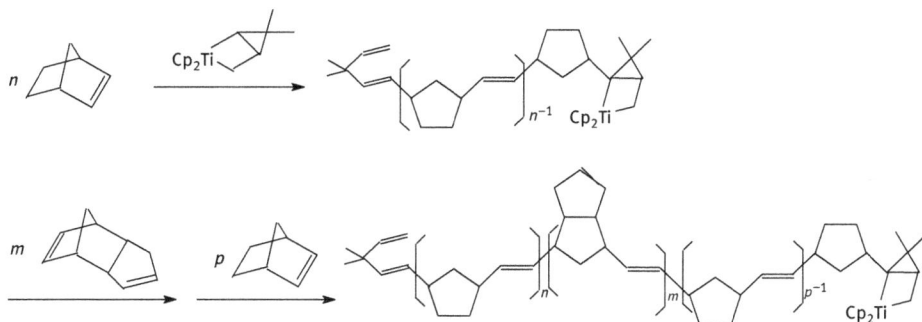

Fig. 5.14: Synthesis of norbornene-dicyclopentadiene-norbornene triblock copolymer.

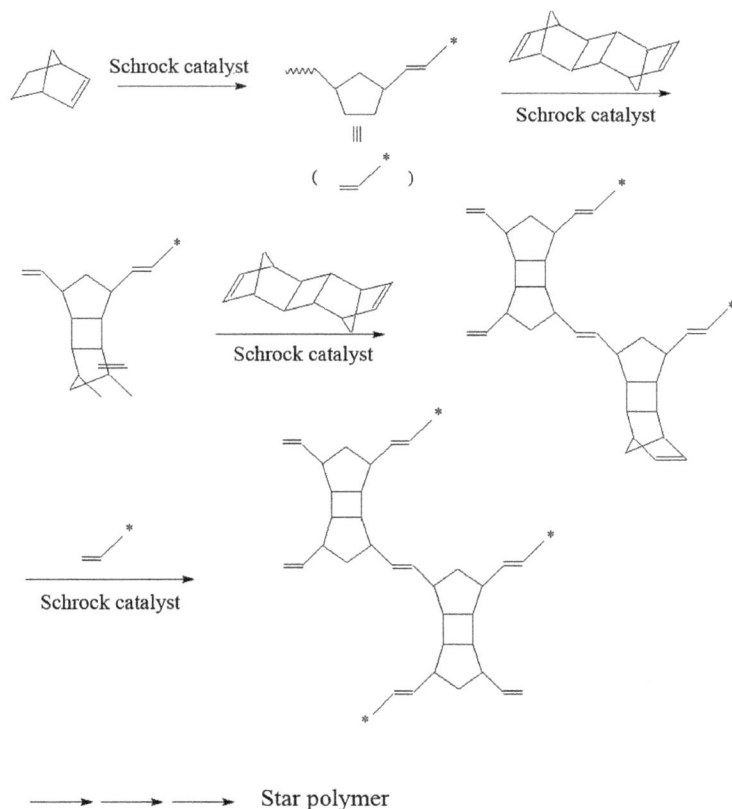

Fig. 5.15: Schematic diagram of the synthesis of norbornene star polymer. Reprinted with permission [28].

linked polydicyclopentadiene. This polymer is excellent in mechanical properties and is a high impact polymer. Figure 5.16 shows the polymerization formula.

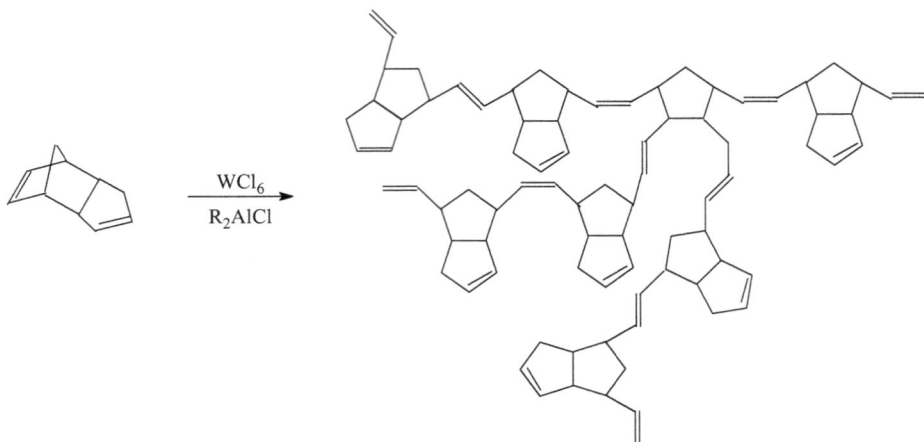

Fig. 5.16: Intent of polymerization of cross-linked polydicyclopentadiene. Reprinted with permission [65]. Copyright, Wiley, 1985.

5.3.3.3 Active ring-opening disproportionation polymerization to another polymerization to prepare block copolymer

The conversion of one living polymerization to another type of living polymerization to prepare a graft or block copolymer is a commonly used means in recent years. As with other living polymerizations, grafting and block copolymers can also be prepared by the polymerization of active sites or reactive groups on the living polymerization product of norbornene and its derivatives into other polymerization mechanisms.

For example, a ring-opening deuteration polymerization of norbornene is first initiated using a titanocene cyclobutane compound as a catalyst to obtain a homopolymer having a relative molecular mass of 1×10^4 and a relative molecular mass distribution $M_w/M_n = 1.15$. Then, it is reacted with an equimolar amount of methanol under mild conditions to convert the terminal portion of the titanocene cyclobutane on the polymer into a metallocene methoxy titanium, and then catalyze the polymerization of ethylene in the presence of EtAlCl$_2$ to obtain Polynorbornene-block-polyethylene copolymer. In the same way, the titanocene cyclobutane end group carried by the polycyclonordene obtained by active ring-opening deuteration polymerization is converted into an alkoxytitanium, and then the ethylene and propylene are coordinated in the presence of an EtAlCl$_2$ promoter. By copolymerization, a PNBE-b-poly(ethylene-co-propylene) block copolymer can be synthesized.

The active ring-opening deuteration polymerization of norbornene is carried out using the same catalyst, and then the PNBE bearing the terminal phase of the titanocene cyclobutane is reacted with the previously synthesized polyether ketone. A graft copolymer having a polyether ketone as a main chain and a PNBE as a branch can be obtained. Figure 5.17 is a schematic diagram of the polymerization reaction.

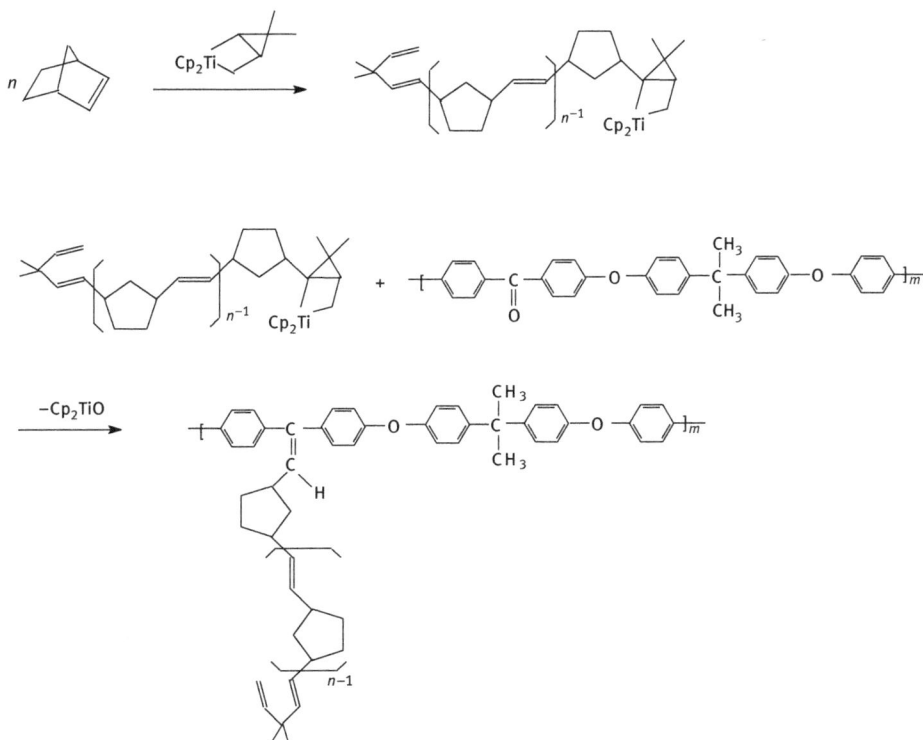

Fig. 5.17: Synthesis of polyether ketone-polynorbornene graft copolymer.

By utilizing the characteristics that the cyclic olefin can be deuterated with the double bond of the linear olefin, a preformed polymer with unsaturated side groups can be deuterated with a cyclic olefin to obtain a graft copolymer, as shown in Fig. 5 18. In this way, cyclooctene has been grafted onto the natural rubber backbone, cyclopentene has been grafted onto the 1,2-polybutadiene backbone, and 1,5-cyclooctadiene has been grafted to ethylene – a propylene-diene terpolymer backbone.

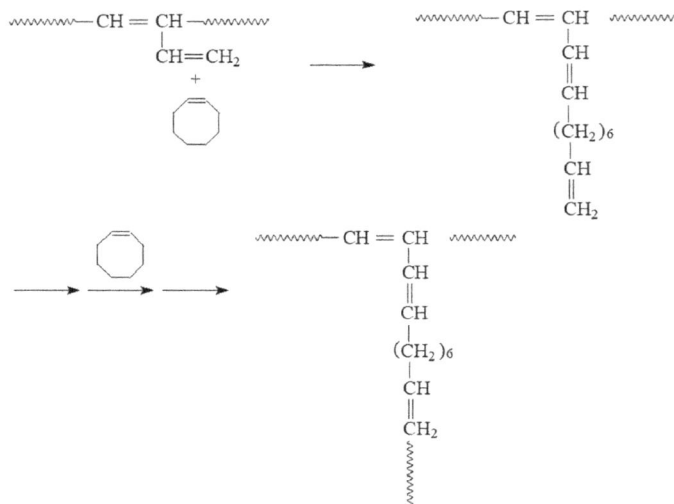

Fig. 5.18: Schematic diagram of preparing graft copolymer by deuteration reaction of unsaturated pendant polymer with cycloolefin.

5.4 Application of ring-opening deuteration polymerization in polymer molecular design

5.4.1 Synthesis of constant ratio copolymer

So far, it has been very difficult to prepare a constant ratio copolymer by radical polymerization or other chain polymerization because only a few pairs of comonomers have a reactivity ratio satisfying the condition of $r_1 = r_2 = 1$.

The ring-opening deuteration polymerization of DCPD and 2-acetoxy-5-norbornene with WF_6 as catalyst showed that the reactivity of these two monomers was $r_1 = r_2 = 1$, and the reaction was constant copolymerization, as shown in Fig. 5.19.

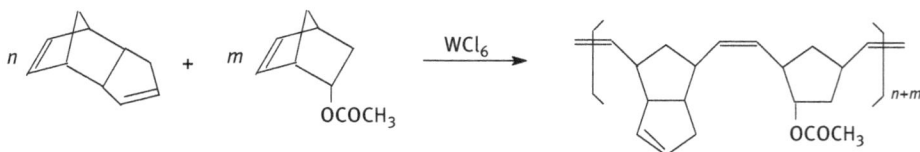

Fig. 5.19: Constant ratio ring-opening deuteration polymerization of dicyclopentadiene and 2-acetoxy-5-norbornene.

The earlier constant ratio copolymerization reaction rate is moderate, and more importantly, the monomer's copolymerization rate does not depend on the type of substituent on the norbornene ring. Therefore, it is possible to synthesize various copolymers having a predetermined composition by adjusting the ratio of the two monomers and to prepare functionalized polymers having different properties by changing the kinds of substituents on the norbornene.

5.4.2 Synthesis of ideal alternating copolymers

The synthesis of ideal alternating copolymers has important theoretical and practical significance. However, it is impossible to synthesize an ideal ternary alternating copolymer by a general chain polymerization method because there are nine chain-growth reactions in the copolymerization reaction involving six colloidal ratios. However, an ideal ternary alternating copolymer can be synthesized by ring-opening deuteration polymerization. A typical example is shown in Fig. 5.2. The obtained ternary alternating copolymer is a functional polymer whose functional groups are evenly distributed along the molecular chain and has important significance both in theory and in practical applications.

5.4.3 Synthesis of all-cis and all-trans polymers

The cyclic olefin ring-opening deuterated polymerization product has a double bond in its molecular backbone, and thus a cis or trans structure may exist. Since the molecular chain aggregation state of the cis–trans structural polymer is greatly different and the physical properties of the polymer are obviously affected, it is of great theoretical and practical significance to synthesize a single-ring open-chain deuterated polymer having a double-bond configuration of the main chain. Studies have shown that different catalysts can be used for ring-opening deuteration polymerization of cyclic olefins to form polymers of different configurations.

For example, using $RuCl_3$ as a catalyst, the polymer produced by the ring-opening polymerization of norbornene has almost the all-trans configuration (the content of the trans configuration is >95%). This is a thermoplastic polymer with excellent shock absorption properties.

When the two isomers of DCPD are subjected to ring-opening deuteration polymerization using different catalysts, polymers of different configurations can be obtained, as shown in Figs. 5.8 and5.9. It can be noted from Figs 5.8 and5.9 that the same stereogenic monomer (exo-DCPD) has different stereo configurations when using different catalysts ($RuCl_3$ and $ReCl_5$), which fully indicates the structure control effect of catalyst on the product of ring opening disproportionation polymerization.

5.4.4 Synthesis of isotactic all-cis and syndiotactic all-trans polymers

The controlled polymerization of stereoregular polymers is of great importance both in theoretical research and in the development of new materials with excellent properties. A controlled stereoregular polymer can be prepared by ring-opening deuteration polymerization of a cyclic olefin.

Using Schrock-type molybdenum carbene complex Mo(CHCMe$_2$Ph)(NAr)[BIPH (t-Bu)$_4$] (Ar is 2,6-Me$_2$C$_6$H$_3$, BIPH is biphenyl) as catalyst, 2,3-disubstituted norbornene diene as a monomer for ring-opening deuteration polymerization, an isotactic full-cis polymer can be synthesized. And the relative molecular mass is 19,500, polydispersity index is 1.05, and good controllable characteristics are also achieved. The reaction process is shown in Fig. 5.20.

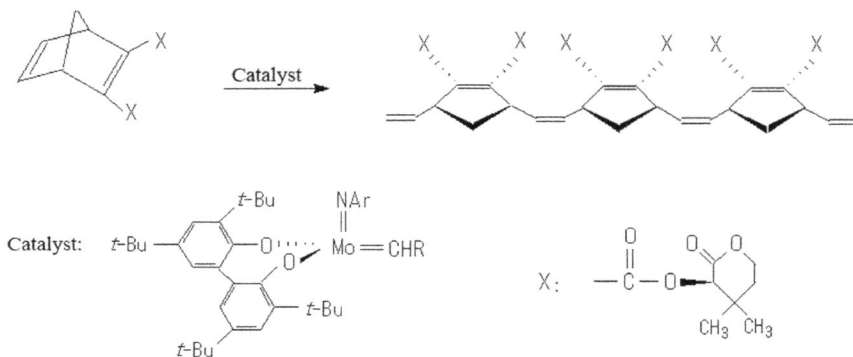

Fig. 5.20: Schematic diagram of isotactic cis-polymer synthesis.

Monomers bearing different X groups can be polymerized into isotactic full cis polymers using the same or similar catalysts. For example, the ring-opening deuteration polymerization of an alkene, utilizing a molybdenum carbene complex Mo(CHCMe$_2$Ph) (NAr)[BINO(t-Bu)$_4$] (Ar is 2,6-Me$_2$C$_6$H$_3$, BINO is biphenol) as a catalyst for 2,3-dimethoxycarbonyl norbornazole, gives an isotactic cis-functionalized polymer having an ester group.

Using molybdenum carbene complex Mo(CHCMe$_2$Ph)(NAr)(Ot-Bu)$_4$ (Ar is 2,6-diisopropylphenyl) as catalyst and 2,3-disubstituted norbornadiene as monomer to carry out ring-opening deuteration polymerization, a product having a syndiotactic total trans-polymer content of >94% can be obtained with a relative molecular mass of 16,300 and a polydispersity index of 1.03. The reaction formula is shown in Fig. 5.21.

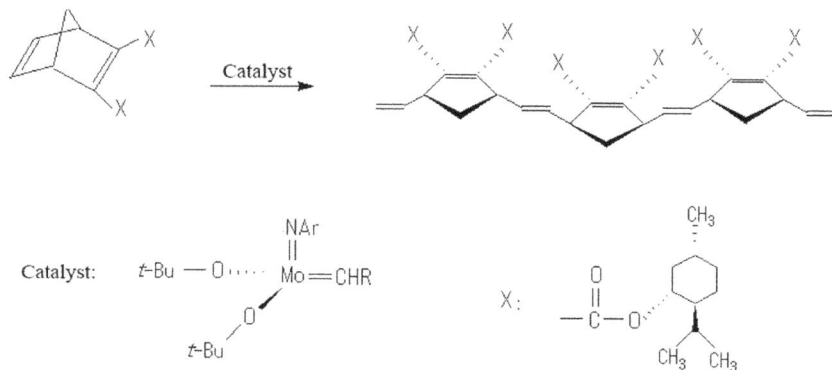

Fig. 5.21: Schematic diagram of syndiotactic all-trans polymer synthesis.

5.4.5 Synthesis of block and graft copolymers

Active controlled ring-opening deuteration polymerization can be carried out by using a suitable catalyst, which provides a powerful means for molecular structure design and tailoring of polymer materials. Block copolymers having the desired composition and structure can be synthesized by active controlled ring-opening polymerization.

The foregoing Figs. 5.6 and 5.14 show examples of the preparation of block co-polymers by living polymerization of norbornene. In recent years, great progress has been made in this area of work. For example, a reactive ring-opening deutera-tion polymerization can be combined with other living polymerizations to prepare a block copolymer having a specific structure. Figure 5.22 is an example of combining active ring-opening deuteration polymerization with reactive group transfer poly-merization. In this example, a living polymer of norbornene is first synthesized and then reacted with terephthalaldehyde to give an aldehyde-terminated polymer. The diblock copolymer can be obtained by initiating the aldol group transfer polymeri-zation of tert-butyldimethylvinylsilyl ether with aldehyde-based polymer by using $ZnCl_2$ as a catalyst.

After the resulting block copolymer being reduced by $NaBH_4$ and treated with Bu4NF, and the silyl group can be converted into a hydroxyl group to form a PNBE/ polyvinyl alcohol amphiphilic block copolymer. By changing the amount of two monomers charged, controlling the length of the two segments, the amphiphilicity of the polymer can be adjusted. An amphiphilic block copolymer of another struc-ture can be prepared by substituting DCPD for norbornene. These amphiphilic block copolymers have important applications in flocculants, wetting agents, par-ticulate surface modifiers, and polymer blend compatibilizers.

The coupling reaction of a ring-opening deuterated living polymer with other polymers containing specific functional groups is also an important method for pre-paring block copolymers. For example, a Witting coupling reaction is carried out

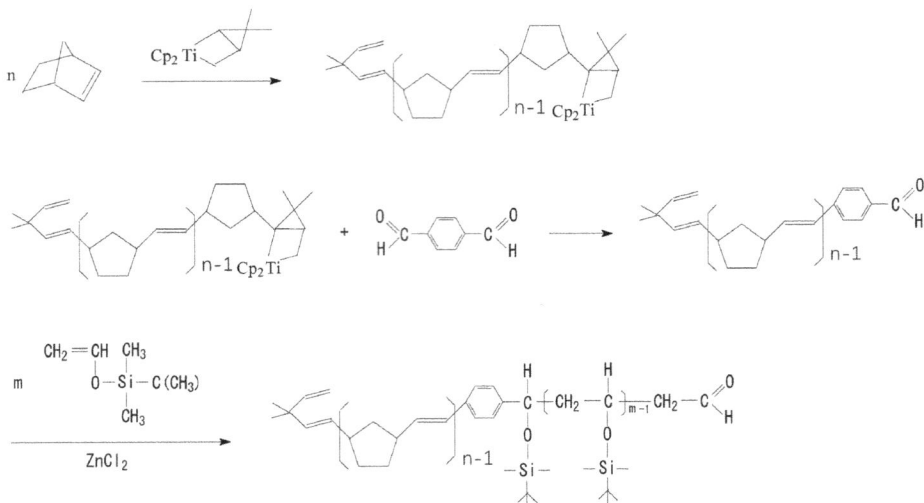

Fig. 5.22: Schematic diagram of preparation of block copolymer by active ring-opening deuteration polymerization and reactive group transfer polymerization.

between a reactive norbornene and a carbonyl-containing polymer to obtain a tri-block copolymer. The reaction process is shown in Fig. 5.23.

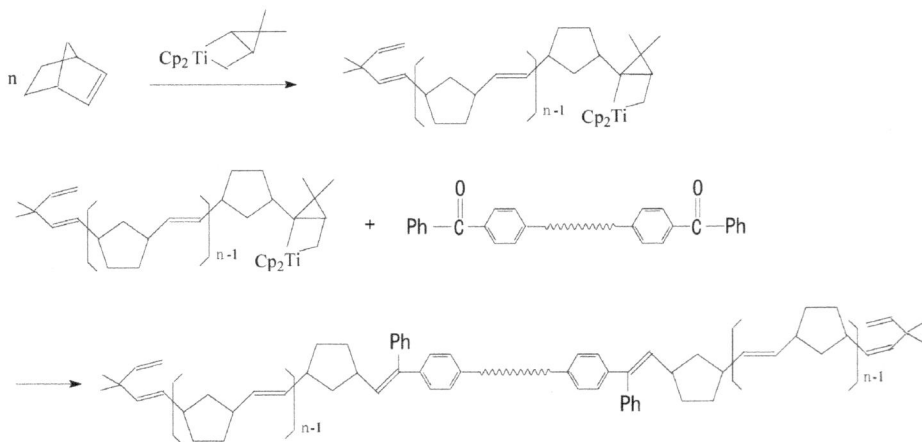

Fig. 5.23: Schematic diagram of preparation of triblock copolymer by coupling reaction of active ring-opening deuterated polymer.

The preparation of block copolymers by combining active ring-opening deuteration polymerization with atom transfer radical polymerization is another novel method.

For example, the molybdenum carbene complex Mo(CHCPhMe$_2$)(NAr)(Ot-Bu)$_2$ (Ar is 2,6-Me$_2$C$_6$H$_3$) is used as a catalyst to carry out active ring-opening deuteration of cyclic olefins such as norbornene and DCPD polymerization. The resulting living polymer is then reacted with *p*-bromomethylbenzaldehyde to convert the polymer end groups to the benzyl bromide group. The benzyl bromide-containing polymer can initiate the active controlled atom transfer radical polymerization of monomers such as styrene and methacrylate under the catalysis of CuCl$_2$/bipyridine, and finally obtain the AB-type two-end copolymer. For example, the reaction for preparing a PNBE-b-polystyrene (PS) diblock copolymer by this method is shown in Fig 5.24. Similar methods have been used to synthesize diblock copolymers such as PNBE-b-polymethyl methacrylate (PMMA), polycyclopentadiene (PDCPD)-b-PS and PDCPD-b-PMMA.

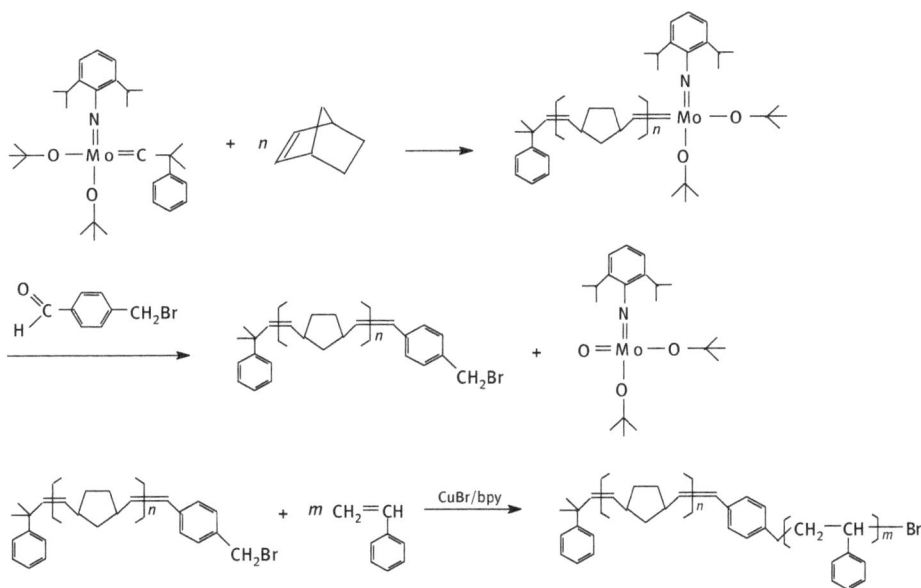

Fig. 5.24: Schematic diagram of preparation of block copolymer by active ring-opening deuteration polymerization combined with atom transfer radical polymerization.

Active ring-opening deuteration polymerization also has a lot of work in the preparation of graft copolymers, especially the active ring-opening deuteration polymerization is combined with other living polymerization methods to become an important method for preparing graft copolymers. The synthesis of the polyether ketone-PNBE graft copolymer shown in Fig. 5.16 is a typical example.

Obviously, this method of combining active ring-opening deuteration polymerization with other living polymerization to prepare a block or graft copolymer greatly broadens the preparation route of the block or graft copolymer and provides a more molecular design for the polymer. Strong means having broad application prospects.

Active ring-opening deuteration polymerization also has many effects in the preparation of graft copolymers; it is an important method for preparing graft copolymers combining active ring-opening deuteration polymerization with other living polymerization methods. The synthesis of the polyether ketone-PNBE graft copolymer shown in Fig. 5.16 is a typical example.

Obviously, this method of combining active ring-opening deuteration polymerization with other living polymerization to prepare a block or graft copolymer greatly broadens the preparation route of the block or graft copolymer, provides a stronger means for the polymer molecular design, and have broad application prospects.

5.4.6 Synthesis of comb and star copolymers

Comb polymers have high intramolecular order and tend to exhibit very strange physical and chemical properties. Therefore, they have received extensive attention in polymer science theory and practical applications. The active ring-opening deuteration polymerization is combined with other living polymerization reactions to synthesize a comb polymer having a regular structure and a relatively narrow molecular weight distribution. For example, a method for preparing a comb polymer having a main chain of PNBE and a side chain of PS is as follows:

The living anionic polymerization of styrene is first carried out using butyllithium as an initiator, and the resulting active polymer is converted into an oxyanion at the end of the polymer when reacted with propylene oxide. Then, it is reacted with 2,3-*trans*-carbonyl chloride-5-norbornene to prepare a macromonomer 2,3-bispolystyrene alkoxycarbonyl-5-norbornene. Finally, the earlier macromonomer was actively opened by using Schrock-type molybdenum carbene complex Mo(CH-t-Bu)(NAr)(Ot-Bu) (Ar is 2,6-diisopropylphenyl) as a catalyst. By deuteration polymerization, a comb polymer can be obtained. The reaction process is shown in Fig. 5.25.

The comb polymer synthesized by this method had a relative molecular mass of 1,200 for each "comb" and a dispersion index of 1.06; the comb polymer had a relative molecular mass of 17,520 and a dispersion index of 1.09. That is, both the macromonomer and the comb polymer are nearly monodisperse, with almost equal lengths of "combs" and "combs," like a batch of combs made in the same mold.

The "comb" of the earlier comb polymer is attached to the PNBE main chain via an ester bond and is poor in acid resistance and alkalinity. Another method for synthesizing a comb polymer is to prepare a comb polymer in which a "comb" is bonded to a polymer backbone via a carbon–carbon bond, as shown in Fig. 5.26. The relative mass of the "comb" in the product was 2,600 and the dispersion index was 1.05; the relative molecular mass of the comb polymer was 30,000, and the dispersion index was 1.08.

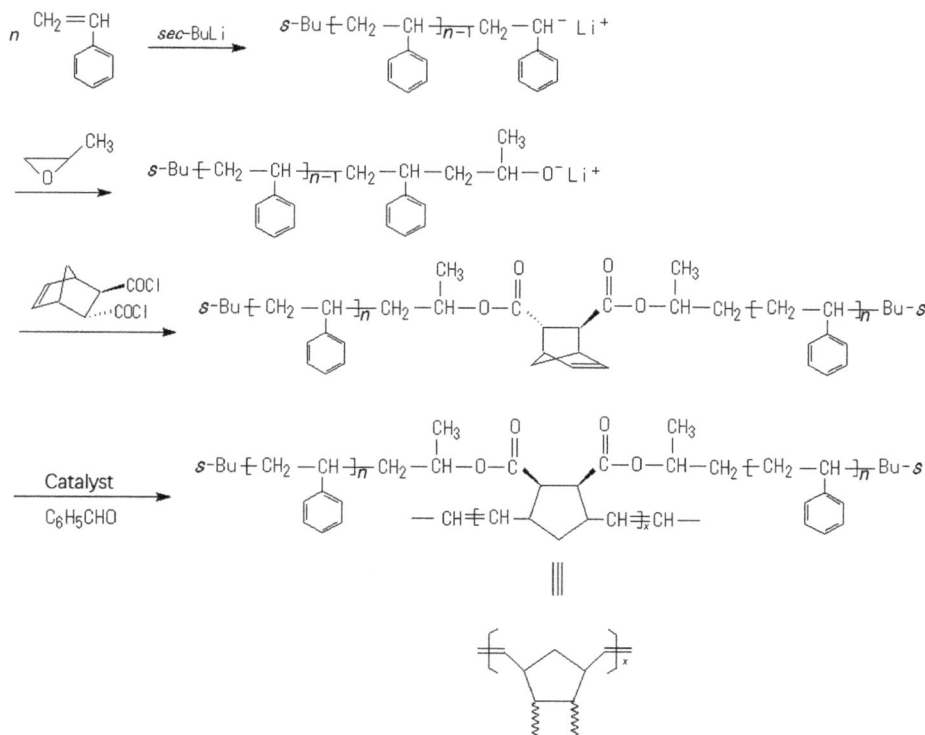

Fig. 5.25: Schematic diagram of the synthesis of comb polymer.

Star-shaped polymers can be prepared by active ring-opening deuteration polymerization. For example, a Schrock-type molybdenum carbene complex is used as a catalyst to carry out active ring-opening polymerization of norbornene, and then to initiate polymerization of dimerized norbornene to obtain a star-shaped polymer. The synthesis of this star polymer has been described in Fig. 5.15.

Star-shaped polymers with functional groups can also be prepared by active ring-opening deuteration polymerization. The active ring-opening disproportionation polymerization of 2,3-dimethoxycarbonylnorbornadiene was carried out using molybdenum carbene complex Mo(CHR)(NAr)(O-t-Bu)$_2$ as a catalyst to obtain a living polymer. Further addition of norbornene for active ring-opening disproportionation polymerization gives a living block copolymer. Finally, it is reacted with dimerized norbornadiene to obtain a star-shaped polymer having an ester group. Figure 5.27 is a schematic diagram of this synthetic process.

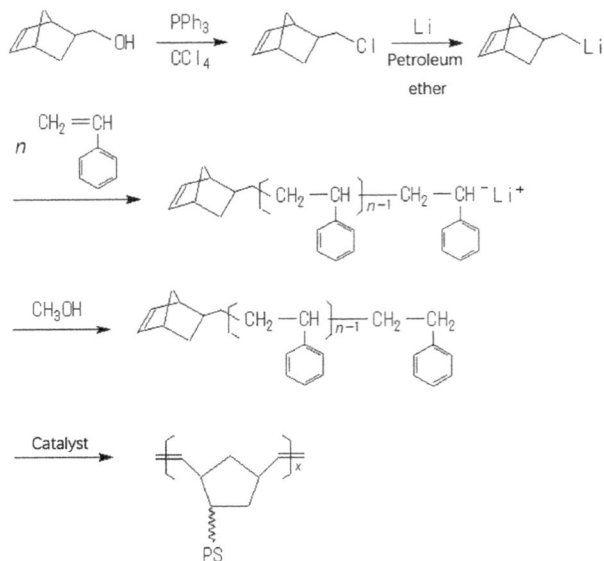

Fig. 5.26: Schematic diagram of the synthesis of a comb polymer with a carbon–carbon bond attached to the polymer backbone.

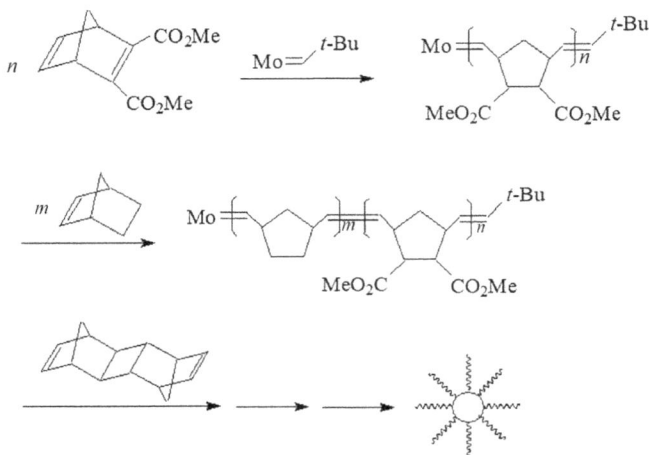

Fig. 5.27: Schematic diagram of the synthesis of a star-shaped polymer with functional groups.

5.4.7 Synthesis of telechelic polymers

Telechelic polymer plays an important role in the preparation and application of functional polymers. Telechelic polymers can be synthesized by ring-opening disproportionation polymerization. For example, (ArO)2WC4-SnMe4 is used as a catalyst for ring-opening disproportionation polymerization of DCPD. In the polymerization, methyl 3-hexene-1,6-dicarboxylate is used as a chain transfer agent, and the product has ester groups at both ends. Then, the telechelic polymer with hydroxyl groups at both end can be obtained by LiAlH4 reduction. If the polymer is hydrolyzed with an acidic substance, a telechelic polymer having a carboxyl group at both ends can be obtained. A similar double-ended hydroxyl telechelic polymer can also be synthesized using norbornene as a monomer. Figure 5.28 is a schematic illustration of the preparation of a telechelic polymer by open-loop disproportionation polymerization.

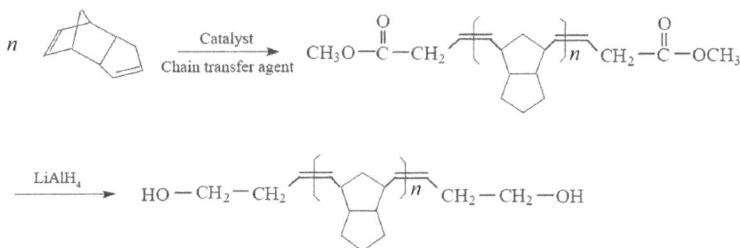

Fig. 5.28: Schematic diagram of preparation of telechelic polymer by open-loop disproportionation polymerization. Reprinted with permission from [52].

5.4.8 Preparation of conductive polymer

Polyacetylene is a kind of conductive polymer material with excellent performance, but its solubility is not good and its processing performance is very poor, which limits its practical use. The establishment of a ring-opening deuteration polymerization technique for cyclic olefins has led to a new method for preparing polyacetylene. For example, WCl_6-Me$_4$Sn is used as a catalyst to carry out ring-opening disproportionation polymerization of 7,8-bis(trifluoromethyl)tricyclo[2.2.0]-3,7,9-nonanetriene, which can obtain a soluble and easy to be processed polymer P. Figure 5.29 shows schematic views of the preparation of polyacetylene by this method. After the latter is processed

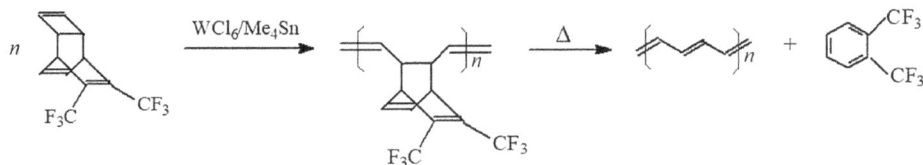

Fig. 5.29: Schematic diagram of preparation for polyacetylene by ring-opening polymerization.

into a desired shape (usually a film), the volatile *ortho*-trifluoromethylbenzene is removed by heating to obtain a polyacetylene. This polyacetylene is an amorphous all-trans structure, easy to process, and has many uses. For example, the polymer P is coated on a silicon crystal, and then heated and doped to form a transistor; the polymer P is made into a sandwich sandwiched with an insulator (P-insulator-P), which can be heated as a capacitor and the polymer P can be first formed into a film, which can then be processed into a flexible wire in a computer, and so on. However, the polyacetylene prepared by this method has a low conductivity because the fluoroaromatic hydrocarbon is difficult to remove. Figure 5.30 is an improved method.

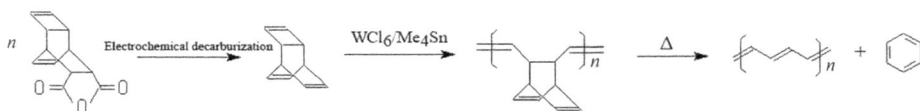

Fig. 5.30: Schematic diagram of an improved method for preparing polyacetylene by ring-opening deuteration polymerization.

Another method for preparing polyacetylene by ring-opening deuteration polymerization is to use potent benzene as a monomer and a tungsten carbene complex W $(CH$-t-$Bu)(NAr)(O$-t-$Bu)_2$ as a catalyst. Polymerization gives a polymer having a full cis main chain double bond structure, which is then treated with $HgCl_2$ to isomerize the polymer to polyacetylene. The reaction process is shown in Fig. 5.31.

Fig. 5.31: Schematic diagram of preparing polyacetylene using pot benzene as monomer.

The intermediate polymer of polyacetylene prepared by this method is easy to process and can be made into fibers and films having high strength. In addition, the bicyclobutane structure with high tension in the intermediate polymer molecule is extremely unstable and has a high internal energy (250.8 kJ/mol), which is expected to be used as a solid rocket fuel. However, this method also has disadvantages because $HgCl_2$ in the product is not easily removed, resulting in affinity cross-linking of the polymer and loss of partial double bonds, resulting in a decrease in electrical conductivity.

Recently, the direct preparation of polyacetylene using cyclooctanetetraene ring-opening deuteration polymerization has been successful. For example, using Schrock-type tungsten carbene complex $W(CH$-t-$Bu)(NAr)[OCMe(CF_3)]$ as a catalyst, ring-opening polymerization of cyclooctatetraene can be carried out into a *cis*-structured polyacetylene, and then at a high temperature. The process is transformed into *trans*-polyacetylene (see Fig. 5.32). The conductivity of the polyacetylene after doping with iodine is 50–350 S/cm.

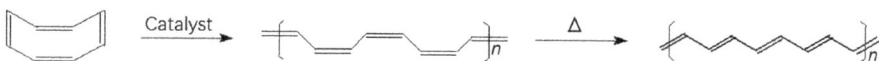

Fig. 5.32: Schematic diagram of polyacetylene prepared by ring-opening deuteration polymerization of cyclooctate.

In addition to the earlier polyacetylene-based conductive polymers, many other forms of conductive polymers have been prepared by ring-opening deuteration polymerization in recent years.

5.4.9 Preparation of ion exchange resin

Active ring-opening disproportionation polymerization was carried out using molybdenum carbene complex $Mo(CHCMe_2Ph)(NAr)[OCMe(CF_3)_2]$ as catalyst and 2,3-carboxylic anhydride-5-norbornene as monomer. After all the monomers were polymerized, dimerized norbornadiene was added for block copolymerization. The active chain is terminated with benzaldehyde, and after hydrolysis and acidification, a functionalized block copolymer having a carboxyl group is obtained. This is a low exchange capacity, weakly acidic cation exchange resin with an exchange capacity of up to 10 mol/g resin. It swells in water and polar solvents and can be used to extract amines, pyridine alkaline substances, and purify air from water. It has high practicability. The reaction process and structural formula of this ion exchange resin are shown in Fig. 5.33.

5.4.10 Synthesis of special performance polymers

Norbornene can be obtained from the petrochemical by-product cyclopentadiene and ethylene by Diels–Alder reaction. After open-loop disproportionation polymerization, a thermoplastic PNBE having a relative molecular mass of 2×10^6 or more is obtained, and its main chain is a trans structure. Industrialization was carried out in 1976 under the trade name "Norsorex." This is a highly oil-absorbing resin that absorbs about 10 times its own weight of oil. It can be used as sound insulation materials, shock absorbing materials, damping materials, and sealing materials (Fig. 5.34).

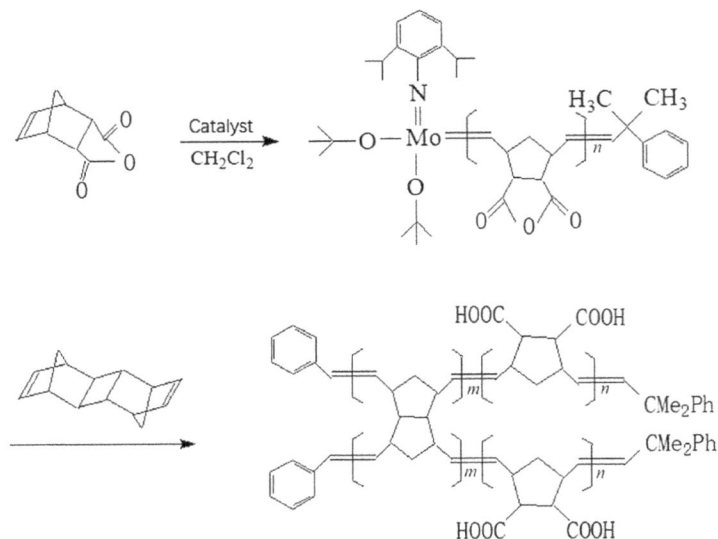

Fig. 5.33: Schematic diagram of preparation of ion exchange resin by active ring-opening disproportionation polymerization.

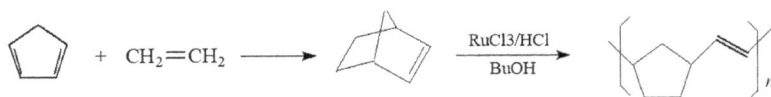

Fig. 5.34: Preparation of *trans*-polynorbornene.

The ring-opening disproportionation polymerization of cyclooctene under the catalysis of a tungsten catalyst can obtain a polymer mainly composed of a trans-main chain structure, and the relative molecular mass is 10^5 or more. *Trans*-polycyclooctene has been industrialized in 1980. These products are excellent rubber compounding agents, which can significantly improve the processing fluidity of rubber and improve the elasticity of vulcanized rubber (Fig. 5.35).

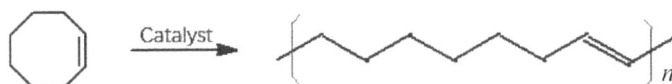

Fig. 5.35: Preparation of *trans*-polycyclooctene.

Cyclopentene can form a polymer with two main chains of trans or cis under the action of different catalysts. Both polycyclopentenes are excellent elastomers (Fig. 5.36). The former is close to natural rubber, while the latter is like natural rubber owing excellent low-temperature properties.

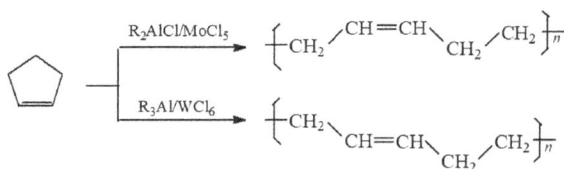

Fig. 5.36: Preparation of polycyclopentene.

In the process of petrochemical cracking to produce ethylene, there are many C_5 fraction by-products (about 15–17% of ethylene production), which contain a large amount of DCPD. In the past, it was mainly burned as fuel, which pollutes the environment and wastes precious chemical resources. Since the advent of open-loop disproportionation polymerization, this problem has been solved to some extent.

The DCPD can undergo ring-opening disproportionation polymerization under the action of a tungsten catalyst WCl_6-Et_2AlC to form a cross-linked polymer. This polymer has high impact, tensile, and flexural strength and is a new type of high impact plastic that can be used in the manufacture of automotive parts and sports equipment and is also used in the machine industry. In the 1980s, reactive injection molding of polydicyclopentadiene was also industrialized. The synthesis of this polymer has been described in Fig. 5.16. A disadvantage of the WCl_6-Et_2AlC catalytic system is that the polymer yield is not high enough (94.6%) and the mechanical strength of the polymer needs to be improved. Li Hong et al. of Nankai University used polymer-supported catalytic system to carry out ring-opening disproportionation polymerization of DCPD, which not only improved the polymer yield (97.9%), but also greatly improved the mechanical properties of the polymer.

Since the advent of the open-loop disproportionation polymerization in the mid-1960s, research and development work in this field has continued to develop. The focus is mainly on the research of the basic theory of active ring-opening disproportionation polymerization, the development of new high-efficiency catalysts, and the development of new materials for polycycloolefins. With the deepening of active ring-opening disproportionation polymerization, this method has become an effective means for molecular structure design, tailoring, and assembly of polymer materials. At present, there are still many unclear points in the catalyst for the open-loop disproportionation polymerization catalyst system, so the research work in this field is expected to be further deepened. In addition, the combination of active ring-opening disproportionation polymerization with other living polymerization methods will lead to greater achievements in the development of new materials for polycycloolefins.

References

[1]   Anderson, AW, Mercking, NG. Polymeric bicyclo-(2, 2, 1)-2-heptene. United States patent 2721189, 1955.
[2]   Elcuterio, HS. Polymerization of cyclic olefins. United States patent 3074918, 1963.
[3]   Truett, WL, Johnson, DR, Robinson, IM, Montague, BA. Polynorbornene by coördination polymerization[1]. J Am Chem Soc. 1960;82(9):2337–2340.
[4]   Calderon, N, Chen, HY, Scott, KW. Olefin metathesis-A novel reaction for skeletal transformations of unsaturated hydrocarbons. Tetrahedro Lett. 1967;8(34):3327–3329.
[5]   Kůdela, V, Stoy, A, Urbanova, R. Electrochemical properties of hydrophilic weak-acid membranes from hydrolysed polyacrylonitrile – I. Effect of cross-linking and chemical composition. European Polym J. 1974;10(10):905–910.
[6]   Dolgoposk, BA, Oreshkin, IA, Smimov, SA. Generation of transition metal carbene complex es and their relative activity in chain processes of cyclo-olefin ring-opening polymerization and olefin metathesis. European Polym J. 1979;15(3):237–240.
[7]   Dolgoplosk, BA. The nature of active centers and mechanism of cyclo-olefin polymerization and olefin metathesis. J Polym Sci, Polym Symp. 1980;67(1):99–114.
[8]   Schaverien, CJ, Dewan, JC, Schrock, RR. Multiple metal-carbon bonds. 43. Well-characterized, highly active, Lewis acid free olefin metathesis catalysts. J Am Chem Soc. 1986;108(10):2771–2773.
[9]   Schrock, RR, Feldman, J, Cannizzo, LF, et al. Ring-opening polymerization of norbornene by a living tungsten alkylidene complex. Macromolecules. 1987;20(5):1169–1172.
[10]   Knoll, K, Krouse, SA, Schrock, RR. Preparation and isolation of polyenes containing up to 15 double bonds. J Am Chem Soc. 1988;110(13):4424–4425.
[11]   Grubbs, RH. The development of functional group tolerant ROMP catalysts. J Macromol Sci, Part A. 1994;31(11):1829–1933.
[12]   Hamilton, JG, Ivin, KJ, Rooney, JJ. Ring-opening polymerization of endo and exo-dicyclopentadiene and their 7, 8-dihydro derivatives. J Mol Catal. 1986;36(l-2):115–125.
[13]   Cannizzo, LF, Grubbs, RH. Block copolymers containing monodisperse segments produced by ring-opening metathesis of cyclic olefins. Macromolecules. 1988;21(7):1961–1967.
[14]   Risse, W, Grubbs, RH., Risse, W, Grubbs, RH. Polynorbornene and poly (exo-dicyclopentadiene) with aldehyde end groups. Makromol Chem, Rapid Commun. 1989;10(2):73–78.
[15]   Tadatomi, N, Jiro, U, Kenkichi, M, Takashi, I. Study of polymeric photosensitizer. 2. Syntheses of multifunctional polymeric photosensitizers containing a pendant nitroaryl group and a quaternary phosphonium salt and their application to the photochemical reaction of potassium cinnamate. Macromolecules. 1988;21(6):1583–1589.
[16]   Risse, W, Wheeler, DR, Cannizzo, LF, et al. Di-and tetrafunctional initiators for the living ring-opening olefin metathesis polymerization of strained cyclic olefins. Macromolecules. 1989;22(8):3205–3210.
[17]   Craig, GSW, Cohen, RE, Schroek, RR, et al. Synthesis and nonlinear optical analysis of triblock copolymers containing discrete polyenes. Macromolecules. 1994;27(7):1875–1878.
[18]   Coca, S, Paik, H, Matyjaszewski, K. Block copolymers by transformation of living ring-opening metathesis polymerization into controlled "living" atom transfer radical polymerization. Macromolecules. 1997;30(21):6513–6516.

[19] Heroguez, V, Brenunig, S, Gnanou, Y, et al. Synthesis of α-norbornenylpoly (ethylene oxide) macromonomers and their ring-opening metathesis polymerization. Macromolecules. 1996;29(13):4459–4464.

[20] Heroguez, V, Gnamou, Y, Fontanille, M. Synthesis of α-norbornenyl polystyrene macromonomers and their ring-opening metathesis polymerization. Macromolecular rapid communications. Macromol Rapid Commun. 1996;17(2):137–142.

[21] Percec, V, Schlueter, D. Mechanistic investigations on the formation of supramolecular cylindrical shaped oligomers and polymers by living ring opening metathesis polymerization of a 7-oxanorbornene monomer substituted with two tapered monodendrons. Macromolecules. 1997;30(19):5783–5790.

[22] Edwards, JH, Feast, WJ. Heat On, Heat Off: A Synthesis of Polyacetylene Revisited. Polymer. 1980;21:595–596.

[23] Edwards, JH, Feast, WJ. New routes to conjugated polymers: 1. A two step route to polyacetylene. Polymer. 1984;25(3):395–398.

[24] Plate, NA, Shibaeu, VP Comb-Shaped Polymers and Liquid Crystals. Springer Science & Business Media. 2012.

[25] Feast, W J, Gibson, V C, Johnson, A F. Well-defined graft copolymers via coupled living anionic and living ring opening metathesis polymerization. J Mol Catal A: Chem. 1997;115(1):37–42.

[26] Brenunig, S, Heroguez, V, Gnamou, Y, et al. Ring-opening metathesis polymerization of ω-norbornenyl polystyrene macromonomers and characterization of the corresponding structures. Macromol Symp. 1995;95(1):151–166.

[27] Bazan, GC, Schlock, RR. Synthesis of star block copolymers by controlled ring-opening metathesis polymerization. Macromolecules. 1991;24(4):817–823.

[28] Stelzer, F, Brunthaler, JK, Leising, G, et al. Synthesis and polymerization of tricyclo [4.2. 2.02, 5] deca-3, 7, 9-triene and tricyclo [4.2. 2.02, 5] deca-3, 9-diene-7, 8-dicarboxylic anhydride. J Mol Catal. 1986;36(1–2):135–143.

[29] Burroughes, J H, Jones, C A, Friend, R H. New semiconductor device physics in polymer diodes and transistors. Nature. 1988;335(6186):137–141.

[30] Swager, TM, Dougherty, DA, Grubbs, RH. Strained rings as a source of unsaturation: polybenzvalene, a new soluble polyacetylene precursor. J Am Chem Soc. 1988;110(9):2973–2974.

[31] Swager, TM, Grubbs, RH. New morphologies of polyacetylene from the precursor polymer polybenzvalene. J Am Chem Soc. 1989;111(12):4413–4422.

[32] Klavett, FL, Grubbs, RH. Polycyclooctatetraene (polyacetylene): synthesis and properties. J Am Chem Soc. 1988;110(23):7807–7813.

[33] Hamilton, JG, Marquess, DG, O'Neill, TJ, et al. Metathesis polymerization of 7-oxa-2,3-bismethoxycarbonylbicyclo [2.2.1] hepta-2,5-diene: synthesis of novel conjugated polyenes, Chem. Commun. 1990;2:119.

[34] Buchmeiser, M R, Atzl, N, Bonn, G K. Ring-opening-metathesis polymerization for the preparation of carboxylic-acid-functionalized, high-capacity polymers for use in separation techniques. J Am Chem Soc. 1997;119(39):9166–9174.

[35] Novak, B M, Grubbs, R H. The ring opening metathesis polymerization of 7-oxabicyclo [2.2.1] hept-5-ene derivatives: a new acyclic polymeric ionophore. J Am Chem Soc. 1988;110(3): 960–961.

[36] Sailor, MJ, Klavetter, FL, Gruubbs, RH, et al. Electronic properties of junctions between silicon and organic conducting polymers. Nature. 1990;346(6280):155–157.

[37] Gorman, CB, Ginsburg, EJ, Sailor, MJ, et al. Substituted polyacetylenes through the ring-opening metathesis polymerization (ROMP) of substituted cyclooctatetraenes: A route into soluble polyacetylene. Synth Met. 1991;41(3):1033–1038.

[38] Grubbs, RH, Gorman, CB, Ginsburg, EJ, et al. New Polymeric materials with cubic optical nonlinearities derived from ring-opening metathesis polymerization. ACS Symp Ser. 1991; 455:672–682.

[39] Sailor, MJ, Ginsburg, EJ, Gorman, CB, et al. Thin films of n-Si/poly-(CH3) 3Si-cyclooctatetraene: conducting-polymer solar cells and layered structures. Science. 1990;249 (4973):1146–1149.

[40] Marder, SR, Perry, JW, Klavetter, FL, et al. Third-order susceptibilities of soluble polymers derived from the ring opening metathesis copolymerization of cyclooctatetraene and 1, 5-cyclooctadiene. Chem Mater. 1989;1(2):171–173.

[41] Swager, TM, Grubbs, RH. Synthesis and properties of a novel cross-conjugated conductive polymer precursor: poly (3, 4-diisopropylidenecyclobutene). J Am Chem Soc. 1987;109(3): 894–896.

[42] Moore, JS, Gorman, CB, Grubbs, RH. Soluble, chiral polyacetylenes: syntheses and investigation of their solution conformation. J Am Chem Soc. 1991;113(5):1704–1712.

[43] Ginsburg, EJ, Gorman, CB, Marder, SR, et al. Poly(trimethylsilylcyclooctatetraene): a soluble conjugated polyacetylene via olefin metathesis. J Am Chem Soc. 1989;111(19):7621–7622.

[44] Steinhäusler, T, Stelzer, F, Zenkl, E. Optically active polymers via ring-opening metathesis polymerization: 1. Polymers from enantiomerically pure 2-acyloxybicyclo [2.2. 1] hept-5-enes. Polymer. 1994;35(3):616–621.

[45] Boutarfa, D, Paillet, C, Leconte, M, et al. Applications of chloro-aryloxide complexes of tungsten in olefin metathesis: The ring-opening metathesis polymerization of dicyclopentadiene. J Mol Catal. 1991;69(2):157–169.

[46] Heroguez, V, Soum, A, Fontanille, M, Fontanille, M Functional oligomerization of dicyclopentadiene. Polymer. 1992;33(15):3302–3304.

[47] Carmail, H, Fontanille, M, Soum, A. Functional oligomers of norbornene: Part 1. Oligomerization by ring-opening metathesis polymerization in the presence of unsaturated diesters. J Mol Catal. 1991;65(1–2):193–203.

[48] Streck, R. Some applications of the olefin metathesis reaction to polymer synthesis. J Mol Catal. 1982;15(1–2):3–19.

[49] Wen-Qi, C, Pang, De-Ren, Yu-Ming, Z, et al. Highly effective activat ors-polyhalophenols in the ring-opening polymerization of cyclopentene. Acta Polym Sin. 1980;4:226–230.

[50] Biagini, SCG, Gibson, VC, Giles, MR, et al. Synthesis of penicillin derived polymers utilising ring-opening metathesis polymerisation methodology. Chem Commun. 1997;12:1097–1098.

[51] Streck, R. Some applications of the olefin metathesis reaction to polymer synthesis. J Mol Catal. 1982;15(1–2):3–19.

[52] Streck, R. Olefin metathesis in industry-a status report. J Mol Catal. 1988;46(1–3):305–316.

[53] Schneider, WA, Muller, MF. Crystallinity of trans-polyoctenamer: characterization and influence of sample history. J Mol Catal. 1988;46(1–3):395–403.

[54] Anon. Plastics World, 1985, 43: 22

[55] Graulich, W, Swodenk, W, Theisen, D. Make new TPR[trans-1,5-polypentenamer] rubber from C5's. Hydrocarbon Process. 1972;51:71–75.

[56] Breslow, DS. How we made neat stuff. Chemtech. 1990;20(9):540–544.

[57] Amass, AJ. Trans-polypentenamer: a general purpose rubber from a novel reaction. Brit Polym J. 1972;4(4):327–341.

[58] Mateijka, L, Houtman, C, Macosko, CW, et al. Polymerization of dicyclopentadiene: A new reaction injection molding system. J Appl Polym Sci. 1985;30(7):2787–2803.

[59] Balcar, H, Dosedlova, A, Perusova, L. Ring-opening metathesis polymerization of dicyclopentadiene by unicomponent catalysts derived from WCl₆. J Mol Catal. 1992;77(3): 289–295.

Exercises

1. **What is open-loop deuteration polymerization?**
   The process in which a cyclic olefin is cleaved in a molecule in the presence of a catalyst and joined to a macromolecule in a head-to-tail manner is called a ring-opening deuteration polymerization.

2. **Please introduce some ring-opening deuteration polymerization catalysts briefly.**
   At present, the most important open-loop deuteration polymerization catalysts have three major categories: The first is a traditional catalyst. In such a catalyst, the main catalyst is MX_n (wherein M = W, Mo, Re, Os, etc. transition metals, X = Cl, Br, etc. halogen atoms). The second type is a water-soluble catalyst, which can be used for the ring-opening deuteration polymerization of an aqueous solution system, and mainly represents $K_2RuCl_3 \cdot H_2O$. The third type is a carbene-type or alkylene-type catalyst developed in recent years. The catalyst has clear initiation mechanism, high catalytic activity, and strong stereoselectivity of the product. Therefore, it is the most important open-loop enthalpy polymerization catalyst of research in development.

3. **In the polymerization process, in addition to the earlier polymerization reaction, chain transfer reactions may occur. Please briefly describe them.**
   1) Self-biting transfer. The growing macromolecular metal carbene complex chain end reverses the double bond on the self-polymer chain, resulting in the formation of a cyclic polymer.

   2) The macromolecular metal carbene complex undergoes a cross-chemical reaction with other macromolecules in the system.

$$P_1 \text{---} CH\text{---}M + P_2 \text{---} CH\!=\!CH \text{---} P_3 \longrightarrow P_1 \text{---} CH\!=\!CH \text{---} P_3 + P_2 \text{---} CH\!=\!M$$

   3) The macromolecular carbene complex undergoes a cross-chemical reaction with the added acyclic olefin in the system.

$$P \text{---} CH\!=\!M + R_1\text{---}CH\!=\!CH\text{---}R_2 \longrightarrow P \text{---} CH\!=\!CH\text{---}R_1 + R_2\text{---}CH\!=\!M$$

4. **Please briefly describe the possible chain termination reactions.**
   1) The α-hydrogen transfer reaction occurs at the active chain end of the growing macromolecular carbene complex to form a macromolecule with a terminal double bond and to reduce the metal atom.

$$P \text{---} CH_2\text{---}CH\!=\!M \longrightarrow P \text{---} CH_2\!=\!CH_2 + M$$

   2) The growing macromolecular carbene complex reacts with the terminator to form a metal carbene complex without initiation activity. Commonly used terminators are trimethylvinylsilane, stilbene, vinyl ether, and the like. For example:

$$P \text{---} CH_2\text{---}CH\!=\!M + Me_3Si\text{---}CH\!=\!CH_2 \longrightarrow P \text{---} CH_2\!=\!CH_2 + Me_3Si\text{---}CH\!=\!M$$

3) The growing macromolecular carbene complex terminates with a cyclopropanation reaction with other molecules containing double bonds.

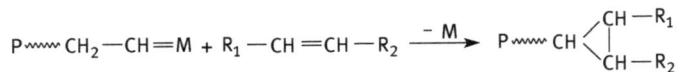

$$P \sim\!\!\sim\!\!\sim CH_2-CH{=}M + R_1-CH{=}CH-R_2 \xrightarrow{\;-\,M\;} P\sim\!\!\sim\!\!\sim CH \begin{array}{c} CH-R_1 \\ | \\ CH-R_2 \end{array}$$

5. **Please list some applications of ring-opening deuteration polymerization in polymer molecular design.**

   1) Synthesis of constant ratio copolymer.
   2) Synthesis of ideal alternating copolymers.
   3) Production of all-cis and all-trans polymers.
   4) Synthesis of isotactic all-cis and syndiotactic all-trans polymers.
   5) Synthesis of block and graft copolymers.
   6) Synthesis of comb and star copolymers.

Chapter 6
Click chemistry

6.1 Introduction

6.1.1 The basic concepts of click chemistry

The development of modern science and technology hopes that people can quickly and easily synthesize the required compounds, and the traditional chemical synthesis methods have become increasingly unsuitable. Therefore, the concept of "click chemistry" came into being.

Click chemistry is also translated as "link chemistry", "dynamic combinatorial chemistry", "speed matching combined chemistry." It is a new method for rapidly synthesizing many compounds proposed by the 2001 Nobel Prize-winning American chemist KB Sharpless in 2001. The main idea is to be fast and reliable through the splicing of small units and complete the chemical synthesis of various molecules. At its core, a series of reliable, modular reactions are used to form heteroatom-containing compounds.

Click chemistry describes the reaction process as simple, efficient, and versatile as clicking a mouse. The name "click" also means that splicing the molecular fragments in these ways is as simple as snapping the two parts of the buckle together. No matter what the buckle is connected to, if the two parts of the buckle come together, they can be combined with each other. And the two-part structure of the buckle determines that they can only be combined with each other. Particularly, it emphasizes the development of new combinatorial chemistry based on carbon-heteroatom bond (C-X-C) synthesis and uses these reactions to obtain diverse molecules simply and efficiently. The concept of click chemistry has contributed greatly to the field of chemical synthesis. It has become one of the most useful and attractive synthetic concepts in the fields of drug development, biomedical materials, and synthesis and preparation of polymer materials.

The concept of click chemistry was first derived from observations of natural products and biosynthetic pathways. It has been found that with more than 20 amino acids and more than 10 primary metabolites, nature can synthesize very complex biomolecules (proteins and polysaccharides) by splicing tens of millions of units of this type (amino acids and monosaccharides). This process has a distinct "modular" feature, and by simply forming a carbon–heteroatom bond by the attachment of certain specific functional groups, the amino acid and the monosaccharide can be assembled into a variety of biomolecules to create a colorful biological world.

https://doi.org/10.1515/9783110597097-006

In 1996, Guida et al. found through computer simulation that a molecule that can be used as a drug contains less than 30 hydrogen atoms with a relative molecular mass of less than 500, from hydrogen, carbon, nitrogen, oxygen, phosphorus, sulfur, and chlorine. It is composed of a bromine atom and it is stable to oxygen and water at room temperature. Then the number of such molecules can be as many as 1,063. However, the actual number of drugs currently used is far less than this number. The reason is that the current synthesis methods are independent and the efficiency is too low. Therefore, mimicking the "modular" synthesis in nature, developing a series of reliable, efficient, and selective click reactions will have revolutionary consequences for drug synthesis. It is also important to develop and synthesize other new materials.

In the history of modern chemistry for more than 150 years, a variety of techniques have been developed to interconnect molecular fragments. A considerable number of them are very delicate and require careful handling of highly reactive reactants under carefully controlled conditions. The "combination chemistry" invented in the 1990s is an important technology in this respect, but this technology has great limitations in the type of structure, and it is more dependent on the functional groups than the traditional synthetic chemistry reaction.

Since the end of the twentieth century, with the growth of new drug demand and the emergence of high-throughput screening methods, the synthesis of many new molecules has become an urgent task for chemical synthesis. The establishment of molecular libraries and the development of molecular diversity have become important issues. With the power of modern technology, new technologies such as chiral technology and high-throughput screening are rapidly improving the quality and speed of development of chemically synthesized drugs.

In 2001, chemists at the Scripps Research Institute in the United States found a technology that was easy to operate and produced high-yield target products. This technology produces products with little or no by-products, works well under many conditions (usually particularly good in water), and is not affected by other functional groups that are linked together. Clicking on the chemistry proposed complies with the requirements for molecular diversity in chemical synthesis.

The click chemical reaction must be modular, wide in application range, have high yield, have harmless by-products and be highly selective of the product. In addition, click chemistry has the following common characteristics:
(1) The reaction conditions are simple.
(2) Raw materials and reagents are readily available.
(3) Do not use solvents or can be carried out in a benign solvent such as water.
(4) It is not sensitive to oxygen and water and the presence of water often plays a role in accelerating the reaction.
(5) The click reaction is generally a fusion process (without by-products) or a condensation process (by-product is water).
(6) Has a higher thermodynamic driving force (> 84 kJ/mol).

(7) The product can be separated by simple crystallization and distillation without complicated separation methods such as chromatography.

Therefore, "click chemistry" is not so much a new chemical synthesis technology but a new chemical synthesis concept. It provides ideas for chemists to open new ways of chemical synthesis.

6.1.2 Click chemistry type

Up to now, there have been four types of click chemical reactions that have been discovered: cycloaddition reactions, especially 1, 3-dipolar cycloaddition reactions, including heterocyclic Diels–Alder reactions; nucleophilic ring-opening reactions, especially the tensioning heterocyclic electrophile is opened; the nonalkal carbonyl reaction; and the carbon–carbon multiple bond addition reaction. Here are a few common types of click chemistry reactions.

6.1.2.1 1, 3-Dipolar cycloaddition reaction
The 1, 3-dipolar cycloaddition reaction of terminal alkyne compounds and azide compounds is known as the "cream of the crop" of click chemistry and is currently the most widely used type of click chemical reaction.

The cycloaddition reaction of azide and acetylene was reported as early as the early twentieth century, but the reaction produced a mixture of 1,4- and 1,5-disubstituted triazoles. Later, Cu(I) catalyst was used, which greatly improved the selectivity of the product. The product was completely single 1,4-triazole, and the yield was as high as 91%, and the reaction time was shortened from 18 to 8 h. The process of this reaction is shown in Fig. 6.1.

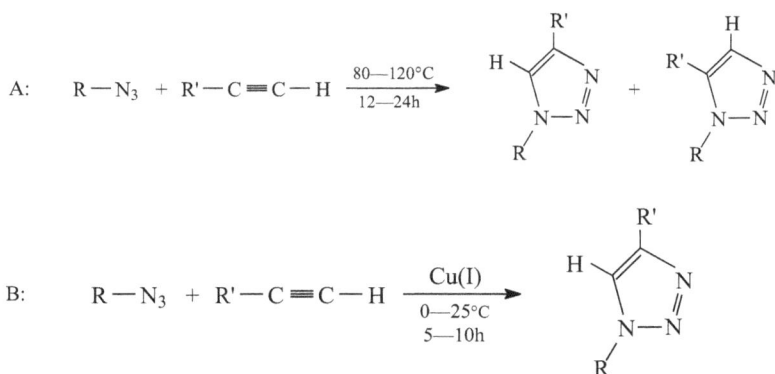

Fig. 6.1: 1,3-Dipolar cycloaddition reaction diagram. Synthesis of poly-5-vinyltetrazole. Reprinted with permission from [3].

6.1.2.2 Nucleophilic ring-opening reaction

The nucleophilic ring-opening reaction is mainly a nucleophilic ring opening of a ternary hetero atomic tension ring, such as an epoxy derivative, aziridine, a cyclic sulfate, a cyclic sulfamide, an aziridine ion, and an arsenazo ion, and many more. Their intrinsic tensile energy is released by the reaction. Among these three-membered heterocyclic compounds, epoxy derivatives and aziridine ions are the most commonly used substrates in click chemical reactions, and various highly selective compounds can be formed by their ring opening.

The epoxide contains a three-membered ring with a large tension, so the ring-opening reaction is a very advantageous process. However, ring opening needs to occur under specific conditions: the nucleophile can only attack one of the carbon atoms along the axis of the C–O bond. Such orbital alignment is detrimental to the elimination reaction competing with the ring-opening reaction, thereby avoiding by-products and obtaining high yields. In addition, the reactivity of the epoxide with water is not high, and the hydrogen bonding ability and polar nature of water are favorable for the ring-opening reaction of the epoxide with other nucleophiles.

The epoxide ring-opening reaction of the amine attack is shown in Fig. 6.2.

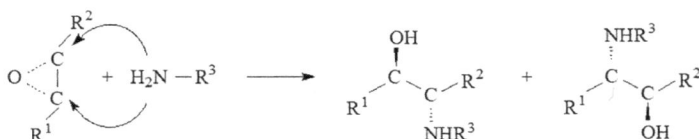

Fig. 6.2: Schematic diagram of epoxide ring-opening reaction of amine attack.

Such a reaction can be carried out in an alcohol/water mixed solvent or without a solvent. For example, di-ethylene oxide and benzylamine can afford a 1,4-diol in a 90% yield in the presence of a protic solvent methanol and in the absence of a solvent, 94% 1,3-diol is obtained. As shown in Fig. 6.3.

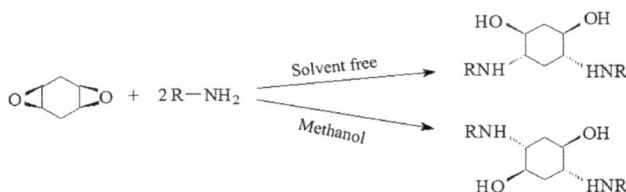

Fig. 6.3: Selective ring-opening reaction of dioxirane and benzylamine.

6.1.2.3 Condensation of carbonyl compounds

Such reliable and widely used reactions include the reaction of aldehydes or ketones with 1,3-diol to form 1,3-epoxypentane; the reaction of aldehydes with hydrazines or hydroxylamine ethers forms hydrazones and hydrazines (oximes); a carbonyl aldehyde, a ketone and an ester are reacted to form a heterocyclic compound or the like.

Figure 6.4 shows that *P*-toluenesulfonic acid is used as a catalyst, *N, N*-Dimethylacetamide is used as a solvent, and cyclohexane is a water-carrying agent. The reaction is carried out by the reaction of a linear saturated aliphatic aldehyde with D-isoascorbic acid, we obtain 3-ethylene pentane ring D-isoascorbic acid acetal reaction diagram.

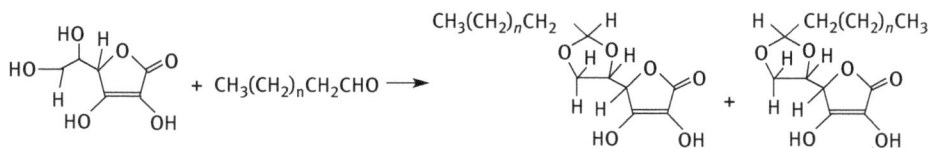

Fig. 6.4: Schematic diagram of aldol reaction.

6.2 "Click chemistry" in materials science

Materials are the foundation of the development of human society. From the history of human development, every important development is closely related to the discovery and utilization of new materials. In recent years, with the development of science and technology, material science and biological science, bioengineering, chemistry, physics, information science and environmental science have been continuously infiltrated, which has greatly promoted the invention and utilization of new materials, but still cannot satisfy people's ever-increasing demand for functionality. Therefore, new technologies and methods are urgently needed to inject new vitality into the research and development of new materials. As an important part of the material field, polymer materials have been the main research content in this field. Click chemistry as a new method developed in recent years has been widely developed in various research directions in polymer science, such as modification of polymer backbone, special structural polymers including block copolymers and star polymers. Click chemistry have found more and more use.

The main applications of chemistry are as follows: (1) synthesis of lead compound libraries, (2) application in new drug research, (3) bioconjugation, and (4) application of polymer chemistry. Among them, in the field of polymer chemistry, some people have successfully synthesized functional materials by means of click chemistry. For example, Sharpless et al. used Cu(I)-catalyzed cycloaddition-alkyne cycloaddition reaction to prepare high-efficiency metal binder materials; Fokin et al. synthesized triazole dendrimers in a similar manner; Löber et al. synthesized functional resin containing a 1,2,3-triazole ring by a chemical method as a carrier for producing a dopaminergic aryl urea compound. Click chemistry is also used for functionalization of material surfaces, such as surface modification of carbon nanotubes.

6.2.1 Click chemistry in the application of functional polymers

Since the click chemistry is a reaction between functional groups and does not involve the chain structure reaction of the polymer, it is possible to introduce a special group or structure into the polymer molecular chain without affecting the structure of the polymer molecular chain itself. This is of great significance for the preparation of functional polymers or special structural polymers.

For example, a linear polymer containing a cinnamate group on a side chain is prepared by performing a click chemical reaction with a small molecule compound containing both a cinnamate group and a terminal alkynyl group and a polymer having an azide group on the side chain.

Sodium alginate is a natural polysaccharide compound with good biocompatibility and is widely used in food and medical fields. Alginic acid is composed of guluronic acid (G unit) and mannuronic acid (M unit). A linear block copolymer composed of two structural units, wherein the G unit can chelate with two valence ions, such as calcium, copper and zinc, and rapidly form hydrogels under mild conditions. Because of its harmlessness, the hydrogel is widely used in drug carriers, tissue engineering scaffolds, cell microencapsulation, and so on. However, when the ionic crosslinked sodium alginate hydrogel is put in place for a period of time, the crosslinked ions will fall off and spread to the surrounding solution, resulting in the decrease of the mechanical strength of the hydrogel. The study shows that the sodium alginate hydrogel can be obtained by cross-linking sodium alginate with bifunctional cross-linking agents. For example, sodium alginate can be oxidized by sodium periodate and crosslinked by adipic acid two hydrazide, and hydrogel with higher mechanical strength can be obtained. However, conventional chemical crosslinking needs to be reacted under more severe conditions, and harmful catalysts may be introduced to reduce the biocompatibility of alginate hydrogels. Sodium alginate hydrogels can be prepared by the chemical reaction of azide and alkyne, which can solve the above problems.

For example, first, using ethanolamine to synthesize 2-azidoethylamine by bromination reaction, azide reaction, and then by 1-(3-dimethylaminopropyl)-3-ethylcarbodiimide hydrochloride and N-hydroxysuccinimide as catalysts, respectively, condense 2-azidoethylamine and propargylamine with sodium alginate to obtain azido sodium alginate and alkyne sodium alginate, and finally use a monovalent copper-catalyzed cycloaddition reaction of an azide group and an alkynyl group, after standing for a few minutes with vigorous stirring, a stable sodium alginate hydrogel can be obtained. The resulting hydrogel needs to be dialyzed against ethylenediaminetetraacetic acid and pure water to remove cuprous ions. Compared with other methods for preparing sodium alginate hydrogel by chemical cross-linking, the method has mild reaction conditions, simple and convenient operation, stable product performance, nontoxic and harmless, and is suitable for food and medical applications. Figure 6.5 is a schematic diagram of the preparation of sodium alginate hydrogel by click chemistry.

By combining click chemistry with other polymer synthesis methods, polymers which cannot be prepared by conventional synthetic methods can be obtained. The preparation of cyclic polymethyl methacrylate by atom transfer radical polymerization (ATRP) in combination with click chemistry is a typical example.

Cyclic polymers exhibit different properties from linear types due to the absence of end groups in the molecular chain, which has attracted the attention of chemists and materials scientists. Most of the methods for synthesizing cyclic polymers are currently formed by a coupling reaction between the double-end groups of linear prepolymers, and the activity of the terminal groups is often a key factor affecting the ring-forming efficiency. The combination of ATRP and click reaction to prepare a cyclic polymer can avoid the influence of end-group activity and greatly improve the efficiency of ring formation. According to the principle of ATRP, methyl methacrylate (MMA) polymerization is initiated by using an organic halide containing a terminal alkynyl group with cuprous bromide (CuBr) and pentamethyldiethylenetriamine as an initiation system. One end of the product is an alkynyl group, and the other end is a halogen atom. The halogen atom can be easily converted into an azide group, whereby the click reaction can be used to ring the polymer. The process of preparing ring-shaped polymethyl methacrylate (PMMA) by click chemistry is shown in Fig. 6.6.

It has been reported that the use of this ATRP in combination with a click reaction can also produce dendritic star polymers. The specific method is as follows: first, α-bromoisobutyryl is reacted with pentaerythritol to obtain a tetrafunctional bromine-containing initiator for initiating ATRP reaction of styrene, and the product is bromine-containing polystyrene $(PS-Br)_4$. The PS-Br was further treated with NaN_3, that is, converted to $(PS-N_3)_4$. Then, $(PS-N_3)_4$ is reacted with 2,2-bis(2-bromoisobutyryloxy) methylpropionic acid propargyl ester (click chemical reaction) to obtain polybromobenzene containing dibromo ethylene $(PS-Br_2)_4$. The ATRP reaction is carried out by using this polymer as an initiator to initiate MMA polymerization, and a dendritic star polymer $[(PMMA)_2PS]_4$ can be obtained. The reaction process is shown in Fig. 6.7.

Similarly, reversible addition-fragmentation chain transfer radical polymerization (RAFT) in combination with click chemistry can also be used to prepare block copolymers. Wu Qi et al. reported the preparation of poly(styrene-b-p-chloromethylstyrene) diblock copolymer (PS-b-PCVB) by RAFT living radical polymerization, and then using PS-b-PCVB by click chemistry. The chloromethylene group is converted into a triazole group to obtain a novel diblock copolymer Poly [styrene-block-(1-(4-Vinylbenzyl)-1H-1,2,3-triazol-4-yl)methanol](PS-b-PVBTM), and the reaction process is

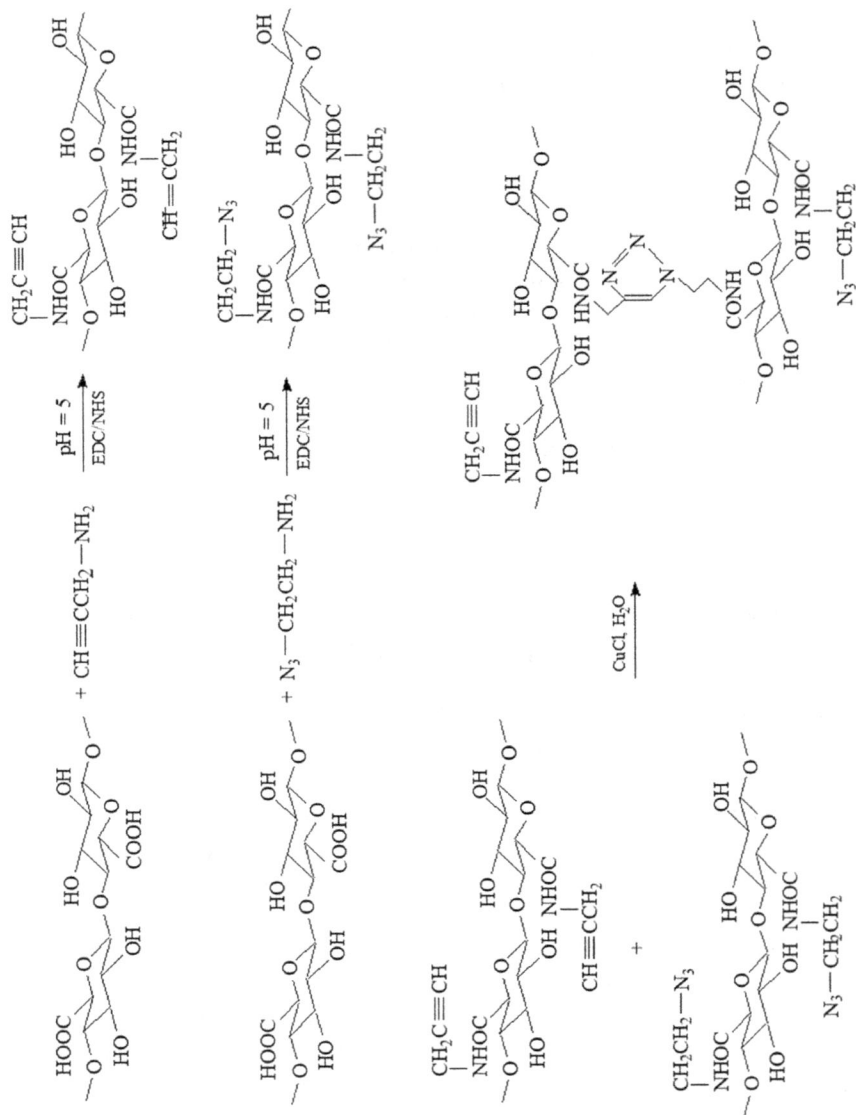

Fig. 6.5: Schematic diagram of preparation of sodium alginate hydrogel by click chemistry.

$CH \equiv CCH_2 - OH + Br - \overset{\overset{O}{\|}}{C} - \overset{\overset{CH_3}{|}}{\underset{|}{C}} - Br \longrightarrow CH \equiv CCH_2 - O\overset{\overset{O}{\|}}{C} - \overset{\overset{CH_3}{|}}{\underset{|}{C}} - Br$
$\qquad\qquad\qquad\qquad\quad CH_3 \qquad\qquad\qquad\qquad\qquad CH_3$

$\xrightarrow[\text{CuBr, PMDETA}]{\text{MMA}} CH \equiv CCH_2 - O\overset{\overset{O}{\|}}{C} - \overset{\overset{CH_3}{|}}{\underset{|}{C}} \!\left(\!CH_2 - \overset{\overset{CH_3}{|}}{\underset{|}{C}}\!\right)_{\!\!n}\!\! Br$
$\qquad\qquad\qquad\qquad\qquad CH_3 \qquad COOCH_3$

$\xrightarrow[\text{NaN}_3,\ \text{DMF}]{} CH \equiv CCH_2 - O\overset{\overset{O}{\|}}{C} - \overset{\overset{CH_3}{|}}{\underset{|}{C}} \!\left(\!CH_2 - \overset{\overset{CH_3}{|}}{\underset{|}{C}}\!\right)_{\!\!n}\!\! N_3$
$\qquad\qquad\qquad\qquad\qquad CH_3 \qquad COOCH_3$

$\xrightarrow[\text{CuBr, PMDETA, DMF}]{}$ (cyclic structure)

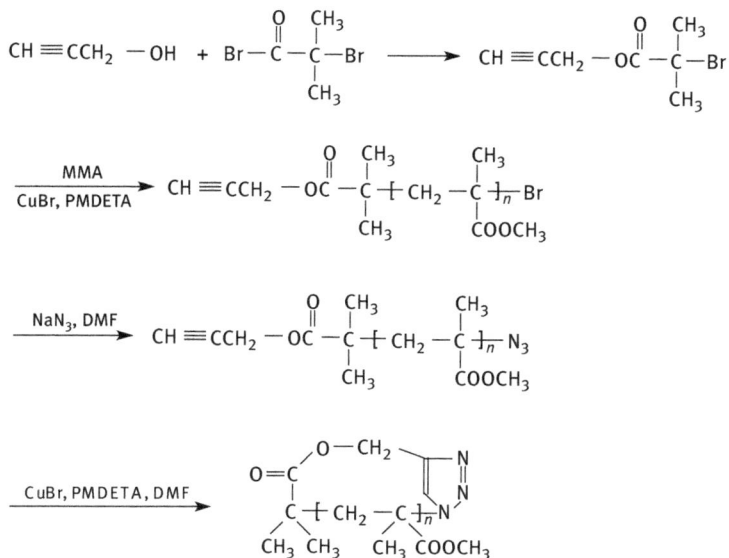

Fig. 6.6: Schematic diagram of preparation of cyclic PMMA by ATRP combined with click chemistry. Reprinted with permission from [9].

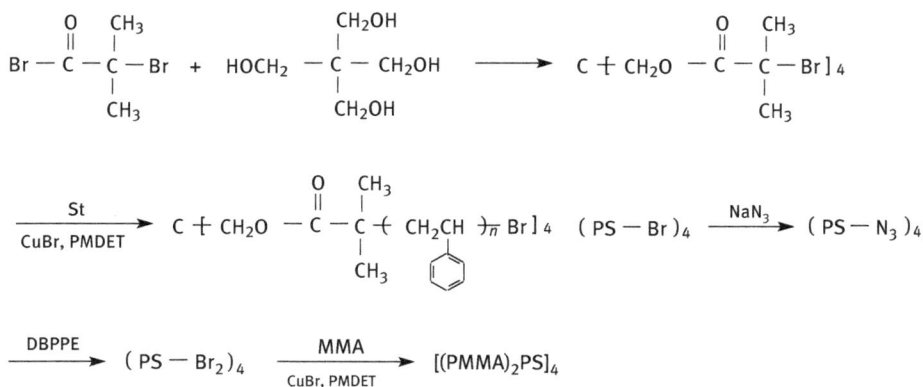

$Br - \overset{\overset{O}{\|}}{C} - \overset{\overset{CH_3}{|}}{\underset{CH_3}{C}} - Br + HOCH_2 - \overset{\overset{CH_2OH}{|}}{\underset{CH_2OH}{C}} - CH_2OH \longrightarrow C\!\left(\!CH_2O - \overset{\overset{O}{\|}}{C} - \overset{\overset{CH_3}{|}}{\underset{CH_3}{C}} - Br\right]_{\!4}$

$\xrightarrow[\text{CuBr, PMDET}]{\text{St}} C\!\left(\!CH_2O - \overset{\overset{O}{\|}}{C} - \overset{\overset{CH_3}{|}}{\underset{CH_3}{C}} \!\left(\!CH_2CH\right)_{\!\!n}\! Br\right]_{\!4} \quad (PS-Br)_4 \xrightarrow[]{\text{NaN}_3} (PS-N_3)_4$

$\xrightarrow[]{\text{DBPPE}} (PS - Br_2)_4 \xrightarrow[\text{CuBr, PMDET}]{\text{MMA}} [(PMMA)_2PS]_4$

Fig. 6.7: Schematic diagram of the preparation of dendritic star polymers by ATRP combined with click chemistry. Reprinted with permission from [10].

shown in Fig. 6.8. The block copolymer can be electrospun to obtain a microphase-separated structural nanofiber in which a nanoparticle and a continuous phase coexist.

The azide-alkyne AB$_2$ type monomer can directly obtain the triazolyl high-nitrogen hyperbranched polymer by click chemistry, and the polymerization process is simple, and the product hyperbranched algebra can be controlled. Figure 6.9 shows the

$$n \ CH_2 = CH \xrightarrow[80°C]{DTCMA, AIBN} +CH_2-CH \}_{\overline{n}} S \diagdown^{S}_{C_{12}H_{25}}$$

$$\xrightarrow[90°C]{CVB, AIBN} + CH_2 - CH \}_{\overline{n}} \{ CH_2 - CH \}_{\overline{m}} S \diagdown^{S}_{C_{12}H_{25}}$$

$$CH_2Cl$$

$$\xrightarrow[80°C]{NaN_3, DMF} + CH_2 - CH \}_{\overline{n}} \{ CH_2 - CH \}_{\overline{m}} S \diagdown^{S}_{C_{12}H_{25}}$$

$$CH_2N_3$$

$$\xrightarrow[CH\equiv CCH_2OH]{CuSO_4, sodium \ ascorbate} + CH_2 - CH \}_{\overline{n}} \{ CH_2 - CH \}_{\overline{m}} S \diagdown^{S}_{C_{12}H_{25}}$$

$$CH_2$$

DTCMA: Dodecyl trithioisobutyrate

AIBN: Azobisisobutyronitrile

CVB: *P*-chloromethylstyrene

$$CH_2OH$$

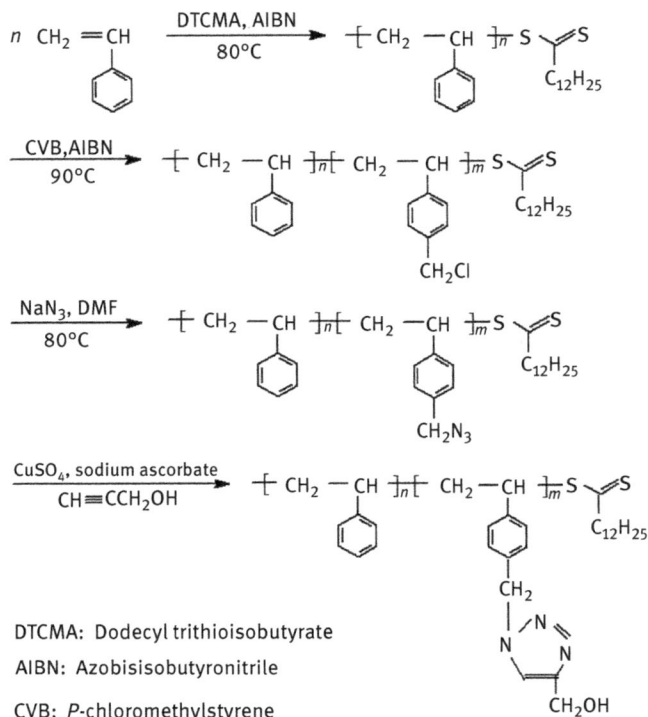

Fig. 6.8: Schematic diagram of preparation of diblock copolymer PS-b-PVBTM by RAFT combined with click chemistry. Reprinted with permission from [14].

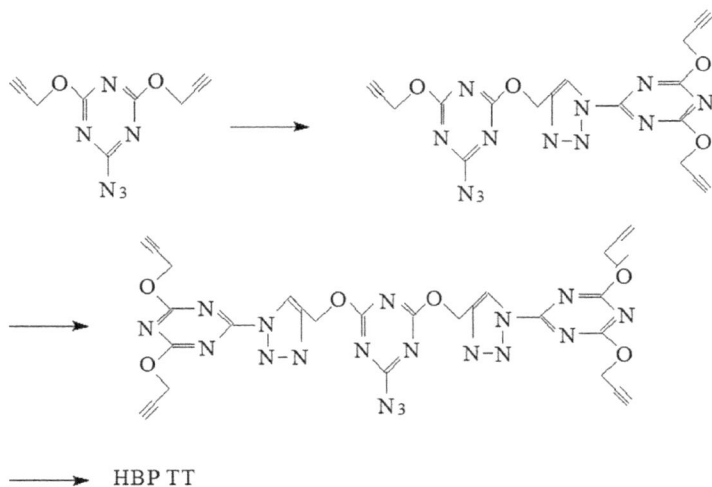

⟶ HBP TT

Fig. 6.9: Schematic diagram of the synthesis of hyperbranched poly(1,2,3-triazole-1,3,5-triazine) (HBP TT).

process of this reaction. The hyperbranched polymer molecule contains a large amount of N–N, C–N, N = N bonds, and has high-energy content, and is expected to be used as an energetic binder.

The sugar-containing polymer refers to a polymer having a sugar-based component on a polymer chain. These materials have good hydrophilicity and biocompatibility and are specifically recognized in living organisms, so they are an important class of biological functional materials. Biochemical modification of polymers by click chemistry is simple, efficient, and selective and does not destroy the original structure of the polymer. Therefore, it has obvious advantages for preparing sugar-containing polymers.

Wang et al., Nankai University, combined RAFT with sulfhydryl-olefin click chemistry to prepare a sugar-containing polymer–polypeptide bioconjugate with a comb structure (see Fig. 6.10). The specific method is as follows: first, the two-block copolymer poly(3-O-methacryloy-α,β-D-glucopyranose)-block-P(2-hydroxyethyl methacrylate) (PMAGlc-b-PHEMA) is prepared by RAFT, and then the acrylate group is bonded to the terminal copolymer by esterification reaction, and the acrylate-modified sugar-containing copolymer is obtained after hydrolysis PMAGlc-b-PHEMA. The reduced-type glutathione and the sugar-containing copolymer are subjected to a click reaction at room temperature to obtain a comb-structured sugar–containing polymer-polypeptide bioconjugate poly(3-O-methacryloy-α,β-D-glucopyranose) -glutathione (PMAGlc-GSH). It was found that the conjugate can self-assemble into spherical micelles in aqueous solution, and the size of the micelle depends on the pH of the environment, and it has specific recognition for concanavalin A.

It is known from the polymerization mechanism that the polymer end prepared by RAFT contains a dithioester group and can be converted into a mercapto group in the presence of a primary amine or a reducing agent. Various modifications can be made to the polymer using click chemistry. Figure 6.11 is a specific example of the preparation of a sugar-containing polymer-polypeptide bioconjugate by the exchange reaction of RAFT with a thiol-disulfide compound.

6.2.2 "Click chemistry" in the application of material surface treatment

6.2.2.1 Application of "click chemistry" on the surface functionalization of nanocarbon materials

Carbon nanotubes have unique structures and properties, and their superior conductivity and high-heat capacity make them have great potential in nanoelectronic components and sensor applications. Surface-functionalized carbon nanotubes offer a wider range of properties, extending their reach to pharmaceuticals, biomaterials, and composites. The surface functionalization of carbon nanotubes can be conveniently achieved using click chemistry.

In order to obtain a hydrophilic carbon nanotube, the water-soluble polymer polyethylene glycol is used to modify the carbon nanotube by a Huisgen cycloaddition

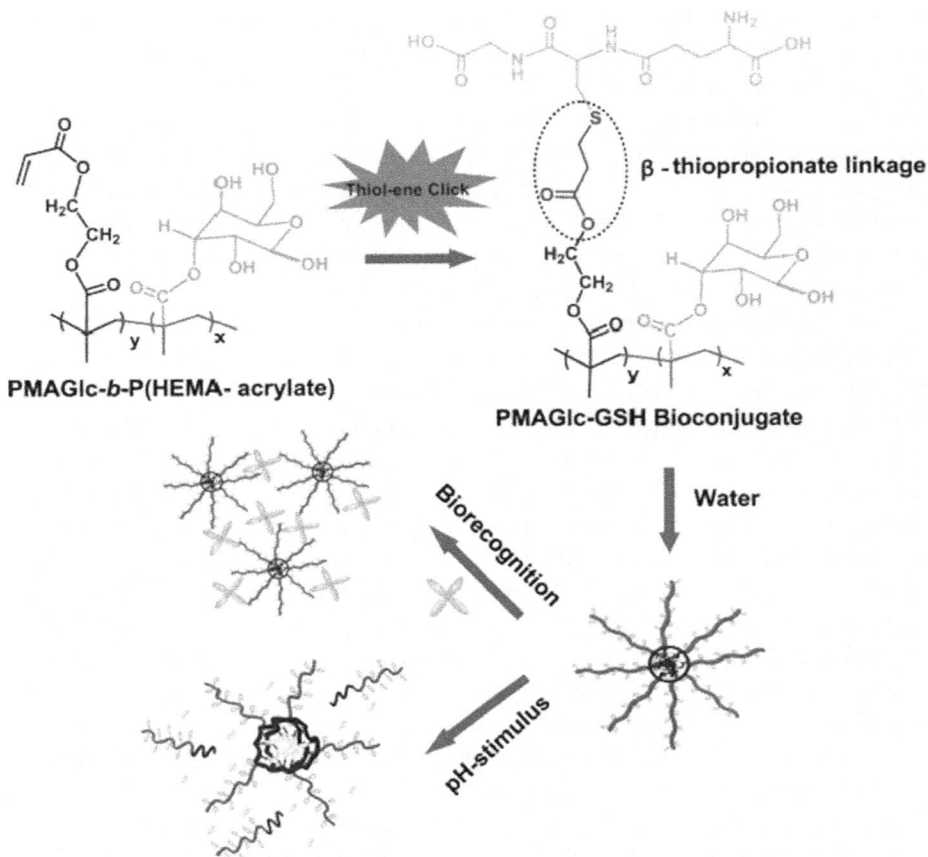

β-thiopropionate linkage

PMAGlc-*b*-P(HEMA- acrylate)

PMAGlc-GSH Bioconjugate

Fig. 6.10: Schematic diagram of preparation of comb-type sugar-containing polymer–polypeptide bioconjugate by RAFT combined with click chemistry. Reprinted with permission from [26].

reaction, thereby obtaining a high-graft density. The specific method is as follows: Introducing an alkynyl group by reaction of an aryl diazonium salt with a single-walled carbon nanotube. Then, it is reacted with polyethylene glycol (PEG) which has been subjected to end-group functionalization with an azide group, thereby obtaining a water-soluble polymer PEG surface-modified single-walled carbon nanotube. It has been experimentally proven that this method can provide relatively high-grafting density, complete determination of polymer molecular weight and structure, and availability of single-walled carbon nanotubes with good water solubility. The reaction process and conditions are shown in Fig. 6.12. As can be seen from the figure, the process is mainly divided into three parts, namely (1) preparation of alkynyl functionalized carbon nanotubes, (2) preparation of terminal azide PEG and (3) making the earlier two products undergo a Huisgen cycloaddition reaction to prepare hydrophilic

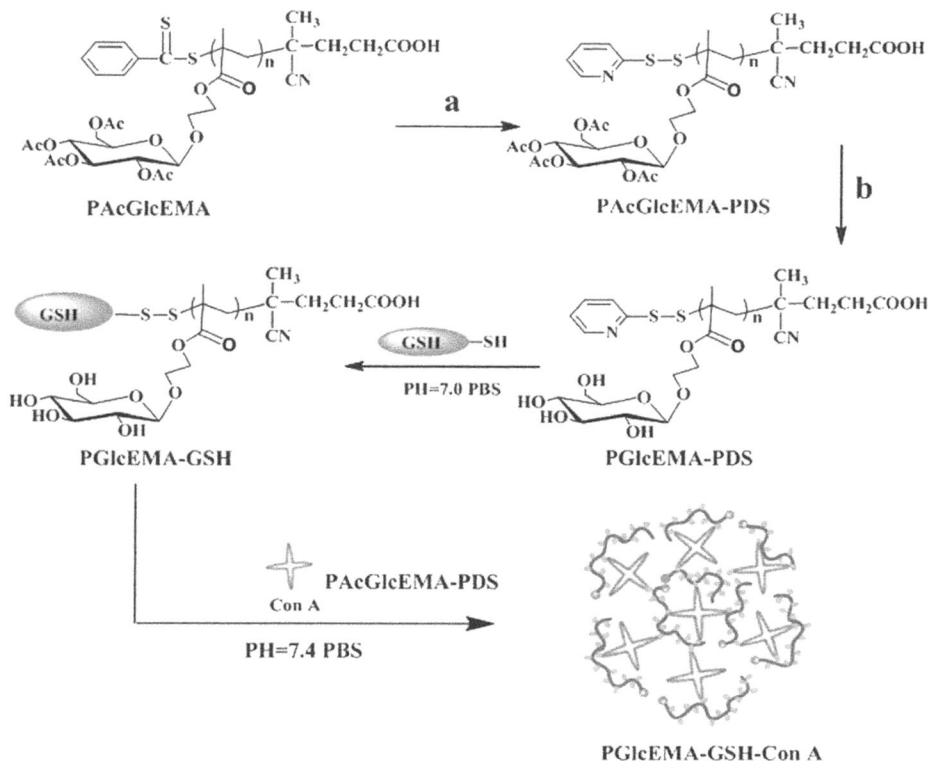

Fig. 6.11: Schematic diagram of preparation of sugar-containing polymer-polypeptide bioconjugate by RAFT combined with thiol-dithio exchange reaction. Reprinted with permission from [16].

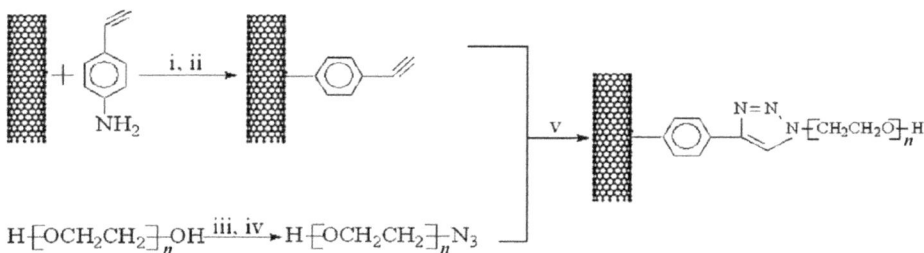

Fig. 6.12: Reaction process and conditions of PEG modified by Huisgen cycloaddition reaction. Reprinted with permission from [17].

carbon nanotubes. Figure 6.13 shows the condition in which carbon nanotubes were dispersed in water with different molecular weights of PEG and allowed to stand for 7 days. It is shown in the figure that the single-walled carbon nanotubes modified

Fig. 6.13: Dispersion of PEG-modified SWNTs with different molecular weights in water. Reprinted with permission from [17].

with PEG having a relative molecular mass of 2,000 have the best dispersion effect in water, followed by the PEG-modified carbon nanotubes with a relative molecular mass of 800 and the dispersion effect. The worst is a PEG-modified carbon nanotube having a relative molecular mass of 400.

If a perfluoroaryl azide compound is reacted with an alkynyl functionalized carbon nanotube, a superhydrophobic carbon nanotube can be obtained.

A similar work was carried out by reacting a terminal azide poly (tert-butyl acrylate)-styrene block copolymer PS-b-PtBA-N$_3$ with an alkynyl functionalized C$_{60}$, and the resulting product PS-b-PtBA-C$_{60}$ was acid hydrolyzed and then polymerized. The tert-butyl acrylate is embedded into a polyacrylic acid (PAA). A PS/PAA block copolymer-modified C$_{60}$ (PS-b-PAA-C$_{60}$) was obtained. PS-b-PAA-C$_{60}$ can form self-assembled vesicles in water, in which C$_{60}$ are regularly arranged in a double corona shape, exhibiting good photoconductivity and can be directly used as a photoconductive material.

6.2.2.2 Application of "click chemistry" in surface modification of other materials

Click chemistry has also made great achievements in the modification of metal surfaces.

For example, self-assembly on a gold surface is achieved by using alkenyl mercaptans of various lengths and terminal azide thioundecane, and then 1-acetylenyl-2-nitrobenzene is covalently grafted onto the surface of gold by click chemistry. The electron transfer resistance (2.5×10^5 Ω) of the gold surface is much larger than that of the bare gold electrode surface.

The polyelectrolyte poly-4-styrenesulfonic acid-maleic acid was bonded to the gold nanorods, and the azide was functionalized and subjected to a click chemical reaction with the alkynylated trypsin. Activity experiments demonstrated that the activity of the bioconjugated enzyme was maintained at 57%, higher than the activity of carbodiimide coupling (20%) and electrostatic adsorption enzymes.

Zhang Tao et al. used a combination of click chemistry and RAFT to prepare gold nanoparticles with dual temperature and pH responses, which realized the simplification and controllability of gold nanoparticle modification. The specific method is as follows: first, PEG-modified gold nanoparticles are prepared by surface adsorption, and azide is used to obtain AuPEG-N3 which can be used for a click reaction. Then, the poly(N-isopropylacrylamide) and poly-4-vinylpyridine containing alkynyl groups at the end were simultaneously modified onto AuPEG-N$_3$ by click reaction, thereby obtaining gold nanoparticles with double response temperature and pH value.

Clay nanoparticles have a very significant effect on the overall properties of polymers, especially mechanical properties, heat resistance and flame retardancy. Therefore, the research of polymer/clay nanoparticle composites has attracted wide attention. Due to the large surface area of nanomaterials, the extremely high-surface energy and easy agglomeration, the dispersion of clay nanoparticles is a key issue in this study.

By clicking on the chemically synthesized diblock copolymer PS-b-PVBTM (see Section 2.1), the triazole group can be better inserted into the lamellar structure of montmorillonite, thus effectively montmorillonite tablets layer is peeled off to achieve the purpose of dispersion. Experiments show that most of the montmorillonite in the formed PS-b-PVBTM/montmorillonite composite loses its accumulation structure and has a flaky structure. The thermal and mechanical properties of the product are greatly improved. It has been studied to carry out the click chemical reaction of alkynyl-functionalized polycyclosiloxane-caprolactone with azide montmorillonite. The results show that the clay layer in the product is exfoliated and randomly dispersed in the polymer substrate. Thereby, the thermal stability and flame retardancy of the nano-composite are better improved.

6.2.3 Application of "click chemistry" reaction in energetic materials

1,3-Dipolar cycloaddition is one of the important types of click chemical reactions. Through such a reaction, a triazole or tetrazole high nitrogen compound can be pro-duced. The cyclic structure of the azole compound contains N–N, C–N, N = N bonds with high-bond energy, and so on, so that the compound has a high positive en-thalpy; the nitrogen atom in the molecule is beneficial to increase the density of the material; More gas is produced during combustion, and the flame temperature is low; the large conjugate system is beneficial to improve the stability of the molecule. These characteristics are required for a new generation of energetic materials. At present, azole compounds have attracted attention at home and abroad as a fourth-generation energetic material with high energy density and good safety performance. This pro-vides a useful place for the application of click chemistry in energetic materials.

In the preparation of solid propellants, the hydroxyl groups of the binder are often reacted with isocyanate for curing. However, the use of isocyanate as a cross-linking agent has the following disadvantages: isocyanate is more toxic, the overall energy of the material is reduced, the reaction conditions are harsh (requires abso-lute water avoidance), and it has poor capacity with green energetic materials such as ammonium amide and HNF nitrox oxime.

The formation of a triazole ring by click chemistry to cure the adhesive not only overcomes the above disadvantages but also has the following new advantages: the same effect is achieved with less cross-linker than isocyanate. the amount of reactants does not require precise stoichiometry, the product morphology can be controlled by controlling the degree of cross-linking, the characteristics of low-pressure self-extinguishing can be improved or even overcome, and the burning rate of the propellant can be improved, by adjusting the type of azide polymer, the mass fraction of azide and the cross-linking time products with different hardness distribution and combustion performance can be obtained.

For example, to prepare energetic materials, 3 M company uses different specifications of polyazide glycidyl ether (GAP) and poly 3,3-bis(azidomethyl)butylene oxide and other azide polymers and pentaerythritol triacrylate to react with unsaturated carbon–carbon double bonds and carbon–carbon three-bond cross-linking agents such as 1,6-hexanediol diacrylate and 1,6-hexanediol dipropynyl ester. The obtained material can be used for rocket fuel and explosives and can be stably burned under normal temperature and normal pressure, and the burning rate is 13.97 mm/s.

In order to improve the combustion properties of energetic materials, it is often necessary to add a combustion performance modifier to the formulation. The metal salt of the azide compound can be used as a combustion performance modifier, but such a catalyst has a high mechanical sensitivity. The tetrazole catalyst has similar chemical properties to the azide compound but has a lower mechanical sensitivity. Therefore, using the 1, 3-dipolar cycloaddition of nitrile group and sodium azide for click chemical reaction, 5-phenyltetrazole and 5-methyleneditetrazole were synthesized, and the corresponding copper salt, lead salt and strontium were further prepared (Fig. 6.14). The results show that these tetrazolium metal salts can improve the burning rate of energetic materials and have good catalytic effects. Inspired by this, a polymer-based tetrazolium salt was synthesized by reacting polyacrylonitrile with sodium azide, which also has good combustion catalysis.

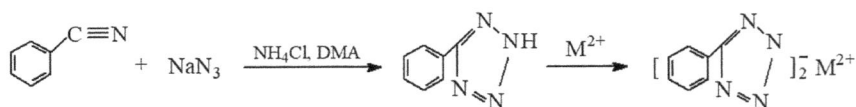

Fig. 6.14: Preparation of 5-phenyltetrazolium salt. Reprinted with permission from [20].

The triazole ring and the tetrazole ring have a higher positive enthalpy (+ 272 kJ/mol), so polymers having a triazole and tetrazole structure are a class of energetic polymers.

The triazole glycidyl ether series polymer can be obtained in a high yield by reacting phenylacetylene with different functional groups and azide GAP (Fig. 6.15). Poly-5-vinyltetrazole (PVT) can be obtained by using a polyacrylonitrile as a raw material and performing a click chemical reaction by reacting a side nitrile group with sodium azide or ammonium chloride (Fig. 6.16).

PVT is a high-nitrogen thermoplastic polymer with high nitrogen content (mass fraction of 5.83%) and high formation (about 2,000 kJ/kg), high density (about 1.45 g/cm^3), high decomposition temperature (about 200 °C), low mechanical sensitivity,

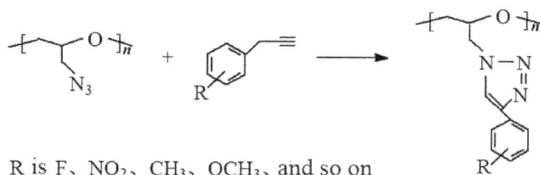

R is F、NO$_2$、CH$_3$、OCH$_3$、and so on

Fig. 6.15: Synthesis of triazole glycidyl ether series polymer. Reprinted with permission from [20].

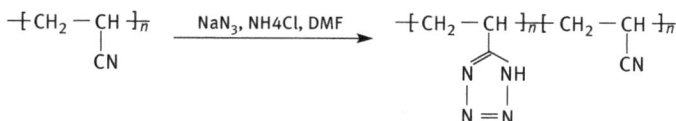

Fig. 6.16: Synthesis of poly-5-vinyltetrazole. Reprinted with permission from [23].

and low combustion temperature. It can be used in the fields of explosives, energetic materials, gas generating agents, and the like. PVT has a strong acidity and reacts with caustic or ammonia to form a water-soluble salt. When the concentration of the salt is increased, it becomes a gel. The aqueous alkali metal salt solution of the polymer can also form a gel by cross-linking with high-valent metal ions, so it can be used as a reactive binder for water-filled composite explosives and energetic materials.

References

[1] Finn, MG, Kolb, HC, Fokin, VV, Sharpless, K B. Click chemistry – interpretation and objectives. Process Chem. 2008;20(1):1.
[2] David, DD, Punna, S, Holzer, P, et al. Click chemistry in materials synthesis. 1. Adhesive polymers from copper-catalyzed azide-alkyne cycloaddition. J Polym Sci, Part A: Polym Chem. 2004;42(17):4392–4403.
[3] Wu, P, Feldman, A K, Nugent, A K, et al. Efficiency and fidelity in a click-chemistry route to triazole dendrimers by the copper (I)-catalyzed ligation of azides and alkynes. Angew Chem, Int Ed. 2004;43(30):3928–3932.
[4] Löber, S, Rodriguez, L P, Gmeiner, P. Click linker: efficient and high-yielding synthesis of a new family of SPOS resins by 1, 3-dipolar cycloaddition. Org Lett. 2003;5(10):1753–1755.
[5] Meng, J C, Averbuj, C, Lewis, W G, et al. Cleavable linkers for porous silicon-based mass spectrometry. Angew Chem, Int Ed. 2004;43(10):1255–1260.
[6] Fazio, F, Bryan, MC, Blixt, O, et al. Synthesis of sugar arrays in microtiter plate. J Am Chem Soc. 2002;124(48):14397–14402.
[7] Juan, Z, Jia-Rui, Z, Xiao-Lin, K. Study on the preparation of linear polymers containing cinnamate groups by "click chemistry". Ion Exch Adsorpt. 2012;28(3):277–281.
[8] Qiao, Z, Guo-Ping, C, Si-Xue, C. Preparation of alginate hydrogels via "click chemistry". J Wuhan Univ of Technol. 2011;33(7):14–16.

[9] Juan, L, Ming, D, Zhong-Wen, F. Combination of ATRP and "click chemistry" to prepare cyclic polymethyl methacrylate. Chem J Chin Univ. 2007;28(6):1197–1199.

[10] Ming, D, Lie-Hui, Z, Juan, L. Combination of ATRP and "click chemistry" to prepare dendritic star polymer. Chem J Chin Univ. 2008;29(10):2118–2120.

[11] Helms, B, Mynar, JL, Hawker, CJ. Dendronized linear polymers via "click chemistry". J Am Chem Soc. 2004;126(46):15020–15021.

[12] Lutz, JF, Bolmer, HG, Weichenhan, K. Combining ATRP and "click" chemistry: A promising platform toward functional biocompatible polymers and polymer bioconjugates. Macromolecules. 2006;39(19):6376–6383.

[13] Thibault, RJ, Takizawa, K, Lowenheilm, P, et al. A versatile new monomer family: functionalized 4-vinyl-1, 2, 3-triazoles via click chemistry. J Am Chem Soc. 2006;128(37):12084–12085.

[14] Qi, W, Hewen, L. Reversible addition-fragmented chain transfer polymerization / click chemistry method for preparing block polymers containing 1,2,3-triazole functional groups and its electrospinning. Chin J Appl Chem. 2010;26(2):29–32.

[15] Boyer, C, Granville, A, Davis, TP, et al. Modification of RAFT-polymers via thiol-ene reactions: A general route to functional polymers and new architectures. J Polym Sci, Part A: Polym Chem. 2009;47(15):3773–3794.

[16] Haiting, S, Li, L, Xiao-Bei, W, Jing-Yi, L, et al. Glycopolymer–peptide bioconjugates with antioxidant activity via RAFT polymerization. Polym Chem. 2012;3(5):1182–1188.

[17] Sizhe, H. Study on hydrophilic modification of carbon nanotubes. Master's thesis of Tongji University, Shanghai, China, 2007.

[18] Tao, Z, Yuan-Peng, W, Zhao-Hui, Z. Preparation of dual responsive gold nanoparticles using click chemistry and living radical polymerization. Chem J Chin Univ. 2010;31(11):2303–2307.

[19] Qiu-Ning, S, Hui-Yi, T, Zhi-Da, L. Application of azides on surface modification of inorganic materials. Sci Technol Chem Ind. 2012;20(2):58–63.

[20] Jun-Jie, C, Yu, X, Xiao-Qin, Z., et al. Applications of 1,3-dipolar cycloaddition reaction in energetic materials. Chem Prop Polym Mater. 2011;9(3):25–28.

[21] Jung, JH, Lim, YG, Lee, KH, et al. Synthesis of glycidyl triazolyl polymers using click chemistry. Tetrahedron Lett. 2007;48(37):6442–6448.

[22] Qing, L, Ning, G, Tao, H, et al. Synthesis and characterization of 5-vinyltetrazole. Chin J Explos Propellants. 2009;32(1):29–31.

[23] Heng-Xin, H, Yi-Lin, C, Zhong-Xiang, S Synthesis and properties of poly(5-vinyltetrazole). Chin J Energ Mater. 2007;15(5):492–495.

[24] Williams, GK, Burns, SP, Mishra, I. Gas generant compositions. United States patent, 20050230017, 2005.

[25] Katritzky, AR, Meher, NK, Hanci, S, et al. Preparation and characterization of 1, 2, 3-triazole-cured polymers from endcapped azides and alkynes. J Polym Sci, Part A: Polym Chem. 2008;46(1):238–256.

[26] Xiao-Bei, W, Li, L, Yan, L, Haiting, S, Jing-Yi, L, Han-Ying, Z. Comb-shaped glycopolymer/peptide bioconjugates by combination of RAFT polymerization and thiol-ene "Click" chemistry. Macromol Biosci. 2012;12(11):1575–1582.

Exercises

1. **How many types of "click chemistry" are there?**

   Up to now, there have been four types of click chemical reactions that have been discovered:
   (1) cycloaddition reactions, especially 1, 3-dipolar cycloaddition reactions, including heterocyclic Diels–Alder reactions
   (2) nucleophilic ring-opening reactions, especially the tensioning heterocyclic electrophile is opened

(3) the nonalkal carbonyl reaction
(4) the carbon–carbon multiple bond addition reaction

2. What is the main idea and core of chemistry?

The main idea is to fast and reliable synthesize various chemical molecules through the splicing of small units. The core is to use a series of reliable, modular reactions to form heteroatom-containing compounds.

3. Please list some characteristics of click chemical reaction

(1) The reaction conditions are simple.
(2) Raw materials and reagents are readily available.
(3) Do not use solvents or can be carried out in a benign solvent such as water.
(4) Not sensitive to oxygen and water; the presence of water often plays a role in accelerating the reaction.
(5) The click reaction is generally a fusion process (without by-products) or a condensation process (by-product is water).
(6) With a higher thermodynamic driving force (>84kJ/mol).
(7) The product can be separated by simple crystallization and distillation without complicated separation methods such as chromatography.

4. What are the applications of click chemistry?

The main applications of chemistry are as follows:

(1) synthesis of lead compound libraries
(2) application in new drug research
(3) bioconjugation
(4) application of polymer chemistry

5. Please introduce one example of the successful application of click chemistry in the field of surface modification.

The main idea is to fast and reliable synthesize various chemical molecules through the splicing of small units. The core is to use a series of reliable, modular reactions to form heteroatom-containing compounds.

The surface functionalization of carbon nanotubes can be conveniently achieved using click chemistry. In order to obtain a hydrophilic carbon nanotube, the water-soluble polymer polyethylene glycol is used to modify the carbon nanotube by a Huisgen cycloaddition reaction, thereby obtaining a high-graft density. The specific method is as follows: Introducing an alkynyl group by reaction of an aryl diazonium salt with a single-walled carbon nanotube. Then, it is reacted with PEG which has been subjected to end-group functionalization with an azide group, thereby obtaining a water-soluble polymer PEG surface-modified single-walled carbon nanotube.

If a perfluoroaryl azide compound is reacted with an alkynyl functionalized carbon nanotube, a superhydrophobic carbon nanotube can be obtained.

Chapter 7
Macromolecular self-assembly

7.1 Introduction

7.1.1 Supramolecular chemistry

In the 1930s, the German Wolf created the term "supramolecular" to describe an ordered system formed by the association between molecules to distinguish it from traditional compounds. In 1978, the French Lehn and others broke through the research field of traditional chemistry and first proposed the concept of "supramolecular chemistry." He pointed out: "The existence of covalent bonds exists in the field of molecular chemistry, based on molecular assemblies and intermolecular bonds." "Supramolecular chemistry" is undoubtedly a major ideological leap. After nearly 20 years of rapid development, supramolecular chemistry has formed a complete scientific system, which is an emerging discipline at the intersection of modern chemistry, materials chemistry, and life sciences. In 1987, the Nobel Prize in Chemistry was awarded to three chemists, C. J. Pedersen, J. M. Lehn, and D. J. Cram, for their pioneering work in the theory of supramolecular chemistry.

Supramolecular chemistry is the chemistry of forming molecular aggregates based on noncovalent interactions between molecules. In the Nobel Prize in Chemistry, Ryan defined supramolecular chemistry as the study of supramolecular chemistry with specific structures and functions that are formed by the interaction of two or more chemical species through intermolecular interactions. science. Unlike traditional molecular chemistry based on atomic building molecules, supramolecular chemical chemistry above the molecular level, it mainly studies two or more molecules through weak interactions between noncovalent bonds between molecules, such as hydrogen bonds. The structure and function of molecular aggregates generated by dipole–dipole interactions, hydrophilic–hydrophobic interactions, and synergies between them. The advent of supramolecular chemistry has enabled scientists to broaden their research from individual molecules to molecular assemblies.

Compared with the classic chemistry of the eighteenth and nineteenth centuries, the distinctive characteristics of modern chemistry are from macroscopic to microscopic, from static research to dynamic research, from individual and meticulous research to interpenetrating and interrelated research, and from atomic arrangement within the molecule to develop the interaction between molecules. In a sense, supramolecular chemistry dilutes the boundaries between organic chemistry, inorganic chemistry, biochemistry, and material chemistry, emphasizing supramolecular chemistry with specific structures and functions, and the four basic chemistries (organic chemistry, inorganic chemistry, analytical chemistry, and physical chemistry) are organically integrated into one.

https://doi.org/10.1515/9783110597097-007

The core issue involved in supramolecular chemistry is how the directionality and selectivity of various weak interactions determine the identification of molecules and the assembly properties of molecules. These include more fundamental scientific issues, such as what is the nature of weak interactions and how synergies between them work. Therefore, the study of self-assembly process in supramolecular chemistry is one of the most important central topics.

At present, supramolecular chemistry has developed into a "supramolecular science system based on molecular recognition, molecular self-assembly as a means, and assembly function," and forms a new discipline with life sciences, information science, materials science, and nanoscience, promoting the development of science and technology. The multidisciplinary intersection will surely become an important source of new concepts and high-tech in the twenty-first century, and it is also of great significance for revealing the mysteries of life phenomena. It is an important direction of chemical development in the twenty-first century.

7.1.2 Molecular self-assembly

Molecular self-assembly is a common phenomenon in nature. Generalized molecular self-assembly refers to the formation of a certain structure and function by the non-covalent bond force between the molecule and the molecule (including Coulomb force, van der Waals force, hydrophobic force, π–π stacking force, and hydrogen bond). The process of the body is spontaneous and does not require external force. Self-assembly phenomena are common in nature and engineering, ranging from atomic or molecular scales to large galaxy scales. For example, protein formation, cell formation and evolution, formation of viral molecules and enzymes, and signal storage and delivery between DNA molecules are achieved through molecular self-assembly. Many biomacromolecules form highly organized, informative, and functional complex structures through self-assembly processes. In the field of chemistry, molecular self-assembly is also ubiquitous, such as crystal growth, liquid crystal formation, self-generation of artificial lipid bilayers, synthesis of metal coordination compounds, and ordered arrangement of molecules on the surface.

The physical nature of molecular self-assembly is the interaction between permanent multipole moments, instantaneous multipole moments, and induced multipole moments. According to the background conditions generated by self-assembly, self-assembly phenomena can be divided into two categories: static self-assembly and dynamic self-assembly. Static self-assembly refers to the self-assembly that the system can complete and maintain without consuming energy under balanced conditions, and does not change with time, such as crystals. Dynamic self-assembly means that the system must consume energy or material flow. Self-assembly is completed and maintained, and the structure often changes with time, such as the birth of cells and the development of organisms. At present, most self-assembly research focused on static self-assembly is a spontaneous process, once the assembly begins, the process is largely controlled by nature.

The formation of a molecular self-assembly system requires two necessary conditions: the driving force of self-assembly and the guiding effect. The weak interaction of noncovalent bonds maintains the structural stability and integrity of the self-assembled system. In general, there are three main steps in creating a molecular self-assembly system: the first step is to form a complex, complete intermediate molecular body through ordered covalent bonds; the second step is to pass the noncovalent bond from the intermediate molecular body. Synergistic action to form a structurally stable macromolecular aggregate; in the third step, one or several molecular aggregates are used as structural units, and multiple repeats are self-organized to form an ordered molecular assembly. The interactions in the supramolecular system are more additive and synergistic, and have certain directionality and selectivity, and their total binding force can be no less than chemical bonds. Molecular recognition is the embodiment of this combination of weak interactions, which is the key to the formation of advanced ordered assemblies. At the same time, most supramolecular embodiments have an additional feature: they have internal adjustment capabilities for error correction (such as the continuous improvement of crystals in the aging process), which is usually not possible with purely covalent systems.

Molecular self-assembly is one of the best methods for manufacturing nanomaterials so far, especially for the fabrication of structurally functional materials, self-assembly has shown its unique superiority. Molecular self-assembly also has a wide range of applications in robotics and manufacturing. Therefore, molecular self-assembly has become an important research direction between research fields such as chemistry, physics, biology, materials, manufacturing, and nanoscience. It is a hot topic in the international scientific and technological circles in recent years. It can be said that molecular self-assembly research not only has important academic significance, but also has a wide range of technical application prospects, thus attracting the attention of many scientists.

7.1.3 Macromolecular self-assembly

In a supramolecular system, ordered, functionalized molecular aggregates and supramolecular assemblies are formed by the synergistic interaction between the substances themselves, the building blocks may be inorganic molecules, small organic molecules, macromolecules, and biological macromolecules. Supramolecular assemblies of various controllable configurations and sizes have been prepared, in which the self-assembly of macromolecules has attracted widespread attention due to its unique structural characteristics and properties.

Macromolecular self-assembly is a cross-research field between supramolecular chemistry and polymer chemistry. It is the study of the interaction between polymers, between polymers and small molecules, between polymers and nanoparticles, or between polymers and substrates. It achieves the science of regular structures on different sizes by noncovalent bonding. Since the 1990s, macromolecular self-assembly has attracted extensive research interest in the international academic community. In the field of supramolecular chemistry in which macromolecules are assembled units, the most intensive research is the micellization of block copolymers in solution and phase separation in bulk. The highly regular structure of the block copolymer, the interaction between the same segments, the mutual repulsion between the blocks, and the chemical linkage between the blocks make them the best self-assembling units, thus creating a variety of macromolecular assembled body. In recent years, in addition to block copolymers, it has been found that homopolymers, oligomers, ionomers, random copolymers, and graft copolymers can be used as assembly units. Under certain conditions, a variety of supermolecular ordered structures are spontaneously formed by various weak interactions (hydrophobic, hydrogen bonding, electrostatic forces, etc.). After the self-assembly is formed, its morphology can be "permanently" maintained by chemical modification. At present, macromolecular self-assembly has been regarded as one of the main ways to construct functional nanomaterials with regular structure. As a kind of "soft matter," polymer nanomaterials have a wide range of potential applications, such as coatings, drug delivery vehicles, nanoreactors, sewage treatment agents, or templates for the synthesis of structured nanomaterials.

7.2 Principle of molecular self-assembly

The basic principle of molecular self-assembly is to use molecular recognition between a molecule and a molecule or a fragment of another molecule to form a molecular aggregate with a specific arrangement sequence by noncovalent interaction. The synergy of the weak interaction forces of molecules spontaneously through numerous noncovalent bonds is the key to self-assembly.

"Weak interaction force" refers to hydrogen bonding, van der Waals force, electrostatic force, hydrophobic force, π–π stacking force, ion–π adsorption, and the like. The weak interaction forces of these noncovalent bonds maintain the structural regularity, integrity, and stability of the self-assembled system.

However, not all molecules are capable of self-assembly. Self-assembly requires two conditions: the driving force of self-assembly and the guiding action.

The driving force of self-assembly refers to the synergy of weak interaction forces between molecules, which provide energy for molecular self-assembly. The guiding effect of self-assembly refers to the complementarity of molecules in space, that is, in order for the self-assembly of molecules to occur, the molecular rearrangement requirements must be met in the size and direction of the space.

For example, the most classical method for obtaining macromolecular self-assemblies is the method where the amphiphilic block copolymer is micellized in a selective solvent, the driving force of which is derived from the hydrophobicity of the segment, and the linear structure of the block copolymer provides the possibility that the macromolecules are arranged close together, as shown in Fig. 7.1.

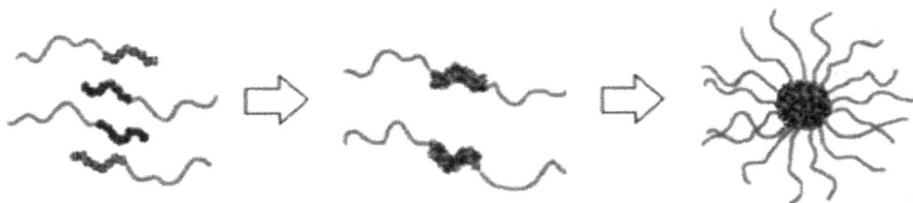

Fig. 7.1: Schematic diagram of self-assembly of block copolymer. Reprinted with permission from [39]. Copyright American Association for the Advancement of Science, 1999.

The development of these work is based on two important basic researches: First, through the development of various living polymerization methods, it is possible to prepare various block copolymers with specific structures and molecular weights as precursors for macromolecular self-assembly; The second is to chemically modify the macromolecular self-assembly to stabilize the structure of the assembly, and is more suitable for future practical applications.

Self-assembled membranes are also the main direction of current research in the field of self-assembly. Self-assembled films are classified into self-assembled monolayer films and layer-by-layer self-assembled films according to their film formation mechanism. As shown in Fig. 7.2, the film formation mechanism of the self-assembled film is a chemically bonded, aligned, tightly oriented two-dimensional ordered monolayer on the substrate through chemical adsorption between the solid–liquid interface, which is nanoscale ultrathin film.

The chemical reaction between the head group of the active molecule and the substrate causes the active molecules to occupy each of the bondable sites on the surface of the substrate and to closely align the adsorbed molecules by intermolecular forces. If the tail group of the active molecule also has a certain reactivity or adsorption capacity, it can continue to react or adsorb with other substances to form a multilayer film, that is, a layer-by-layer self-assembled film.

In addition, depending on the mode of action between the layers, the self-assembled multilayer films can be further divided into two categories, namely, in addition to the chemical adsorption-based self-assembled films described earlier, there is also a self-assembled film which is alternately deposited. The alternately deposited self-assembled film mainly refers to a film formed by chemical adsorption of a polyelectrolyte with oppositely charged groups. This method can control the thickness of the film at the molecular level, and is a composite organic ultrathin film. The preparation method is shown in Fig. 7.3.

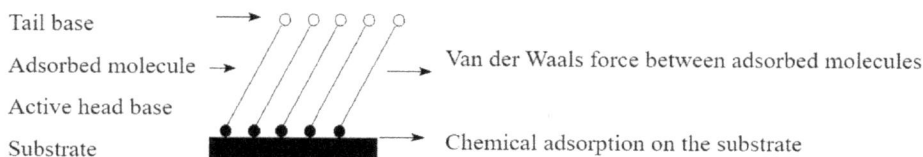

Fig. 7.2: Schematic diagram of self-assembled monolayer film structure.

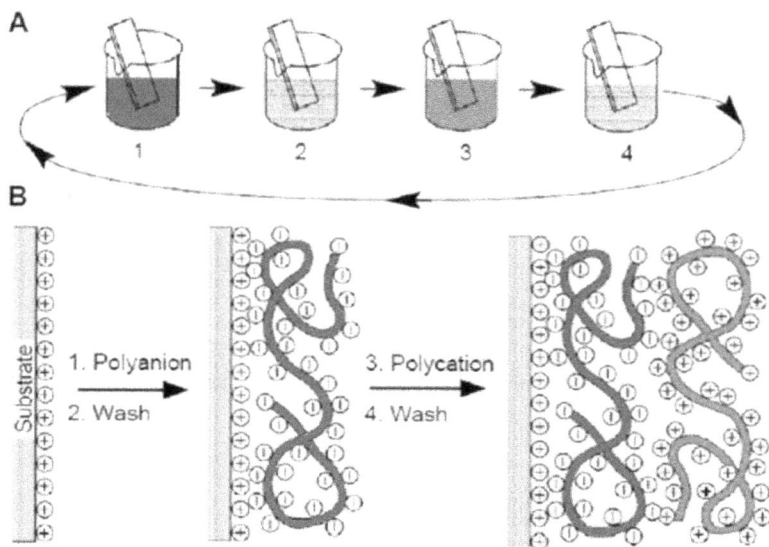

Fig. 7.3: Schematic diagram of preparation of alternately deposited self-assembled films using glass sheets and beakers.

7.3 Solution self-assembly of block copolymer

7.3.1 Self-assembly of amphiphilic block copolymers

7.3.1.1 Principle of self-assembly of amphiphilic block copolymers

The self-assembly of amphiphilic block copolymers is the earliest reported macromolecular self-assembly and the most extensive and in-depth self-assembly. In 1995, Professor L. Zhang and Professor A. Eisenberg of McGill University in Canada first reported in the journal Science that block copolymers self-assembled in solution to form a series of self-assembled bodies of various shapes, which created the first river of the block copolymer self-assembly.

In fact, the understanding of the self-assembly of amphiphilic block copolymers can be inspired by the self-assembly of small molecule surfactants. Small molecule surfactants generally consist of a hydrophilic head group (polar group) and a hydrophobic tail chain (about several to several dozen carbon atoms in length). Due to the repellency of the hydrophobic tail chain in water, the surfactant can spontaneously form micelles in water. Obviously, this is a typical self-assembly phenomenon. The types of polar groups in the surfactant molecule, the length of the hydrophobic tail chain, the type of ions, the concentration of the surfactant, and the ambient temperature directly affect the assembly form of the surfactant in water. For example, sodium lauryl sulfate forms spherical micelles at a lower concentration, and becomes a rod-shaped micelle when the concentration is high, as shown in Fig. 7.4.

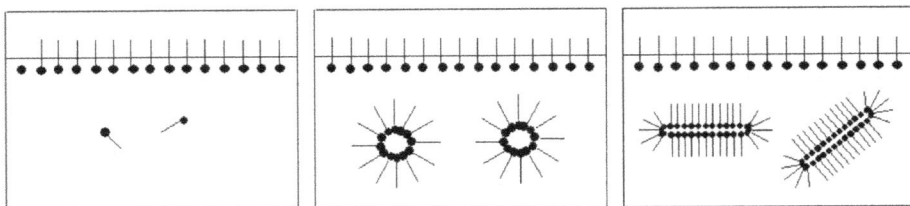

Fig. 7.4: Different assembly forms of surfactants at different concentrations.

Amphiphilic block copolymers can be considered as amplified surfactants, as shown in Fig. 7.5.

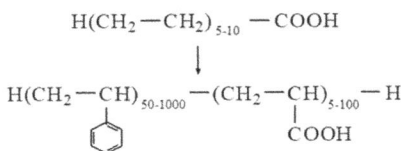

$$H(CH_2-CH_2)_{5-10}-COOH$$
$$\downarrow$$
$$H(CH_2-CH)_{50-1000}-(CH_2-CH)_{5-100}-H$$

Fig. 7.5: Relationship between small molecule surfactants and amphiphilic block copolymers. Reprinted with permission from [2]. Copyright Beijing Science Press, 2006.

On the other hand, the aggregate structure of bulk block copolymers has been a topic of concern. For example, polystyrene/polybutadiene block copolymers having a large molecular weight are incompatible systems and generally exhibit an amphoteric structure in the bulk. It has been observed that its microphase structure includes spheres, hexagonal cylinders, and sheet structures, and varies with the content of one of the blocks (Fig. 7.6).

A ball   A stick   AB layer   B stick   B ball

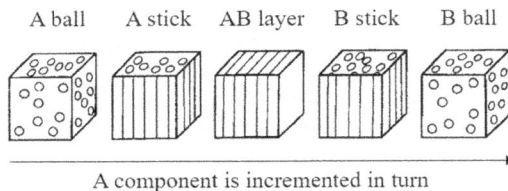

A component is incremented in turn

Fig. 7.6: Schematic diagram of two-phase structure in polystyrene/polybutadiene block copolymer.

Block copolymers typically still maintain incompatibility between their different blocks in solution, especially for amphiphilic block copolymers containing hydrophilic and hydrophobic segments. In aqueous solution, the hydrophobic segments drive the aggregation of the polymer chains. The simplest case is the formation of spherical micelles. The micelle core consists of a hydrophobic segment that forms a shell around the core in a solvated form to maintain the stability of the micelle.

7.3.1.2 Factors affecting the morphology of self-assembled amphiphilic block copolymers

1. Effect of block relative length
By changing the length of a block in the amphiphilic block copolymer, self-assembled bodies of different shapes can be obtained. For example, an amphiphilic block copolymer having a longer hydrophilic segment forms a star-shaped micelle having a small core and a large shell in water, and an amphiphilic block copolymer having a longer hydrophobic segment is aggregated into a large core-shell micelle. Only a variety of morphological changes occur in the flat micelle aggregates.

Since the length of the hydrophobic segment or the hydrophilic segment of the amphiphilic block copolymer can be adjusted by the polymerization process, the aggregate structure is also well tunable.

2. Effect of block copolymer concentration
The aggregation of the amphiphilic block copolymer has a critical micelle concentration (CMC). Below the CMC value, the polymer is dissolved in a solvent in the form of a single molecular chain. The properties of the block copolymer, the length of the block, and the total molecular weight, the differences in the interaction parameters χ between the segments and the solvent all affect the CMC.

For example, the larger the molecular weight of the insoluble homopolymer, the more the number of insoluble units, the more easily the polymer precipitates. Similarly, for block copolymers, the larger the molecular weight of the insoluble segment (the molecular weight of the soluble segment is unchanged), the lower the CMC value.

However, since the insoluble segment is chemically bonded to the soluble segment and the insoluble segment is aggregated, the soluble segment prevents the precipitation from occurring and is replaced by a micellization process.

3. Effect of preparation method
The block copolymer micellization process and its aggregate morphology are thermodynamically driven. However, slow chain motion of the polymer prevents the system

from reaching a thermodynamically stable state. Therefore, the preparation method directly affects the stability of the self-assembly.

Currently, the most common method for preparing amphiphilic block copolymer micelles is to add water to the dissolution system. The entire assembly process is to minimize the Gibbs free energy of the system.

The general method is to first dissolve the block copolymer in a common solvent of two segments (and the cosolvent can be mutually miscible with the precipitant of a certain segment), and prepare a low concentration polymer solution (the mass fraction is usually lower than 2%). For example, tetrahydrofuran is a cosolvent for the PS/PAA block copolymer and aggregates with the PS segment. However, due to the presence of the hydrophilic segment PAA, the enlargement and precipitation of the aggregates are prevented, resulting in the formation of spherical micelles. After the spherical micelles are formed, the cosolvent is dialyzed off with distilled water to obtain a pure aqueous micelle solution.

Different amounts of water can be obtained in different aggregate forms, as shown in Fig. 7.7.

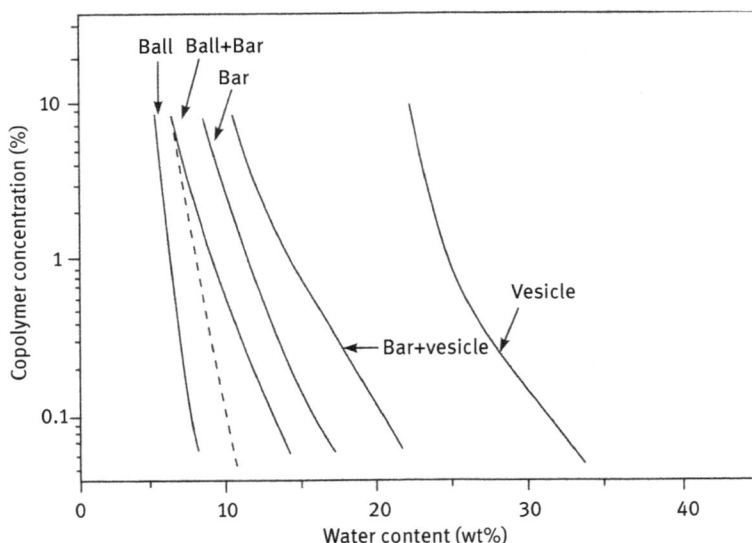

Fig. 7.7: Phase diagram of copolymer PS$_{310}$-b-PAA$_{52}$ dioxane/water-mixed solution. Reprinted with permission from [37]. Copyright American Chemical Society, 1999.

4. Impact of other factors

When the system reaches equilibrium, the Gibbs free energy of the micellization process is composed of various contributions, such as the entropy change of water molecules around the hydrophobic segment and the entropy change of different segments at the aggregate interface. However, only three contributions contribute to the morphology of the aggregate, that is, the interfacial energy between the aggregate core and the shell, the stretched state of the core segment and the repulsion of the shell segment. The amount of water added above affects the morphology of the aggregates due to the effects of these factors.

All factors affecting the balance between these three forces affect the morphology of the aggregates, such as the relative chain length of the block polymer, polymer concentration, solvent composition and properties, additives, and temperature.

Taking the system ionic strength and pH value as an example for the PS_{410}-b-PAA_{25} system microspheres to be rod-shaped (Fig. 7.8).

NaCl was added to the system. When the NaCl concentration was 2.1 mmol/L, the aggregates were spherical, 4.3 mmol/L was rod-shaped, and when it reached 16 mmol/L, it became vesicle-like. When HCl is added to the system, the above change can occur at a lower concentration (the vesicle structure can be formed when the HCl concentration is 10 mmol/L, and the vesicle structure appears when the NaCl concentration is 16 mmol/L). Obviously, the change in aggregate morphology caused by the addition of salt or acid is caused by the change in the repulsive force between the shell sizes.

The PS-b-PAA system is self-assembly in water to form aggregates with PAA as the shell. Adding an electrolyte such as salt, acid, or alkali to the system will change the effective charge of the shell segment, thereby changing the repulsive force between the shell molecular chains. The addition of salt can cause electrostatic shielding of the aggregate shell. The addition of acid can protonate the shell of the aggregate (eliminate the charge), both of which can reduce the repulsive force between the shell molecules and make more molecular chains enter the same aggregation body. An increase in the number of molecular chains in the micelle core leads to an increase in the size of the aggregate and an increase in the elongation of the polymer chain. Therefore, as the amount of salt or acid added increases, the self-assembled body naturally changes from microsphere to rod, and then evolves into a vesicular structure. On the contrary, the addition of alkali causes the shell molecules to be deprotonated, resulting in an increase in the repulsive force between the shell molecular chains, and the molecular weight and the degree of extension of the micelles are decreased, resulting in the opposite effect of adding salt and adding acid.

A 5% PS homopolymer was added to the PS-b-PAA system, and the aggregate morphology was changed from a rod shape to a spherical shape. The reason is that the PS homopolymer can enter the micelle core, resulting in a decrease in the total stretch of the polystyrene segments in the core, which keeps the globular structure stable.

Temperature also affects changes in aggregate morphology. Because temperature changes have an effect on the solubility parameter between the polymer and the solvent. For example, PS386-b-PAA79 can be formed into spherical micelles by heating to 160 °C in butanol, and a mixture of microspheres and rods at 115 °C.

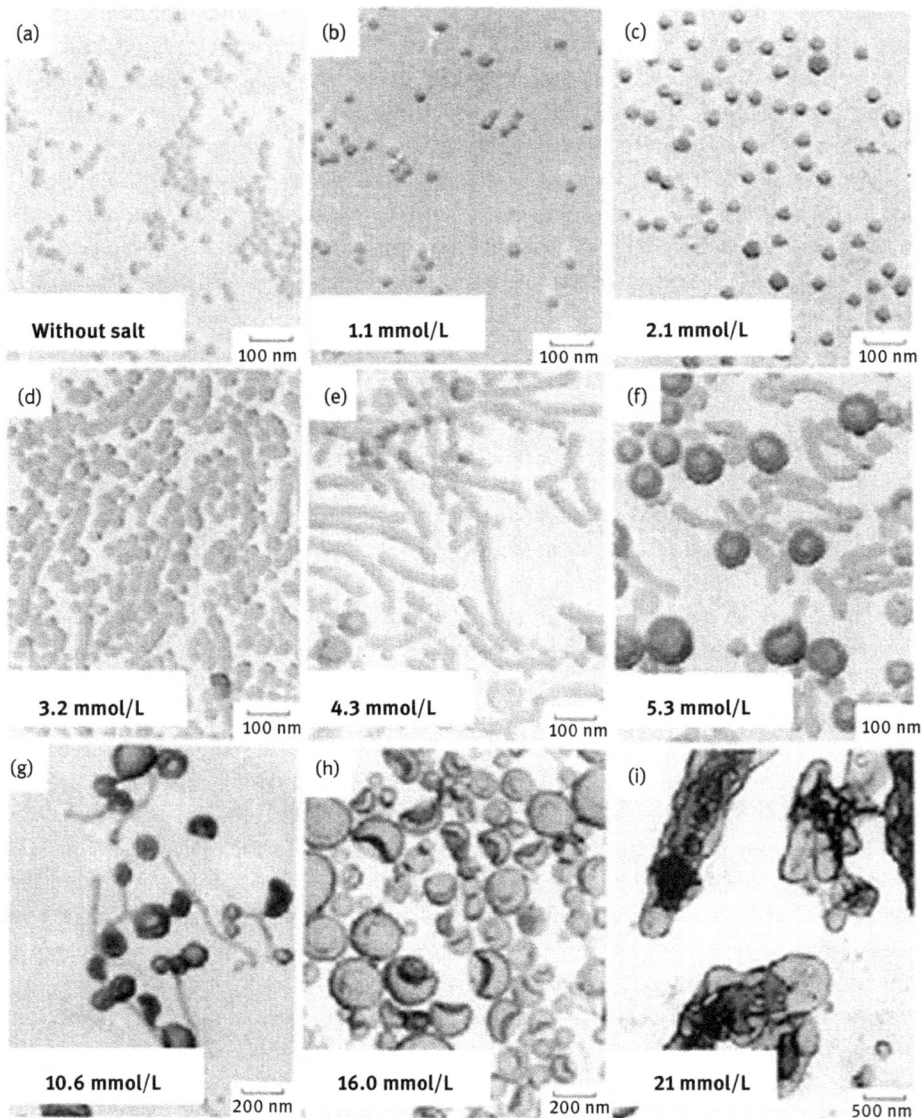

Fig. 7.8: Effect of salt concentration on the morphology of the PS_{410}-b-PAA_{25} system self-assembly. Reprinted with permission from [10]. Copyright American Chemical Society, 1996.

7.3.2 Self-assembly of fully hydrophilic block copolymers

7.3.2.1 The concept of fully hydrophilic block copolymer and its self-assembly

The fully hydrophilic block copolymer refers to a block copolymer composed of two or more water-soluble segments each having a different chemical structure and environmental responsiveness. Under certain conditions, DHBCs are completely soluble in water. When environmental conditions such as pH, temperature changes, or the addition of certain polymers or small molecules that form a complex with a segment, one of the segments becomes water-insoluble, allowing self-assembly in aqueous solution.

Self-assembly by DHBCs does not require the introduction of any organic solvent, and can be carried out in a completely hydrophilic environment, which is incomparable compared to the traditional case where the amphiphilic block copolymer must change the solvent to achieve self-assembly. Therefore, the study of DHBCs and their self-assembly behavior has become a hot topic in the field of polymer self-assembly in recent years.

7.3.2.2 Temperature-induced self-assembly of DHBCs

DHBCs capable of temperature-induced self-assembly should have at least one block with temperature responsiveness and the lowest phase transition temperature in the molecular structure. This LCST should generally be above room temperature, and upon heating, this block changes from water soluble to water insoluble, resulting in micellization self-assembly.

For example, Poly(N-isopmpylacrylamide) (PNIPAM) is a water-soluble polymer having the lowest critical solution temperature in water. When the water temperature is lower than 32 °C, PNIPAM can be dissolved in water, and when the water temperature is higher than 32 °C, PNIPAM is insoluble in water. Therefore, the self-assembly of PEO-g-PNIPAM and PEO-g-P (NIPAM-co-AA) can be successfully achieved by controlling the temperature.

The temperature-sensitive micellization behavior of PEO-g-PNIPAM found that when MPNIPAM/MPEO = 1:3, micellization behavior appeared above 30 °C, while below 30 °C, no aggregation behavior of PNIPAM was observed. The content of PNIPAM reduces the LCST of the copolymer.

The study also found that by copolymerizing PNIPAM with polyacrylamide, LCST can be raised to near body temperature, which has potential applications in drug loading.

Poly(2-methoxyethyl vinyl ether)-block-poly (methyl tri(ethylene glycol) vinyl ether) (PMVE-b-PMTEGVE) is composed of two blocks, each having temperature sensitivity. It has been found that the copolymer forms stable micelles only when the temperature is between the LCST of the two blocks; and when the temperature is higher than the LCST of the two blocks, the copolymer becomes water insoluble and precipitates. Therefore, the temperature range of the micellization of the copolymer can be adjusted by changing the degree of polymerization of the diblock. Studies have shown that when the temperature is between the two blocks of LCST, PMVE is hydrophobic to form a micelle core, and PMTEGVE is still in a water-soluble state, forming a micelle shell.

7.3.2.3 pH-induced self-assembly of DHBCs

DHBCs induced by pH can be divided into two categories: one is a block copolymer containing a polyelectrolyte block, and the protonation and deprotonation of the polyelectrolyte are pH dependent, another block It is a water-soluble nonionic block; the other is a block copolymer composed of two cationic polyelectrolytes or an anionic polyelectrolytes each having a different pH value, and the two blocks exhibit different water solubility at different pH conditions. Both types of block copolymers can achieve self-assembly by pH induction.

Poly(2-vinylpyridine)-block-poly(ethylene oxide) (P2VP-b-PEO) is an example of a first class of pH-induced self-assembled DHBCs. Under acidic conditions, P2VP can be protonated and dissolved, so both blocks are water soluble. However, under alkaline conditions, the P2VP block is neutralized and converted to be insoluble in water, thus forming micelles.

A fully hydrophilic block copolymer composed of two anionic polyacids or two cationic polybase blocks can also form an assembly by pH induction. For example, after acidification of sodium polystyrene sulfonate-b-polyvinylbenzoic acid (PSSNa-b-PV-BA) containing two polyacid blocks, the solubility of the weakly acidic PVBA block is reduced, thus forming micelles. This is due to the difference in pH between the two blocks. When the weakly acidic PVBA is protonated and becomes insoluble, the strongly acidic PSSNa remains ionized and water soluble.

If the fully hydrophilic block copolymer consists of a length of anionic polyacid and a cationic base of polycations, different forms of assembly will be formed at different pH conditions. For example, A forms a micelle with PVBA as a core at pH = 2 and 20 °C, and at pH = 10 and 20 °C, a reverse micelle with PDEAEMA as a core is formed.

7.3.2.4 Self-assembly of DHBCs induced by ionic strength changes

Most ionic strength sensitive DHBCs are also pH sensitive. For example, poly(4-vinyl benzoic acid-block-2(diethylamino)ethyl methacrylate) (PVBA-b-PDEAEMA), in which the weakly basic block PMEMA is completely soluble under acidic conditions, precipitates easily upon addition of the salt. The block copolymer is characterized by the formation of PDMAEMA-nuclear micelles and PMEMA-nuclear reverse micelles at pH adjustment and suitable salt concentrations. When pH = 4–8.5, PDMAEMA is formed into a core micelle, and after adding a certain amount of Na_2SO_4, at pH = 6–6.7, a reverse micelle with PMEMA as the core is formed.

7.3.2.5 Complexation-induced self-assembly of DHBCs

When a block copolymer composed of a polyelectrolyte is mixed with a block copolymer or a homopolymer having a counter ion, phase separation occurs due to formation of a complex, thereby realizing self-assembly.

A mixture of poly(ethylene oxide)-b-poly(lysine) (PEO-b-PLys) and poly(ethylene oxide)-b-poly(aspartic acid) (PEO-b-PAsp) can form stable micelles in water. It was found that when the PLys segment and the PAsp segment have the same degree of polymerization, the spherical micelles formed are not only monodisperse but also relatively stable, indicating that the complexation between the two oppositely charged polyelectrolytes has a chain length recognition effect. In addition, the study also found that PEO-b-PAsp and the polyelectrolyte homopolymer PLys with an equimolar opposite charge can also form micelles, and the micelles are stable and monodisperse. The surface charge of the micelle particles was found to be electrically neutral, indicating that the "shell" of the micelle consisted of electrically neutral PEO.

The above results indicate that the reverse ionic polymer complex phase separation is the driving force for self-assembly.

7.4 Macromolecular self-assembly induced by induction

7.4.1 Hydrogen bond induced macromolecular self-assembly

In the study of polymer blends, there are two compelling systems: one is the special interaction (such as hydrogen bonding and ion–ion interaction) and the compatibility problem, that is, by introducing special interactions. The incompatible system is transformed into a compatible system; the second is that the polymer pair with strong interaction forms a polymer complex when mixed, accompanied by a series of changes in physical properties.

Obviously, the abovementioned driving forces for compatibilization and complexation are special interactions between polymers. The study found that the system can undergo an "incompatible–compatible complexation" transition as the density of specific interactions in the system increases. Self-assembly of polymers can also be achieved by this route.

The research group of Professor Jiang Ming of Fudan University prepared a hydrogen bond graft copolymer by means of localization of special action units, and further realized the self-assembly of the block copolymer.

The specific method is: introducing a proton donor unit on the end group of the polymer (A chain), when it is dissolved in the common solvent with the proton acceptor unit polymer (B chain, the receptor unit is randomly distributed in the chain) It is possible to form a "hydrogen bond graft copolymer" by interaction of the A chain end group with the B chain proton acceptor unit. When the hydrogen bond graft copolymer is placed in a selective solvent, it is possible to obtain a micelle structure, as shown in Fig. 7.9.

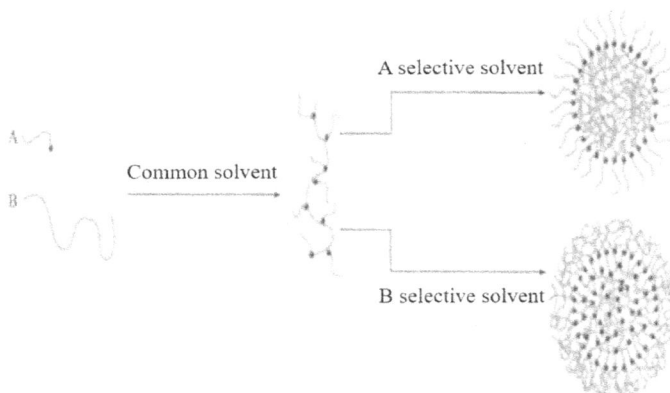

Fig. 7.9: Schematic diagram of hydrogen-bonded graft copolymer self-assembly. Reprinted with permission from [38]. Copyright American Chemical Society, 2005.

For example, Liu et al. prepared a single-end carboxyl styrene (MCPS, Mn = 1, 800–5,500) by anionic polymerization and CO_2 termination reaction. Then, it was dissolved in a common solution of chloroform with poly(4-vinylpyridine) (P4VP, Mn = 1.4×10^5, containing 1,330 repeating units per chain) to obtain a hydrogen bond graft copolymer. The solvent is then switched to a selective solvent. Toluene dissolves MCPS, but it is a precipitant for P4VP. Toluene was added to the mixed solution of P4VP and PS in chloroform, and the solution remained transparent and opalescent, indicating that the P4VP chain had been dissolved into toluene by MCPS to form micelles. Studies have shown that the hydrodynamic radius of micelles increases with the increase in the molar mass of MCP (Fig. 7.10).

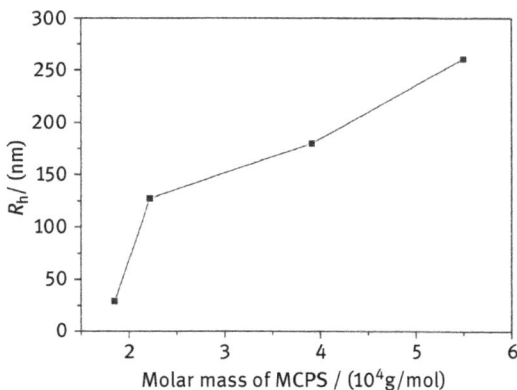

Fig. 7.10: Effect of MCPS molar mass on micellar size. Reprinted with permission from [9]. Copyright Elsevier, 2000.

In the above polymer micelles, the P4VP chain is a core and the MCPS is a shell. Unlike the micelles prepared by using the block copolymer, the core and the shell are connected by hydrogen bonding instead of being chemically bonded. Jiang Ming's research group called this micelle "noncovalently connected micelles, NCCM."

Hydrogen bond graft copolymers and NCCM are also obtained using terminal carboxyl polybutadiene (MCPB) and P4VP. For example, using n-hexane as a selective solvent for MCPB, nitromethane as a selective solvent for P4VP, added to the chloroform solution of MCPB/P4VP, respectively, when the amount of n-hexane and nitromethane added is about 30% by volume, emulsion appears, indicating the formation of micelles. The core molecules of these two micelles are exactly opposite to the shell molecules, so they are called positive micelles and reverse micelles, respectively.

Since such micelles have only hydrogen bonds and no chemical bonds, their stability has attracted attention. The hydrodynamic size of the above positive micelles and reverse micelles was observed by light scattering method as a function of time. It was found that (MCPB)-P4VP was placed for 3 days and 30 days, and (P4VP)-MCPB was placed for 3 days and 150 days. The micelle size did not change.

In the above method of preparing NCCM, it is emphasized that the point of action on the molecular chain is limited to the end of the molecular chain. In fact, this limitation is not required. If the functional groups on the complementary two macromolecules are randomly distributed in the chain, the hydrogen bonding group between the components can be limitedly contacted by the effective control of the mixing process, thereby preventing precipitation and forming regular-rule assembly.

For example, there are polymers A and B which have complementary interaction points and are randomly distributed over the molecular chain. The solution A and the solution B are separately prepared, wherein the solvent of B should be a precipitant of A. When the A solution is added dropwise to the B solution, the polymer A aggregates to rapidly form nanoparticles. However, due to the hydrogen bond between A and B, the B polymer is caused to aggregate around the A particles, thereby preventing the occurrence of macroscopic precipitation.

NCCM can also be prepared by in situ polymerization. The specific process is as follows: (1) First, a solution of the initiator AIBN and the polymer PCL in dimethylformamide (DMF) was added dropwise to the water to form stable PCL nanoparticles. (2) Then the monomer N-isopropylacrylamide (NIPAM) and the cross-linking agent N,N-methylenebis(acrylamide) are added to the above solution. (3) The temperature was raised to 76 °C to initiate NIPAM polymerization. Since the initiator is enriched around the PCL particles and the reaction temperature is much higher than the lowest critical solution temperature (32 °C) of PNIPAM, once formed, PNIPAM is adsorbed on the PCL particles due to hydrophobicity. After the formation of the first layer of PNIPAM, the more hydrophobic monomer can capture more monomer and cross-linker from the aqueous phase, and the polymerization reaction is continued, so that the shell layer is continuously grown to obtain the NCCM of (PCL)-PNIPAM.

(PCL)-PNIPAM micelle size and core-shell thickness can be controlled by adjusting the preparation process, such as the concentration of PCL/DMF solution, the amount of monomer and cross-linker, and PCL/NIPAM ratio.

7.4.2 Small molecule surfactant-induced macromolecular self-assembly

The amphiphilic block copolymers of such assembly systems typically contain a block which is complexed with a polar hydrophilic group of a small molecule surfactant, typically a polyelectrolyte or ionomer, or a repeat structural unit contains a polymer chain of pyridine, amino group or the like.

When the block copolymer and the surfactant coexist in water, the polar group of the surfactant and the block are complexed by hydrogen bonding or ion–ion interaction, and the long hydrophobic tail chains of the surfactant aggregate with each other. The block is induced to aggregate to form a micellar core, while the other hydrophilic segment of the block copolymer forms a shell of micelles and allows the micelles to be stably present.

If both segments of the block copolymer are hydrophilic, it corresponds to a small molecule surfactant with a hydrophobic end for hydrophobic modification of one of the hydrophilic blocks to convert it into a hydrophobic block, and the other hydrophilic block remains hydrophilic, thereby inducing micellization of the block copolymer.

For example, the use of trimethylhexadecylamine bromide to induce polyoxyethylene-b-polymethyl methacrylate to achieve micellization in water provides stable dispersion of polymersomes. Similar to small molecule surfactants, self-assembly of the block copolymer can also be achieved with suitable small molecule linear fatty acids.

7.4.3 Electrostatic induction of macromolecular self-assembly

The polyelectrolyte block copolymer combines the structural characteristics of polyelectrolyte, block copolymer, and surfactant, and can form a regular structure with nanometer size through interaction of electrostatic, hydrophobic, hydrogen bonding, and the like in an aqueous solution. Micellar, microparticle, or hollow microcapsules. Therefore, polyelectrolyte block copolymers have become a hot spot in the field of macromolecular self-assembly in recent years.

Polyelectrolyte self-assembly is very rich, including self-assembly of block copolymers and self-assembly between two homopolymers: small molecule-induced self-assembly and inorganic nanoparticle-induced polyelectrolyte self-assembly.

7.4.3.1 Self-assembly of polyelectrolyte block copolymers and homopolymers

Self-assembly with a polyelectrolyte block copolymer as a precursor has the following two cases:

(1) If the two blocks of the polyelectrolyte block copolymer have significant hydrophilic differences, their self-assembly behavior in aqueous solution is similar to that of ordinary amphiphilic block copolymers.

(2) If the polyelectrolyte block copolymer has both weak acid and strong acid segments, such as P2VP-b-PAA two-block polyelectrolyte, it can self-assemble to form micelles in a certain pH range.

When two polyelectrolyte diblock copolymers are used as precursors, if the two polyelectrolyte copolymers have the same water-soluble nonionic segment, and the ionic segments of the two have opposite charges, they can be realized in an aqueous solution. Self-assembly forms spherical micelles. Wherein, the water-soluble nonionic segment forms a shell of the micelle, and the oppositely charged ionic segment aggregates into the core. Generally, only the polyelectrolyte segments of the exact same number of charges form a particle core, and the unmatched polyelectrolyte block copolymer remains free in water. However, studies have also found that if a polyelectrolyte block copolymer is paired with an oppositely charged polyelectrolyte homopolymer, a nanoself-assembly can be obtained regardless of whether the number of charges is completely the same.

If two polyelectrolyte homopolymers with opposite charges are used as precursors, self-assembly can be achieved by the formation of plums by charge interaction between the two. If the PAA solution is dropped into the polyethyleneimine solution, a polymer vesicle having reversible pH responsiveness can be obtained.

Similarly, the salt formed by the polyelectrolyte acid and the polyelectrolyte base can also be self-assembled, such as polybutadiene-b-polymethacrylate and PS-b-poly (1-methyl-4-vinylpyridine) iodide salt can obtain microphase-separated microcapsules by electrostatic force in tetrahydrofuran.

7.4.3.2 Small molecule-induced polyelectrolyte self-assembly

It has been found that small organic acids can induce self-assembly of polyelectrolytes in aqueous solution by electrostatic interaction. For example, a dilute aqueous solution of a positively charged biomacromolecular PLys is mixed with a dilute aqueous solution of citric acid, and a spherical self-assembly is observed after a few seconds. The driving force of this self-assembly is apparently from the electrostatic interaction between COO^- and NH_4^+.

Similarly, self-assembly of polyelectrolytes can also be achieved with small molecular salt compounds. The process of inducing PEO-b-PAA to form micelles with the water-soluble small molecule dimethylaminopropyl ethylcarbamide (ETC) is as follows: ETC is added to the aqueous solution of PEO-b-PAA, N= in the ETC molecule The C=N structure and the terminal NH_4^+ react with the COO^- of the PAA, respectively, to cause cross-linking between the PAA block molecules, resulting in micellization of the block copolymer.

According to the same principle, small molecule ammonium salt surfactants with different structure of small molecular structure (dodecyldimethylammonium bromide, octadecyldimethylammonium bromide, and octylmethylammonium bromide), can successfully induce PEO-b-PMAA self-assembly to form different forms of micelles.

7.4.3.3 Inorganic nanoparticle-induced polyelectrolyte self-assembly

Murthy et al. used positively charged PLys as a raw material and directly mixed with a negatively charged nano-Au solution to obtain nanoparticles. Further, negatively charged SiO_2 particles were added, and a hollow sphere structure was formed.

Therefore, they proposed a self-assembly "flocculation" mechanism, and believed that the formation of Au/PLys flocs is a key step in determining the structure of the assembled product. Since PLys is excessive relative to Au, the surface of Au/PLys particles has a positive charge. When the negatively charged SiO_2 particles are continuously added, the "vacancy" on the PLys segment continues to bind to the SiO2 particles due to electrostatic adsorption, thereby causing the PLys to have a driving force for outward expansion, resulting in the conversion of the Au/PLys particles into hollow microparticles (shown in Fig. 7.11).

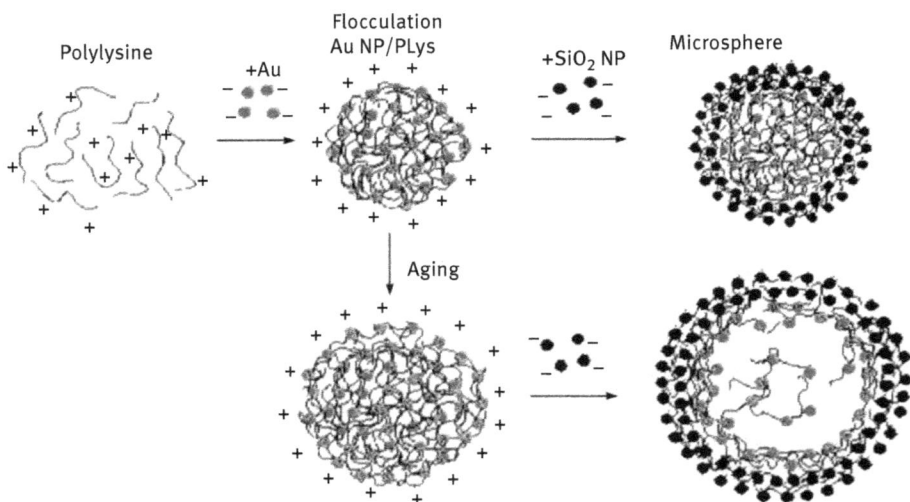

Fig. 7.11: Formation of Au/PLys hollow microcapsules. Reprinted with permission from [40]. Copyright American Chemical Society, 2004.

7.4.4 Rigid-chain-induced macromolecular self-assembly

When rigid-chain macromolecules are used as organizational units, high-density rigid chains usually have regular mutual proximity. The tendency to align in parallel provides a driving force for the formation of micelles and generally induces the formation of hollow structural assemblies.

For example, the amphiphilic flexible-rigid structure two-block copolymer PS-b-polyphenylquinoline (PPQ, the main chain consists entirely of aromatic rings) can quickly form spheres, vesicles, columns, and layers in solution. A variety of self-assemblies. These self-assemblies are large in size, up to the micron scale, and tend to have a hollow structure compared to self-assembled bodies formed from flexible-flexible block copolymers.

According to this finding, Professor Jiang Ming of Fudan University tried to form a "hydrogen bond graft copolymer" with a flexible polymer and a rigid polymer, and then obtained a spherical assembly of a hollow structure in a selective solvent. However, when they were prepared by dissolving rigid polyimide (PI) with carboxyl groups at both ends and flexible P4VP in the common solvent DMF, it was immediately observed that the solution exhibited a pale blue opalescence indicating that the self-assembly had formed.

They believe that by the hydrogen bond between the carboxyl group on the PI and the pyridyl group on the P_4VP, 102 to 103 PI rigid chains can be attached to each P4VP chain, so the concentration of the PI rigid chain around the P4VP chain is higher than that in the solution. The average concentration is much higher. Since the rigid chains usually have a tendency to regularly lie close to each other, it provides a driving force for forming hollow grievances. The process of PI and P4VP forming an assembly is shown in Fig. 7.12.

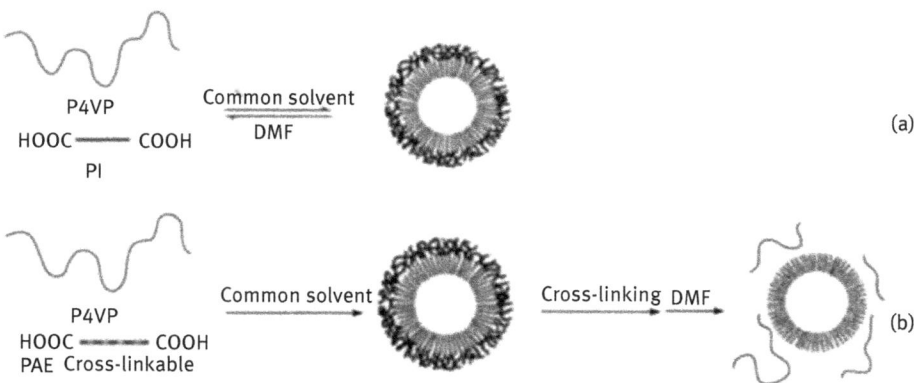

Fig. 7.12: Schematic diagram of the process of PI and P4VP self-assembly into hollow spheres (a), the self-assembling and structural fixation process of cross-linkable PAE and P4VP (b). Reprinted with permission from [12]. Copyright Elsevier, 2004.

7.5 Self-assembly of dendrimers and hyperbranched polymers

7.5.1 Self-assembly of dendrimers

The macromolecular self-assembly described above is based on block copolymers. As research progresses, many macromolecules with more complex structures are introduced into the field of self-assembly research. In these novel building units, dendrimers have been given special attention in the field of self-assembly due to their unconventional, refined structure and precise control of their molecular size, shape, and function. More and more experiments have shown that dendrimers can lead to the formation of self-assemblies through hydrogen bonding, electrostatic interaction, and metal ligand chelation, and exhibit many different behavioral characteristics from linear polymer self-assembly.

In 1996, Zimmerman et al. proposed self-assembly of dendrimers with hydrogen bonding, followed by in-depth studies. Studies have shown that self-assembly of carboxylic acid dimers in nonpolar solvents such as chloroform gives hexagonal dendritic aggregates. In the polar solvent, the tree structure cannot be self-assembled because the polar solvent easily acts on the acid substituent and hinders the formation of hydrogen bonds. Using the fourth-generation diterpene ether branch self-assembly, GPC verification, finally obtained a tree-shaped macromolecule with a molecular weight of 3.4×10^4; in addition, it was found to have a similar dynamic radius with the same size in the valence-type tree-shaped macromolecule. The formation of self-assembled structures and a certain degree of stability have been strongly proved.

In addition, Zimmerman et al. studied the self-assembly of dendrimers with a small molecule of a functional group by hydrogen bonding into a double-tree macromolecule. The molar ratio of dendrimer to small molecule is 2:1, which is small. Molecules, like bridges, organically link two species of branches by hydrogen bonding. The NMR test shows that the hydrogen bond strength of the three-molecule self-assembly is independent of the branch end group size (algebra), and the small molecules in the middle have enough space to form hydrogen bonds.

Frechet et al. also studied the ordered self-assembly of dibenzyl ether dendrimers by hydrogen bonding. They functionalized melamine and cyanuric acid with a second-generation dendrimer that is unsubstituted at the end or has a parasubstituted bromine atom. Studies have shown that due to the existence of hydrogen bonds, a "six-ring"-like self-assembled structure is finally formed, as shown in Fig. 7.13.

Metal-ligand interaction is another noncovalent interaction. For example, a dibenzyl ether dendrimer substituted with a carboxyl group at the center is a shell, and a trivalent lanthanide metal (such as Er, Eu, and Tb) is a core, and a metal-ligand chelate can form a new one. Dendritic structure is shown in Fig. 7.14.

Dendrimers can also be assembled into more complex macromolecules based on electrostatic interactions. For example, a polyarylene ether dendrimer having a central point of cationic ammonium is assembled around the porphyrin by electrostatic

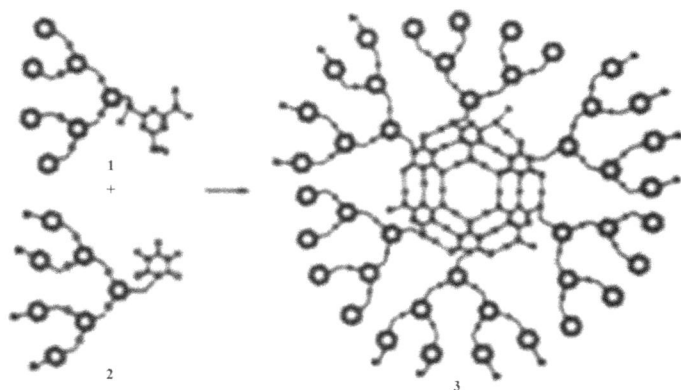

Fig. 7.13: Schematic diagram of the formation of a "six-ring" dendritic self-assembled structure by hydrogen bonding. Reprinted with permission from [26]. Copyright American Chemical Society, 1997.

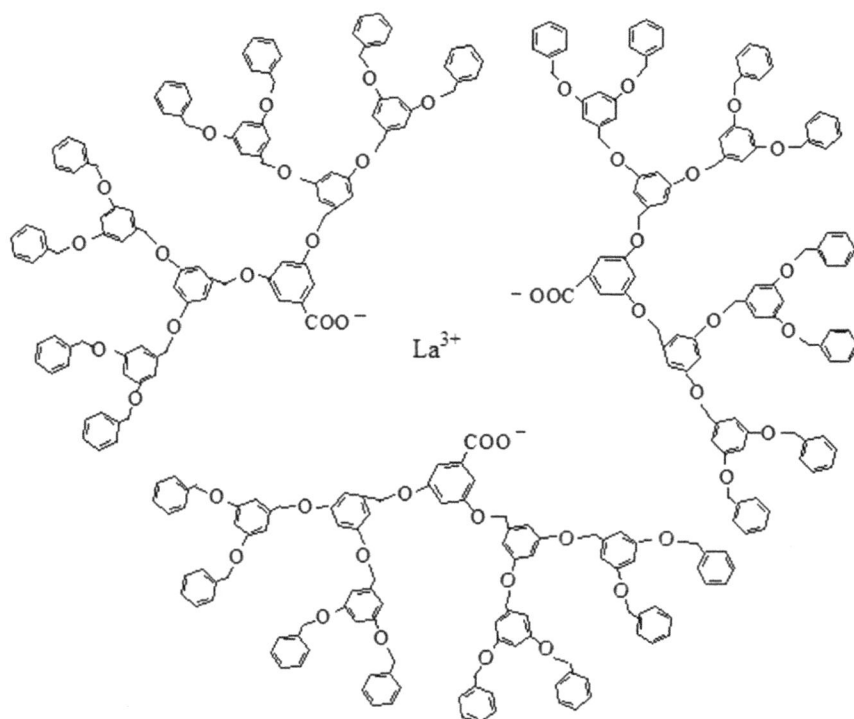

Fig. 7.14: Schematic diagram of dendritic self-assembly structure by metal-ligand chelation. Reprinted with permission from [19]. Copyright Editorial Department of Journal of Functional Polymers, 2003.

action. This cation can electrostatically interact with the four sulfite anion groups on the periphery of the porphyrin to achieve classical self-assembly. The synthesis and assembly process are shown in Fig. 7.15.

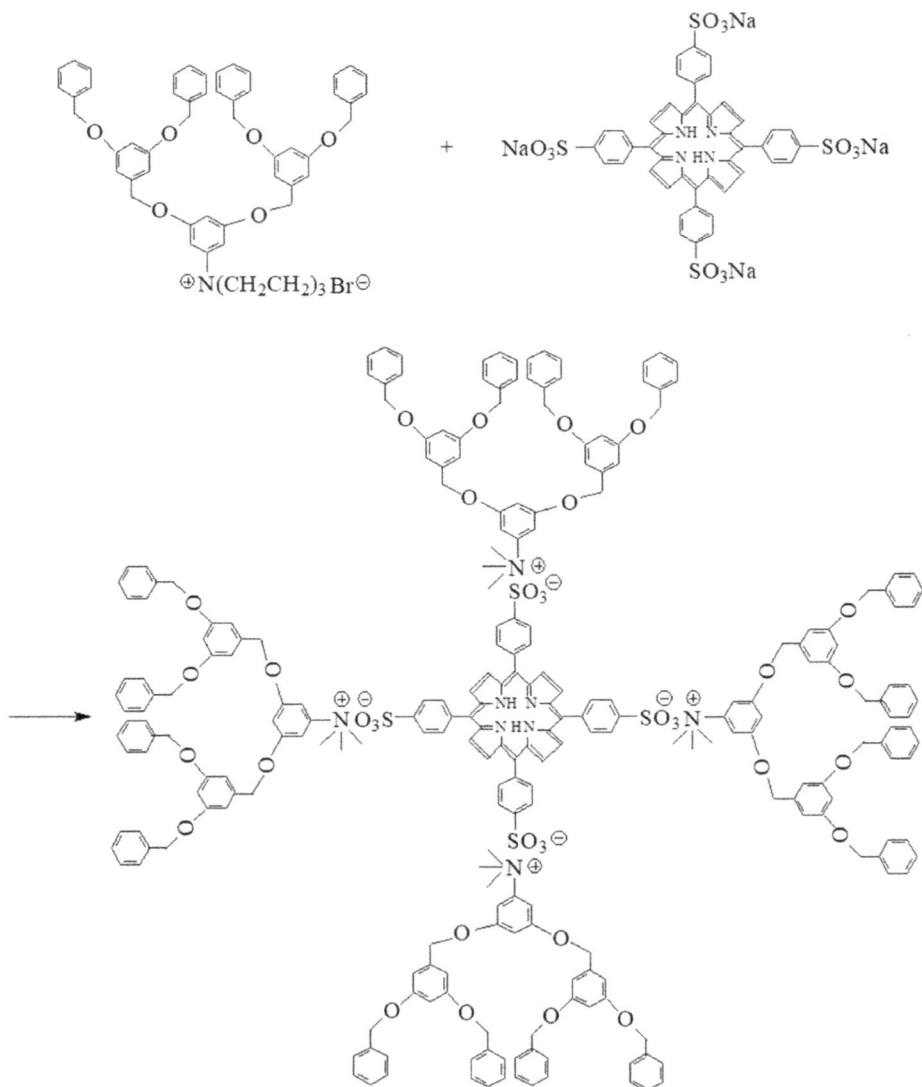

Fig. 7.15: Self-assembly of dendrimers based on electrostatic interaction. Reprinted with permission from [27]. Copyright Editorial office of Chemical Journal of Chinese Universities, 1997.

7.5.2 Self-assembly of hyperbranched polymers

Hyperbranched polymers are a class of polymers with irregular structures. Compared with ordinary block copolymers, hyperbranched polymers have an ellipsoidal structure and degree of branching, and contain a large number of terminal functional groups, which can be further modified; compared with dendritic polymers, hyperbranched polymer structures are not so regular, but its synthesis process is very gradual, can be prepared by "one-pot method." In particular, current synthetic techniques have made it possible to prepare hyperbranched polymers having a hydrophobic hyperbranched core and a large number of linear hydrophilic arms or a hydrophilic core and a large number of linear hydrophobic arms. These unique advantages lay the foundation for the development of hyperbranched polymers in the field of self-assembly research.

7.5.2.1 Self-assembly of amphiphilic hyperbranched polymers

The formation of self-assemblies in water by hyperbranched polymers having a hydrophobic hyperbranched core and a large number of linear hydrophilic arms is representative of such self-assembly. Professor Yan Deyue from Shanghai Jiaotong University took the lead in carrying out work in this field and achieved remarkable results. They successfully self-assembled in water using amphiphilic hyperbranched multiarm copolyether HBPO-star-PEO to obtain self-assembled vesicles. And self-assembly in acetone to obtain macromultiwalled spiral tubes of centimeter length and millimeter diameter, the first time in the laboratory to achieve macroscopic self-assembly of molecules, thus expanding the molecular self-assembly research to the macroscale. Prior to this, the size of self-assembled bodies generally ranged from microscopic (nanoscale) or mesoscopic (micron), which greatly expanded the research content and scope of supramolecular self-assembly, and also proved that irregular molecules can also be assembled into a self-assembled body with a regular structure.

In addition, they can also be self-assembled in water using the amphiphilic hyperbranched multiarm copolymer HBPO-star-PDMAEMA, and found that the polymer can form spherical large micelles with a size exceeding 100 nm in water. Further research found that large micelles assembly mechanism involves a hierarchical self-assembly process. Figure 7.16 is a schematic diagram of the molecular structure of HBPO-star-PDMAEMA and the hierarchical self-assembly process of the large composite micelle (LCM) formed. Studies have shown that at lower than the CMC, the polymer exists as a single molecule micelle; and above CMC, the single molecule micelles interact through the micelles to form a multimolecular spherical large micelles (Fig. 7.16(b)). Therefore, such large micelles belong to LCMs. LCMs are common in assembly studies of linear block copolymers, but were first discovered for highly branched polymers.

As already mentioned in the previous section, Eisenberg advocated a class of linear block copolymers with shorter hydrophilic heads and found a very rich self-assembly behavior. He called the self-assembly of this type of polymer collectively

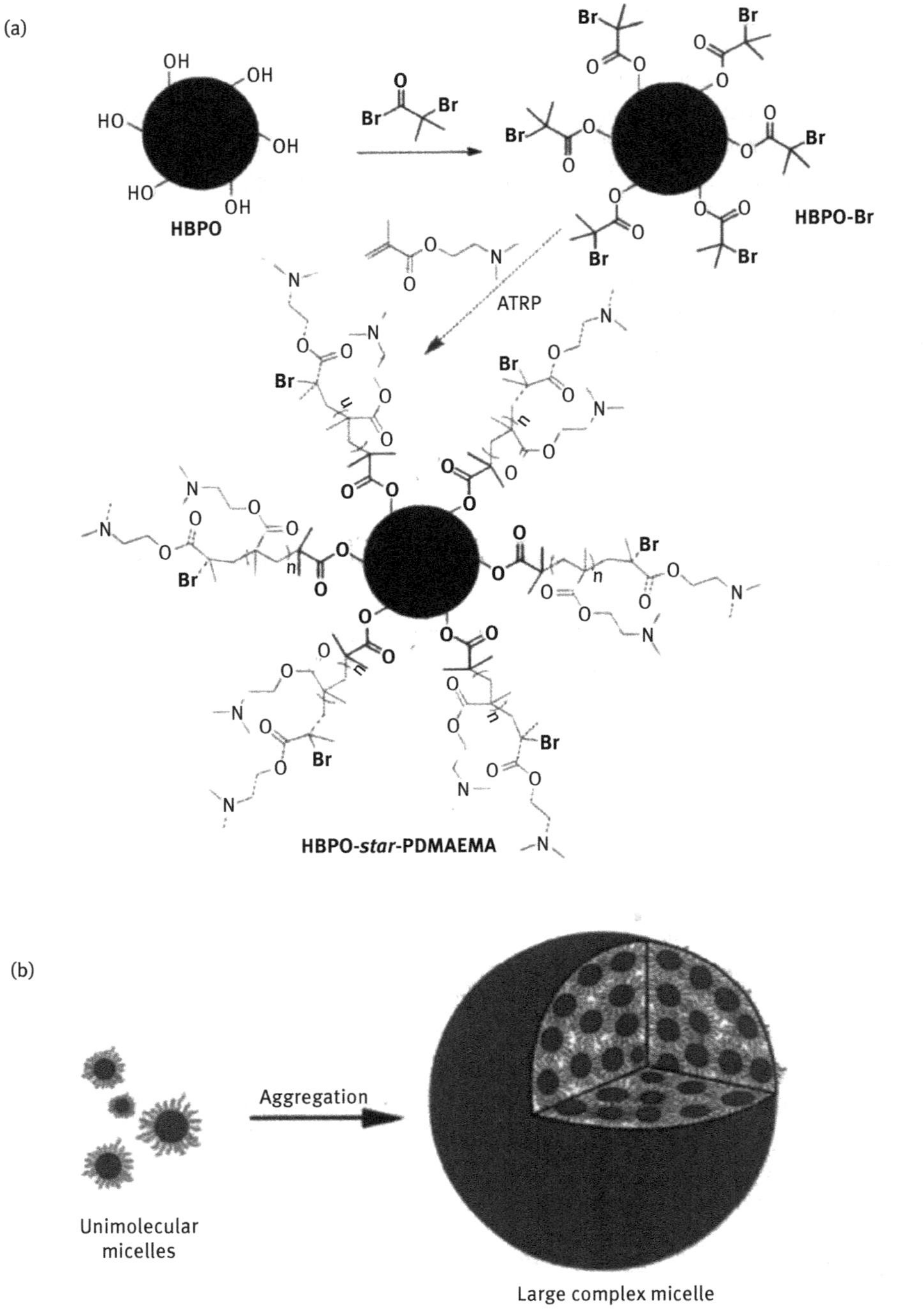

(a)

HBPO

HBPO-Br

ATRP

HBPO-*star*-PDMAEMA

(b)

Unimolecular
micelles

Aggregation

Large complex micelle

Fig. 7.16: Schematic diagram of the molecular structure and hierarchical self-assembly process of HBPO-star-PDMAEMA. Reprinted with permission from [34]. Copyright Wiley-VCH Verlag GmbH & Co. KGaA, Weinheim, 2007.

micellar. Inspired by the work of Eisenberg, the hydrophilic arms of hyperbranched polymers were shortened, and it was found that such flat-headed hyperbranched co-polyethers also exhibited significant hierarchical self-assembly in water. Figure 7.17 shows a real-time assembly process. In Fig. 7.17(b), the black circle shows the vesicles being fused, and the illustration in the upper right corner of Fig. 7.17(d) is the model of the large composite vesicle (LCV).

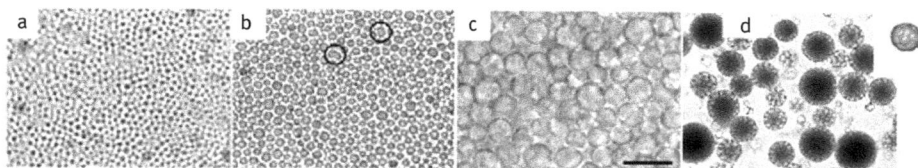

Fig. 7.17: Real-time hierarchical self-assembly process of LCV. Reprinted with permission from [35]. Copyright Wiley-VCH Verlag GmbH & Co. KGaA, Weinheim, 2007.

It can be seen from Fig. 7.17 that the polymer first phase separates in water and self-assembles to form small vesicles; then, as the hydration time increases, the number of small vesicles increases and fuses with each other, transforming into tightly packed large vesicles. The bubble is called a three-dimensional vesicle heap; finally, under the action of external field shearing force, the vesicle heap further self-assembles to form a spherical LCV of about 40 μm, which is composed of inter-connected large vesicles. The greater the number of associated vesicles, the greater the LCV. The results show that the LCV itself is actually in a metastable state, and the vesicles that make up the LCV will fuse with each other, so the number of vesicles will gradually decrease, and finally the entire LCV will be transformed into a single giant vesicle and eventually rupture forming a precipitate. Further studies have found that the stability of LCV is related to the number of vesicles that make up LCV. The more the number of associated vesicles, the more stable the LCV.

7.5.2.2 Electrostatic adsorption self-assembly of hyperbranched polymers

Electrostatic action plays an important role in self-assembly research. Self-assembly of linear polymers and dendrimers has been achieved by electrostatic interaction. Also, an assembly can be formed by electrostatic adsorption self-assembly techniques.

Tang Liming of Tsinghua University and other polyelectrolyte multilayer self-assembled films with specific nanostructures were prepared by using terminal car-boxyl hyperbranched polyester (PMPP) as polyanion and terminal acrylate-based hyperbranched polyesteramine (PEAC) as polycation.

Studies have shown that PEAC and PMPP are assembled into membranes in different pH aqueous solutions, and the membrane obtained between pH = 6–7 is the thickest. As the pH increases from 3.5 to 6.5, the hydrophobicity of PEAC molecules increases gradually and can deposit on the surface of the membrane as larger aggregates. As the number of PEAC molecules on the surface of the membrane increases, the adsorption of PMPP molecules is also promoted. The thickness gradually increases. Since the PEAC molecule has a highly branched structure and a large number of terminal acrylate groups, it gradually cross-links during the illumination process to form a photocured film.

7.5.2.3 pH-induced hyperbranched polymer self-assembly

In the foregoing, it has been described that amphiphilic and fully hydrophilic block copolymers can achieve self-assembly by pH induction. By introducing this idea into highly branched polymers, self-assembly of amphiphilic hyperbranched polymers can also be achieved by pH induction.

Zhong Ling et al. introduced a disulfide structure in hyperbranched polyester to obtain a hyperbranched dithioester chain-transfer agent which can be used as a reversible addition-fragmentation chain-transfer radical polymerization (RAFT) chain-transfer agent, and further prepared to overspend. The self-assembly behavior of the polyester in the aqueous solution and the polyacrylic acid as the amphiphilic hyperbranched copolymer H20-star-PAA was studied. The preparation process of H20-star-PAA is shown in Fig. 7.18.

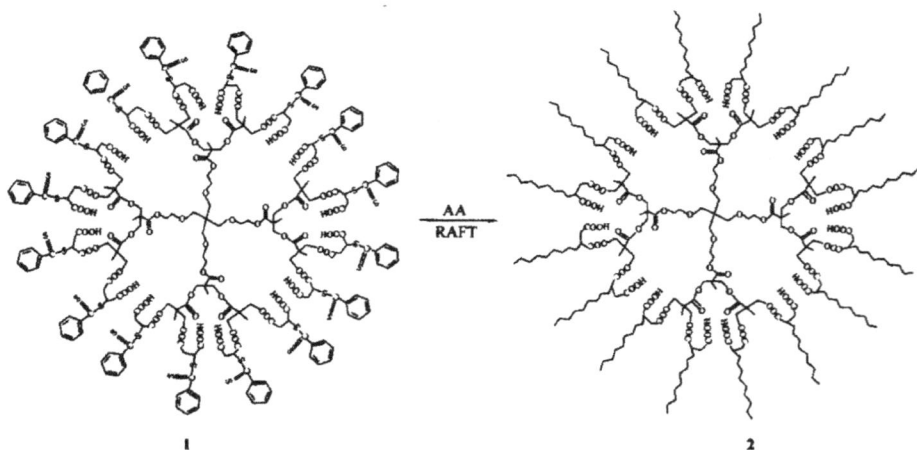

Fig. 7.18: Schematic diagram of preparation of H20-star-PAA. Reprinted with permission from [36]. Copyright Chinese Academy of Sciences, 2007.

The study found that the pH phase transition process of the hyperbranched polymer H20-star-PAA has two mechanisms. In a dilute solution system, the polymer is present as a monomolecular micelle. Under alkaline conditions, the carboxylate of H20-star-PAA exists in the form of sodium carboxylate, the molecular segments repel each other, and the entire molecule is in an extended state. As the pH decreases, the sodium carboxylate on the polymer chain gradually protons into a collapsed state, resulting in a phase transition (Fig. 7.19). Conversely, in a more concentrated solution system, as the pH decreases, the intermolecular hydrogen bonds increase, and the polymer aggregates from the monomolecular micelles into multimolecular micelles (Fig. 7.20).

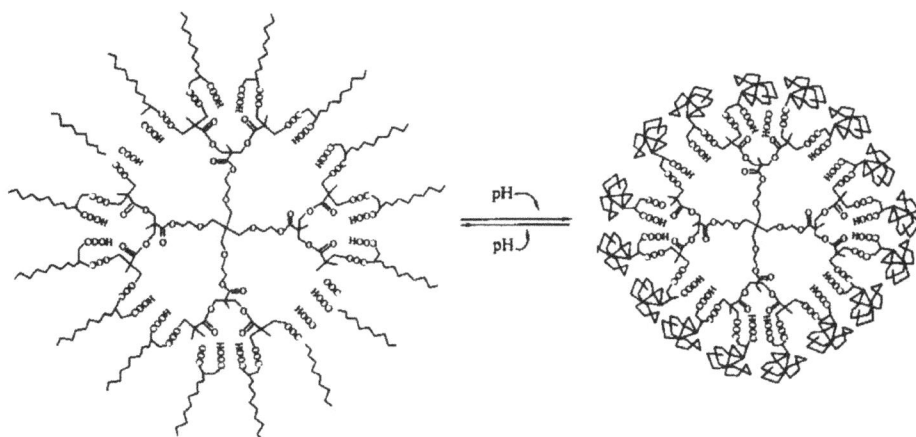

Fig. 7.19: Effect of pH on the morphology of H20-star-PAA single molecule micelles. Reprinted with permission from [36]. Copyright Chinese Academy of Sciences, 2007.

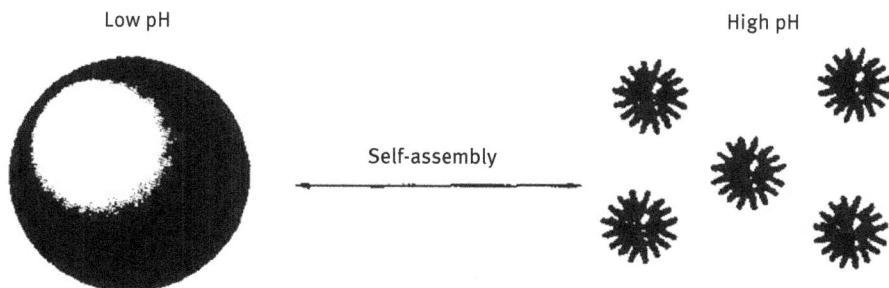

Fig. 7.20: Schematic diagram of phase transition of H20-star-PAA in concentrated solution. Reprinted with permission from [36]. Copyright Chinese Academy of Sciences, 2007.

This phase transition behavior of H2O-star-PAA is expected to find application in the research of intelligent drug carriers.

In fact, due to the structural diversity of hyperbranched polymers, they can exhibit different phase transition characteristics in different environments, such as ion responsiveness, temperature responsiveness, and solvent responsiveness. Research in these areas is further deepening.

References

[1]   Lehn, JM. Perspectives in supramolecular chemistry – from molecular recognition towards molecular information processing and self-organization. Angew Chem, Int Edit. 1990;29(11): 1304–1319.

[2]   Jiang, M, Eisenberg, A. Macromolecules self-assembly. Beijing: Science Press, 2006.

[3]   Huang, HY, Remsen, EE, Kowalewski, T, et al. Nanocages derived from shell cross-linked micelle templates. J Am Chem Soc. 1999;121(15):3805–3806.

[4]   Zhang, L, Eisenberg, A. Multiple morphologies and characteristics of crew-out micelle-like aggregates of polystyrene-b-poly(acrylic acid) diblock copolymers in solution. J Am Chem Soc. 1996;118(13):3168–3181.

[5]   Halperin, A, Tirrell, M, Lodge, TP. Tethered chains in polymer microstructures. Adv Polym Sci. 1992;100:31–71.

[6]   Zhang, L, Eisenberg, A. Formation of crew-cut aggregates of various morphologies from amphiphilic block copolymers in solution. Polym Adv Technol. 1998;9(10–11):677–699.

[7]   Liu, SY, Zhang, GZ, Jiang, M. Soluble graft-like complexes consisting of poly(4-vinyl pyridine) and carboxyl-terminated polystyrene. Polymer. 1999;40:5449–5453.

[8]   Liu, SY, Pan, QM, Xie, JW, et al. Intermacromolecular complexes due to specific interactions 12, graft-like hydrogen bonding complexes based on pyridyl-containing polymers and end-functionalized polystyrene oligomers. Polymer. 2000;41:6919–6929.

[9]   Liu, SY, Jiang, M, Liang, HJ. Intermacromolecular complexes due to specific interactions 13, formation of micelle-like complexes from hydrogen-bonding "graft" complexes in selective solvents. Polymer. 2000;41:8697–8902.

[10]   Zhang, L, Eisenberg, A. Morphogenic effect of added ions on crew-out aggregates of polystyrene-b-poly(acrylic acid) block copolymers in solutions. Macromolecules. 1996;29(27): 8805–8815.

[11]   Jenekhe, SA, Chen, XL. Self-assembled aggregates of rod-coil block copolymers and their solubilization and encapsulation of fullerenes. Science. 1998;279:1903–1907.

[12]   Kuang, M, Duan, HW, Wang, J, et al. Structural factors of rogid-coil polymer pairs influencing their self-assembly in common solvent. J Phys Chem B. 2004;108:16023–16029.

[13]   Bronich, TK, Ming, OY, Kabanov, VA, et al. Synthesis of vesicles on polymer template. J Am Chem Soc. 2002;124:11872–11873.

[14]   Kabanov, AV, Bronich, TK, Kabanov, VA, et al. Soluble stoichiometric complexes from poly (N-ethyl-4-vinylpyridinium) cations and poly (ethylene oxide)-block polymath-acrylate anions. Macromolecules. 1996;29:6797–6802.

[15]   Forder, C, Patrickios, CS, Armes, SP, et al. Synthesis and aqueous solution properties of amphiphilic diblock copolymers based on methyl triethylene glycol vinyl ether and benzyl vinyl ether. Macromolecules. 1997;30(19):5758–5762.

[16] Bronich, TK, Popov, AM, Eisenberg, A, et al. Effects of block length and structure of surfactant on self-assembly and solution behavior of block ionomer complexes. Langmuir. 2000;16(2): 481–489.

[17] Dou, HJ, Jiang, M, Peng, HS, et al. pH-dependent self-assembly: Micellization and micelle-hollow-sphere transition of cellulose-based copolymers. Angew Chem, Int Edit. 2003;42(13): 1516–1519.

[18] Harada, A, Kataoka, K. Formation of stable and monodispersive polyion complex micelles in aqueous medium from poly(L-lysine) and poly(ethylene glycol)-poly(aspartic acid) block copolymer. J Macromol Sci, Pure Appl Chem A. 1997;34(10):2119–2133.

[19] Yi, CF, Shen, YH, Xu, ZS. Self-Assembly of Dendritic Structures. J Funct Polym. 2003;16(4): 599–606.

[20] Zimmerman, SC. Self-assembling dendrimers. Science. 1996;271:1095–1098.

[21] Zeng, F, Zimmerman, SC. Dendrimers in supramolecular chemistry: from molecular recognition to self-assembly. Chem Rev. 1997;97:1681–1712.

[22] Zimmerman, SC. Dendrimers in molecular recognition and self-assembly. Curr Opin Colloid Interface Sei. 1997;2(1):89–99.

[23] Hawker, CJ, Frechet, JMJ. Preparation of polymers with controlled molecular architecture. A new convergent approach to dendritic macromolecules. J Am Chem Sec. 1990;112(21): 7638–7647.

[24] Thiyagerajan, P, Zimmerman, SC. SANS investigation of self-assembling dendrimers in organicsolvents. J Mater Chem. 1997;7:1221–1226.

[25] Wang, Y, Zeng, F, Zimmerman, SC. Dendrimers with anthyridine-based hydrogen-bonding units at their cores: synthesis, complexation and self-assembly studies. Tetrahedron Lett. 1997;38(31):54595462.

[26] Freeman, AW, Vreekamp, R, Frechet, JMJ. The self-assembly of convergent dendrimers based on the melamine cyanuric acid lattice. Polym Mater Sci Eng. 1997;77:138–139.

[27] Bo, ZS, Zhang, X, Shen, JG. The self-assembled dendritic supramolecular complex based on electrostatic attraction. Chem J Chin Univ. 1997;18(2):326–328.

[28] Yan, DY, Zhou, YF, Hou, J. Supramolecular self-assembly of macroscopic tubes. Science. 2004;303(5654):65–67.

[29] Zhou, YF, Yan, DY. Supramolecular self-assembly of giant polymer vesicles with controlled sizes. Angew Chem, Int Ed. 2004;43(37):4896–4899.

[30] Zhou, YF, Yan, DY. Real-time membrane fission of giant polymer vesicles. Angew Chem, Int Ed. 2005;44(30):3223–3226.

[31] Zhou, YF, Yan, DY. Real-time membrane fusion of giant polymer vesicles. J Am Chem Soc. 2005;127:10468–10469.

[32] Zhou, YF, Yan, DY, Dong, WY. Temperature-responsive phase transition of polymer vesicles: real-time morphology observation and molecular mechanism. J Phys Chem: B. 2007;111(6): 1262–1270.

[33] Mai, YY, Zhou, YF., Yan, DY. Synthesis and size-controllable self-assembly of novel amphiphilic hyperbranched multiarm copolyether. Macromolecules. 2005;38(21):8679–8686.

[34] Hong, HY., Mai, YY, Zhou, YF, et al. Self-assembly of large multimolecular micelles from hyper-branched star copolymers. Macromol Rapid Commun. 2007;28:591–596.

[35] Mai, YY, Zhou, YF, Yan, DY. Real-time hierarchical self-assembly of large compound vesicles from an amphiphilic hyperbranched multiarm copolymer. Small. 2007;3:1170–1173.

[36] Zhong, L, He, XH, Zhou, YF. Synthesis and self-assembaly of a pH-sensitive hyperbranched muitiarm copolymer. Acta Polym Sin. 2007;(10):986–992.

[37] Shen, H, Eisenberg, A. Morphological phase diagram for a ternary system of block copolymer PS_{310}-b-PAA_{52}/dioxane/H_2O. J Phys Chem B. 1999;103(44):9473–9487.

[38] Chen, DY, Jiang, M. Strategies for constructing polymetric micelles and hollow spheres in solution via specific intermolecular interactions. Acc Chem Res. 2005;38(6):494–502.

[39] Harada, A, Kataoka, K.. Chain length recognition: core-shell supramolecular assembly from oppositely charged block copolymers. Science. 1999;283(5398):65–67.

[40] Murthy, VS, Cha, JN, Stucky, GD, et al. Charge-driven flocculation of poly(L-lysine)-gold nanoparticle assemblies leading to hollow microspheres. J Am Chem Soc. 2004;26(16): 5292–5299.

Exercises

1. **The physical essence of molecular self-assembly.**
   The physical nature of molecular self-assembly is the interaction between permanent multipole moments, instantaneous multipole moments, and induced multipole moments. According to the background conditions generated by self-assembly, self-assembly phenomena can be divided into two categories: static self-assembly and dynamic self-assembly. Static self-assembly refers to the self-assembly that the system can complete and maintain without consuming energy under balanced conditions, and does not change with time, such as crystals. Dynamic self-assembly means that the system must consume energy or material flow. Self-assembly completed and maintained that the structure often changes with time, such as the birth of cells and the development of organisms. At present, most self-assembly research is focused on static self-assembly is a spontaneous process, once the assembly begins, the process is largely controlled by nature.

2. **Two necessary conditions that are needed is the formation of a molecular self-assembly.**
   The formation of a molecular self-assembly system requires two necessary conditions: the driving force of self-assembly and the guiding effect. The weak interaction of noncovalent bonds maintains the structural stability and integrity of the self-assembled system. In general, there are three main steps in creating a molecular self-assembly system: the first step is to form a complex, complete intermediate molecular body through ordered covalent bonds; the second step is to pass the noncovalent bond from the intermediate molecular body. Synergistic action has to form a structurally stable macromolecular aggregate; in the third step, one or several molecular aggregates are used as structural units, and multiple repeats are self-organized to form an ordered molecular assembly. The interactions in the supramolecular system are more additive and synergistic, and have certain directionality and selectivity, and their total binding force can be no less than chemical bonds. Molecular recognition is the embodiment of this combination of weak interactions, which is the key to the formation of advanced ordered assemblies. At the same time, most supramolecular embodiments have an additional feature: they have internal adjustment capabilities for error correction (such as the continuous improvement of crystals in the aging process), which is usually not possible with purely covalent systems.

3. **Factors affecting the morphology of self-assembled amphiphilic block copolymers.**
   Effect of block relative length; effect of block copolymer concentration; effect of preparation method; impact of other factors.

4. **What is self-assembly of fully hydrophilic block copolymers?**
   The fully hydrophilic block copolymer refers to a block copolymer composed of two or more water-soluble segments each having a different chemical structure and environmental responsiveness. Under certain conditions, DHBCs are completely soluble in water. When environmental

conditions such as pH, temperature changes, or the addition of certain polymers or small molecules that form a complex with a segment, one of the segments becomes water-insoluble, allowing self-assembly in aqueous solution.

5. **Mechanisms of the pH phase transition process of the hyperbranched polymer H_{20}-star-PAA**

The study found that the pH phase transition process of the hyperbranched polymer H_{20}-star-PAA has two mechanisms. In a dilute solution system, the polymer is present as a monomolecular micelle. Under alkaline conditions, the carboxylate of H_{20}-star-PAA exists in the form of sodium carboxylate, the molecular segments repel each other, and the entire molecule is in an extended state. As pH decreases, sodium carboxylate on the polymer chain gradually protons and forms a collapsed state, resulting in a phase transition. Conversely, in a more concentrated solution system, as the pH decreases, the intermolecular hydrogen bonds increases, and the polymer aggregates from the monomolecular micelles into multimolecular micelles.

Chapter 8
Carbon nanomaterials and their polymer modifications

8.1 Introduction

In the mid-1980s, following graphite and diamond, a third crystal form of carbon was found, and its molecular formula was C_n. The currently known n value is at most 540. Such carbon compounds are referred to as carbon cage clusters or fullerenes. Among the wide variety of fullerenes, the most in-depth study one is C_{60} because it is one of the most stable. Due to the special structure of fullerenes and the singular properties of light, electricity, and magnetism, this field has become one of the hotspots of scientists in the world. The discovery of C_{60} has brought human understanding of carbon chemistry to a new level. Before this, chemists rarely used elemental carbon as a starting material for synthetic research. In the short span of more than 10 years since the discovery of C_{60}, it has had a tremendous impact on physics, chemistry, materials science, life sciences, and medical science and has shown great potential applications. To this end, the inventors of C_{60}, American scientist R. E. Smalley, R. F. Curl, and British scientist H. Kroto won the Nobel Prize in Chemistry in 1996.

The discovery of fullerenes dates back to 1984, when Richard Smalley, a professor at Rice University in Houston, USA, led a research team exploring the nature of atomic beams. To this end, they invented the laser-ultrasonic atomic beam ray tester, which is a steel vacuum vessel with a holed anvil. The test sample is placed in the anvil, and then rapidly evaporated by a short-wavelength laser wave with a temperature of up to several tens of thousands of degrees. The vaporized material is carried by an inert helium gas stream to another laser beam, which ionizes the atomic beam. The ionized atomic beam is then passed to a mass spectrometer for analysis and its size is measured. At the right time, Harry Kroto of the University of Sussex in the UK was visiting Rice University. His research topic was the origin of long-chain carbon molecules in interstellar space. In order to simulate the possible C_9 molecules in interstellar dust, they performed experiments with the earlier methods, but obtained large atomic beams such as C_{50}, C_{60}, and C_{70}. The analysis found that C_{60} was three times more than other atomic beams. Smallery's graduate student James Heath improved the test method and the C_{60} obtained was 40 times more than other atomic beams.

After studying these products, they were found to be carbon spheroids. For example, C_{60} is a football-like body composed of 12 regular pentagons and 20 regular hexagons. Each vertex represents a carbon atom, and each carbon atom is at the junction of a pentagon and two hexagons. As shown in Fig. 8.1.

https://doi.org/10.1515/9783110597097-008

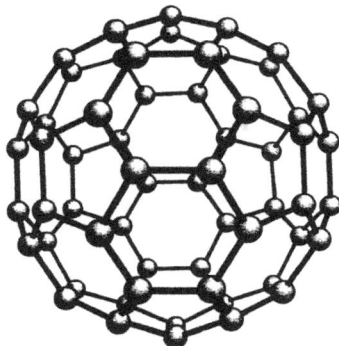

Fig. 8.1: The structure of C_{60} diagram.

The naming process for fullerenes is also very interesting. When Kroto et al. analyzed the molecular structure of C_{60}, they benefited from the 1967 Montreal International Exposition in Canada. Because the US Pavilion was a dome constructed of pentagon and hexagonal components, they proposed the molecular structure of C_{60}. Therefore, they decided to name "Buckminsterfullerene," referred to as "fullerene," named after Buckminster Fuller, the architect of the exhibition hall. The term "ene" is the meaning of olefin, indicating the unsaturation of C_{60}. Later, it was discovered that in addition to C_{60}, there are some such hollow spherical carbon molecules, so fullerene has become a general term for this type of carbon molecule. Since the polyhedral structure of C_{60} carbon molecules is very similar to football, there are many other names, such as Buckyball, Spherene, Soccerene, Carbosoccer, Footballerene, and so on, but Fullerene is the most commonly used name. For a series of fullerene spherical molecules having different numbers of carbon atoms, it is represented by Fullerene C_n (n represents the number of carbon atoms).

On 14 November 1985, Kroto, Smalley, and others published a paper in *Nature*, officially announcing the discovery of C_{60} and its structural model, and won the Nobel Prize in Chemistry.With the deepening of research work, many C_n with larger value of n have been discovered. Moreover, it has been found that as the value of n increases, the fullerene is no longer a football-like circle, but gradually develops toward an ellipse. When n is large enough, the shape of C_n actually becomes a carbon tube with a tubular shape in the middle and a hemispherical shape at both ends, as shown in Fig. 8.2.

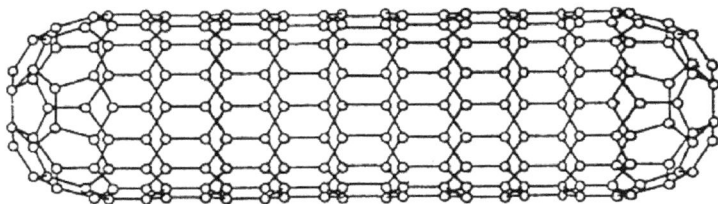

Fig. 8.2: Fullerene when n is large.

Since the diameters of these tubular C_n reach the size of nanometer, they are named after carbon nanotubes (CNTs). It was unexpectedly discovered in 1991 by Iijima of Japan's NEC Corporation when testing fullerenes produced in graphite arc equipment under high-resolution transmission electron microscopy. Since CNTs have many similarities structure to buckyballs, they are also called "bucky tubes."

CNTs are yet another allotrope of carbon found after C_{60}. It is a one-dimensional quantum material with special dismissal (radial size in nanometers and axial dimension in micron order) with typical lamellar hollow structure features. The end cap portion has a polygonal structure composed of pentagon and hexagonal carbon ring. CNTs may have only one wall or multiple walls, which are called single-walled carbon nanotube (SWCNTs) and multi-walled carbon nanotube (MWCNTs) (Fig. 8.3). The distance between the walls is about 0.34 nm and the diameter is generally 2–20 nm. The ratio of length to diameter of CNTs is very large, reaching 10^3–10^6. At present, scientists have been able to synthesize CNTs with a length of millimeters size.

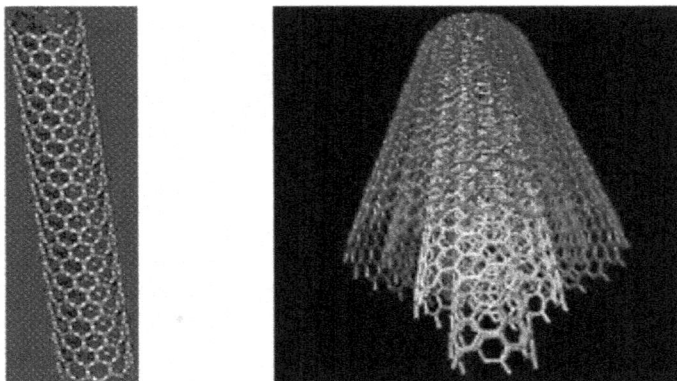

Fig. 8.3: The structure of single-walled nanotubes (left) and multi-walled carbon nanotubes (right).

The discovery of fullerenes and CNTs has led to a further recognition of the link between carbon materials. As long as the environmental conditions change, various carbon materials can be transformed. Based on the hybridization of electrons on the surface of carbon atoms, diamond with stereo structure is sp³ hybridized, while graphite with planar structure of is *sp²* hybridized, and the carbene with one-dimensional linear structure is *sp* hybridized. If we draw a "phase diagram" with the three allotropes of carbon as the apex, the relationship between the various carbon materials can be clearly seen (Fig. 8.4). In this figure, P refers to the five-membered ring structure, and H refers to the six-membered ring.

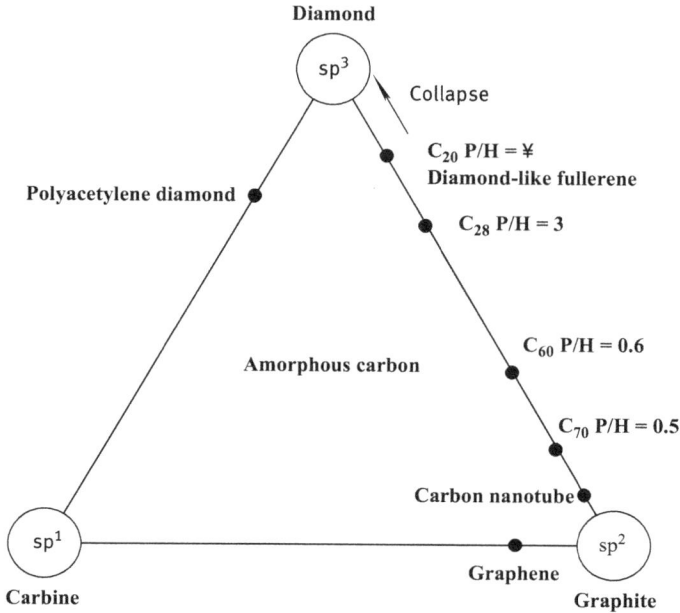

Fig. 8.4: The phase diagram of carbon allotropes.

It is not difficult to deduce from Fig. 8.4 that since spherical fullerene and tubular CNTs can be synthesized by artificial methods, planar graphene should also be artificially prepared.

Graphene is a new material with single-walled sheet structure composed of carbon atoms. It is a two-dimensional material with the thickness of one carbon atom and it is composed of carbon atom with a sp^2 hybrid orbital. Its crystal lattice exhibits a hexagonal honeycomb shape. Graphene had long been considered a hypothetical structure and could not be stably present alone.

In 2004, physicist Andre K. Geim and his colleague Konstantin Novoselov from the University of Manchester used a special plastic tape to glue the sides of the graphite sheet, tearing the tape and splitting the sheet into two. By repeating this process, it is possible to obtain increasingly thinner graphite flakes, and finally obtain graphene composed of only one layer of carbon atoms. Geim and Novoselov used this primitive method to confirm that graphene can exist singly and steadily. The two also won the 2010 Nobel Prize in Physics for their "pioneering test of two-dimensional graphene materials."

Graphene is currently the thinnest and hardest nanomaterial in the world. It can deliver electrons at a speed faster than any known conductor at room temperature. The perfect graphene is two-dimensional, and it only includes hexagons (equal hexagons), as shown in Fig. 8.5. If there are pentagons and heptagons, it will constitute a

Fig. 8.5: The structure of graphene.

defect of graphene. The connection between the carbon atoms inside the graphene is very flexible. When an external force is applied to the graphene, the carbon atom plane is bent and deformed, so that the carbon atoms do not have to be rearranged to adapt to the external force, thereby maintaining the structural stability.

Graphene is the basic unit constituting charcoal, graphite, CNTs, and fullerene allotropes. If it is stacked, it becomes graphite, rolled into a barrel, nanotubes, and if it contains 12 pentagons and 20 hexagonal graphenes, it becomes fullerenes (C_{60}), as shown in Fig. 8.6.

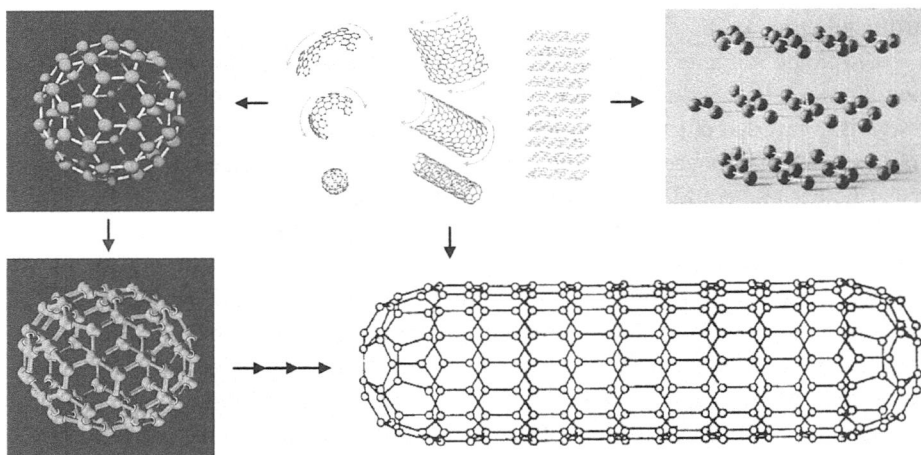

Fig. 8.6: Transition between carbon nanomaterials.

8.2 Structure and properties of fullerenes

8.2.1 Structure and characteristics of C_{60}

C_{60} is a spherical molecule composed of 60 carbon atoms. Each carbon atom is connected to the other three adjacent carbon atoms by σ bond with $sp^{2.28}$ hybridization, forming a highly symmetrical football-like hollow molecule (Fig. 8.1), which is a truncated icosahedron, composed of 12 positive five-membered rings and 20 positive six-membered rings. The 6–6 bond is shorter, 0.1388 nm, close to the double bond, while the 6–5 bond is longer and the bond length is 0.1432 nm, close to a single bond.

The ^{13}C NMR measurement showed that the chemical environment of the 60 carbon atoms was identical. Structural chemistry calculations show that C_{60} has at least 12,500 resonant structures, so its stability is extremely high. Comparing to the planar structure of the usual carbon compound in which the C–C bond sp^2 is hybridized, the cage molecular structure of C_{60} causes tension in the ring, forcing the C–C bond in the molecule to be doped with the sp^3 hybrid component (average $sp^{2.28}$). This structure leads to a low degree of π-electron delocalization in the molecule, and the aromaticity of the molecule is not obvious. The carbon–carbon double bond has an electron-withdrawing induction effect and a strong electrophilic ability. Therefore, the basic chemical property of C_{60} is electron-deficient olefin, which can be used as an electron acceptor. At the same time, the tension of the bond in the molecule is also easy for C_{60} to react with other molecules.

Due to the hollow spherical structure of C_{60}, it can react on both the inner and outer surfaces, thereby obtaining C_{60} derivatives with various structures. Up to now, the chemical reaction of C_{60} has been found which includes the reaction between C_{60} and metal, redox reaction of C_{60}, reaction between C_{60} and radical, addition reaction of C_{60}, and a polymerization reaction of C_{60} itself. These chemical reactions form the basis of C_{60} macromolecularization.

Due to the existence of thermal motion, the intermolecular interactions in the C_{60} crystal are not sufficient to bind the individual electrons, so that even at room temperature, the C_{60} molecule can spin at millions times per second. Each carbon atom provides a p_z electron and a nonplanar three-dimensional conjugate system of π_{60}^{60} is formed inside and outside the sphere. Single crystal X-ray diffraction experiments show that the C_{60} crystal is a face-centered cubic structure at room temperature, the lattice constant is 1.42 nm, and the distance between the two nearest neighbor molecules is 1 nm. This is a typical molecular crystal. Its melting point is higher than 700 °C, insoluble in water, but soluble in toluene and CS_2 and can sublimate without cracking. The pure C_{60} obtained by chromatography is a dark brown solid which is magenta in an organic solvent. C_{70} is a reddish brown solid and shows the color like red wine in an organic solvent.

The density of C_{60} is 1.65 ± 0.05 g/cm^3. The formation of the C_{60} crystal is $\Delta H^0 = 2280 \text{KJ/mol}$; the ionization potential is (7.61 ± 0.02) eV; the electron affinity energy is 2.6–2.8 eV. When a pressure of 15–20 GPa is applied to the solid C_{60}, it is converted into an insulating state C_{60}. When a pressure of 20 GPa is rapidly applied, a large amount of C_{60} is converted into polycrystalline diamond. This transformation has the potential to be used to make diamonds, while it also reflects some relationship among these three elemental carbons materials: diamond, graphite, and C_{60}. The volume compressibility of C_{60} measured by powder X-ray at room temperature is 7.0×10^{-12} cm^3/dyn, which is 40 times that of diamond and 3 times that of graphite. This suggests that C_{60} may be the most elastic among the known solid carbon. The C_{60} can withstand a static pressure of 20 GPa. When shot to a stainless steel plate at 6.7 km/s, C_{60} bounced back without damage. Its impact resistance is also stronger than all the particles so far. Table 8.1 lists the basic structural parameters of these three different forms of carbon crystals: diamond, graphite, and C_{60}.

Tab. 8.1: The basic structural parameters of diamond, graphite, and C_{60}.

Structure properties	Diamond	Graphite	C_{60}
Carbon atom configuration	Tetrahedron	Plane triangle	Football-like
C–C–C average bond angle °	109.5	120	116
Hybrid track type	sp^3	sp^2	$sp^{2.28}$
C–C bond length (nm)	0.1544	0.1418	$(6/6)^*0.1388$ $(6/5)0.1432$
Carbon atom occupied volume ($\times 10^3$) (nm)	50,672	8.744	11.87
Density (g/cm^3)	3.514	2.266	1.678
Crystal type	Atomic crystal	Between the molecule and the atomic crystal	Molecular crystal

Note: * 6/6 indicates the C–C bond shared by two adjacent six-membered rings in the C_{60} molecule. 6/5 indicates the C–C bond shared by a six-membered ring and a five-membered ring.

After determining the structure of fullerenes, the quantum theory can be used to calculate the electronic structure and energy state of fullerenes, and then predict their various spectra. So far, various spectra of C_{60} and C_{70} have been measured, and the theoretical predictions agree well with the experimental results. Table 8.2 shows the main peaks of the infrared spectrum, Raman spectrum, nuclear magnetic resonance, and ultraviolet (UV)-visible spectrum of C_{60} and C_{70}.

As can be seen from Tab. 8.2, the various spectra of C_{60} are relatively simple, especially the NMR spectrum, and the chemical shift has only a single peak at 143.68 ppm. This is mainly due to the extremely symmetrical geometry of C_{60}, which has the same chemical environment for all 60 carbon atoms. This provides great convenience for the structural characterization of C_{60} polymer.

Tab. 8.2: The main peaks of IR, Raman, NMR, and UV-VIS of C_{60} and C_{70}.

Spectral name	Fullerenes	Peak position
Infrared spectrum (cm^{-1})	C_{60}	1,430, 1,183, 577, 527
	C_{70}	1,700, 1,431, 1,414, 1,086, 796, 674, 578, 534
Raman spectrum (cm^{-1})	C_{60}	1,469, 497, 273
Nuclear magnetic resonance (ppm)	C_{60}	142.68 (benzene solution)
	C_{70}	150.7, 148.1, 147.4, 145.4, 130.9
Ultraviolet-visible spectrum (nm)	C_{60}	213 (s), 257 (s), 329 (s), 404 (w), 440–670 (broad)
	C_{70}	214 (s), 236 (s), 331 (mw), 360 (mw), 378 (mw), 440–670 (broad)

8.2.2 Physical properties of fullerenes

People's interest in fullerenes is not only because of its beautiful molecular structure, but also because of its wide range of uses, and it is likely to become a newcomer to the material family in the new century. Recently, the research team led by Professor Yukio Shoji of the Japan Institute of Industrial Science has used C_{60} in solar cell manufacturing for the first time. They made the iron-containing organic ferrocene and C_{60} in order to form a rod-shaped molecule, and then arranged the molecule on a metal electrode to form a film, thereby forming a photosynthesis solar cell. Due to the use of C_{60}, the quantum absorption rate of this battery has increased by 114%, greatly improving the energy conversion rate.

The potential application of fullerenes in many special fields is determined by its structure and the special properties it brings.

8.2.2.1 Solubility of fullerenes

The solubility of fullerenes in aliphatic hydrocarbons increases as the number of carbon atoms in the solvent molecules increases, but generally the solubility is small. It has good solubility in aromatic solvents such as benzene and toluene and has a large solubility in carbon disulfide (CS_2). However, because CS_2 is highly toxic, it is generally not used. Currently, the most common solvent used to dissolve C_{60} is toluene. Table 8.3 lists the solubility parameters and solubility of C_{60} in different solvents.

Tab. 8.3: The solubility parameters and solubility of C_{60} in different solvents.

Solvent	Solubility parameter (J /cm^3)	Solubility (μg/mL)	Solvent	Solubility parameter (J/ cm^3)	Solubility (μg/mL)
Isooctane	14.17	26	Carbon tetrachloride	17.59	447
Pentane	14.52	4	1,3, 5-Trimethylbenzene	18.04	997
Hexane	14.85	40	Toluene	18.20	2,150
Octane	15.45	25	Benzene	18.82	1,440
Decane	15.81	70	Dichloromethane	20.04	254
Dodecane	16.07	91	Dioxane	20.50	41
Tetradecane	16.24	126	Carbon disulfide	20.50	5,160
Cyclohexane	16.77	51			

8.2.3 Superconductivity of fullerenes

C_{60} is a molecular crystal at room temperature, and the energy spectrum calculation shows that the face-centered cubic solid C_{60} is a semiconductor with an energy gap of 1.5 eV. After proper metal doping, it exhibits good electrical conductivity and superconductivity. In March 1991, A.F. Hebard et al. of Bell Labs of the American Telephone and Telegraph Company first reported that K_3C_{60} obtained after potassium doping has superconductivity and its critical temperature is 18 K. Subsequent research on high temperature C_{60} superconductivity has developed extremely rapidly. It has been found that M_3C_{60} series compounds (M is K, Rb, Cs, etc.) are superconductors. M_3C_{60} is formed by a diffusion reaction between C_{60} and a metal. X-ray diffraction (XRD) studies show that M_3C_{60} and C_{60} have the same crystal structure and belong to face-centered cubic structure. Each C_{60} participates in the formation of two tetrahedral voids (with a radius of 0.112 nm) and an octahedral void (with a radius of 0.206 nm). Three M atoms enter the void and form M_3C_{60}. M_3C_{60} is an isotropic nonoxide three-dimensional superconductor. It is different from two-dimensional oxide high-temperature superconductors and one-dimensional organic superconductors. In essence, the oxide superconductor is a kind of ceramic, the texture is very brittle, and the processing is difficult. However, the C_{60} superconductor has a large elasticity, and its volume can be compressed to 60% and can be restored to the original shape, so the processing is much easier. It has been predicted that the doped C_{240} and C_{540} may become superconductors with higher critical temperatures or even room temperatures. Table 8.4 shows the composition of the partial fullerene derivative and their superconducting critical temperature T. Table 8.4 shows that the critical superconducting transition temperature of the iodine-doped C_{60} compound I_xC_{60} is as high as 57 K. Besides, it also has the advantage of avoiding the use of an alkali metal unstable in air, which greatly improves the application possibility of the superconducting material.

Tab. 8.4: Superconductivity of fullerenes and their critical temperatures.

Doping element	Superconductor composition	Critical temperatures T_c (K)	Doping element	Superconductor composition	Critical temperatures T_c (K)
Alkali metal	K_3C_{60}	18–19.3	Alkali and other metals	$Rb_{2.7}Ti_{2.2}C_{60}$	45
	Rb_3C_{60}	28–30		$RbTi_2C_{60}$	48
	Rb_2C_{60}	29	Halogen	I_xC_{60}	57
	Rb_4C_{60}	29		$(IBr)_xC_{60}$	31
	Cs_3C_{60}	30	Other metals	Sm_xC_{60}	37
	$KRbC_{60}$	24.4–26.4		Ca_xC_{60}	8
	K_2RbC_{60}	2,108			
	Rb_2CsC_{60}	31.3			
	$RbCsC_{200}$	33			

8.2.3.1 Optical properties of fullerenes

The peculiar optical properties of fullerenes are found in the beautiful colors of solutions obtained by extracting C_{60} from soot.

The DuPont chemical company test station in Wellington, Delaware, and the US Naval Research Laboratory first discovered that fullerenes have "nonlinear optical properties," meaning that when light penetrates a fullerene sphere, its direction of refraction varies with the intensity of the light. In the future, fullerenes may be used as a switching material for a new generation of optical computers based on this property.

The study also found that the toluene solution of C_{60} and C_{70} transmits light with relatively low light intensity, but can prevent light passing through if it reaches a certain critical intensity. Besides, the C_{60} molecule in the photoexcited state has better absorbance than the C_{60} in the ground state. At low light intensities, the solution of C_{60} follows Lambert-Beer law and the transmittance does not change with increasing light intensity. However, when the light intensity exceeds 100 mJ/cm^2, the transmittance is remarkably lowered and maintained at 65 mJ/m^2. Since the absorption spectrum of the triplet C_{60} is parallel to the ground state, the light confinement effect of C_{60} may be effective over a relatively wide wavelength of light. This nonlinear response of the C_{60} to light can be used to fabricate light limiters, protect optical sensors from damage caused by intense light pulses, and may even facilitate the development of next-generation optoelectronic computers.

Experiments have shown that under light, C_{60} can photochemically react with oxygen to produce a reddish-brown precipitate. Therefore, C_{60} should be stored in vacuum or nitrogen in the dark. Smalley et al. believes that the destruction of C_{60} by UV light in the arc during the preparation is also one of the reasons for the lower yield of C_{60}.

8.2.4 Chemical properties of fullerenes

8.2.4.1 Basic chemical properties of fullerenes
Since the constant preparation of C_{60} and C_{70} and other advanced fullerenes, their various chemical properties have been extensively studied and fruitful results have been achieved. Since C_{60} is the most abundant in various fullerenes, most of the research has focused on the chemical reaction of C_{60}. It can be inferred from the spherical hollow structure of the C_{60} molecule that it should have an aromaticity and can carry out a reaction of general fused aromatic hydrocarbon, such as an alkylation reaction, a reduction reaction to produce a hydride, and the like. It is well known that aromatic hydrocarbons generally exhibit electron-rich reactivity and are susceptible to electrophilic substitution of electrophiles. However, C_{60} is difficult to react with electrophiles, that is, it is easy to react with nucleophiles such as NH3 and metals, exhibiting reactivity of electron-deficient compounds and tends to obtain electrons. Therefore, the chemical behavior of C_{60} is more like an electron-deficient olefin than an aromatic compound. Due to the hollow spherical structure of C_{60}, it can chemically react on both the inner and outer surfaces to obtain various functionalized C_{60} derivatives. Some C_{60} chemical reactions have been well characterized, including the reaction with metals to produce compounds, halogenation reactions, free radical reactions, and the like. These reactions form the basis of surface modification and macromolecularization of fullerenes.

8.2.4.2 Reaction of fullerenes with metals
There are two kinds of reactions of fullerenes with metals: one is that the metal is contained inside the fullerene cage and the other is that the metal reacts with the sphere of the fullerene. In order to distinguish it from the external additive, the inclusion in the cage is indicated by the symbol $M_x@C_n$. In 1991, Chai et al. first gasified graphite-metal rods in a high-temperature furnace filled with helium to obtain $La@C_{2n}$ ($2n = 60, 70, 74, 82$). Later, this technology was used to synthesize $Y@C_{2n}$,

$Y_2@C_{82}$, $U@C_{82}$, $U_2@C_{60}$, and $Ca@C_{60}$. Nowadays, the arc technology and resistance heating technology is widely used in the preparation of metal fullerene inclusions, that is, metal or metal oxide, graphite powder, and binder (asphalt, dextrin) is packed into a graphite rod. After high temperature treatment, this rod acts as a positive electrode on a standard fullerene reactor and discharge which provides a macroscopic amount of fullerene inclusions of various metals. The metals involved include alkali metals, alkaline earth metals, and rare earth metals such as K, Na, Cs, La, Ca, Ba, Sr, U, Y, Ce, Sm, Eu, Gd, Tb, Ho, Th, and so on. During the preparation process, the atom number ratio of metal to carbon in the graphite rod has a direct influence on the number of metals in the metal fullerene. Generally, when the metal atom content is less than a certain value, for example, when La is less than 1%, Eu is less than 0.8%, and Sm is less than 0.9%, only a single metal fullerene can be formed. As the ratio of metal/carbon number increases, the productivity of bimetallic or even trimetalic fullerenes (such as LaC_{80} and Sc_3C_{82}) is observed. Experiments have also shown that metal carbides are better discharge fillers than oxides, which can improve the reaction yield. Therefore, high temperature treatment is often required to increase the degree of carbonization before metal oxide-graphite rods are discharged. When the graphite is evaporated under a $Fe(CO)_5$ atmosphere, a transition metal compound $Fe@C_{60}$ is contained, and the Fe is almost zero valent state.

Comparing the mass spectrum of the direct sublimation of the soot produced by the arc method, the solvent extract, and the extracted residue, it was found that the solubility of the metal fullerene $M_x@C_n$ was worse than C_{60}. Among the single metal fullerene, $M@C_{82}$ has the best solubility and is soluble in most common solvents such as benzene, toluene, pyridine, CS_2, and the like.

The externally-bonded metal fullerene compound which has been prepared is $M_n{}^+C_{60}$ compound containing V, Fe, Co, Ni, Cu, Rh, and La. Compounds such as K_3C_{60} which has superconductivity, as described earlier, may also belong to a C_{60} compound which is a externally-bonded metal.

8.2.4.3 Redox reaction of fullerenes

Using the unsaturation in the spherical structure of the C_{60} carbon cage, the oxidation of olefin by osmium tetroxide is applied to the surface modification of C_{60} to obtain a phthalate ester of C_{60}. By changing the ratio of C_{60} to OsO_4, C_{60} phthalate esters with different degrees of oxidation can be obtained. The reaction formula is as follows:

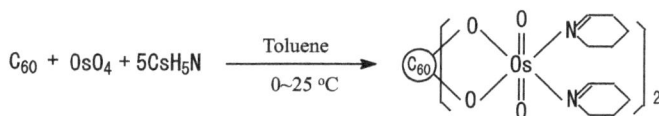

Chemical reaction equation 1

The oxygen-saturated C_{60} benzene solution was irradiated through a quartz lamp at room temperature for 18 h and $C_{60}O$ can be obtained. The reaction mixture was separated on a silica gel column and its yield was 7%. The strongest peak in mass spectrometry is 736, and another strong peak is 720, corresponding to $C_{60}O$ and C_{60}, respectively. Using deuterated benzene as solvent, the ^1H NMR spectrum of $C_{60}O$ gave 16 lines. The chemical shift has a band at 90.18 ppm, and the rest is between 140 and 146 ppm. Under the same conditions, the chemical shift of C_{60} has only one band at 142.68 ppm. This shows that $C_{60}O$ has broken the high symmetry of C_{60}.

C_{60} can be reduced with an aqueous solution of a strong reducing agent lithium to form hydrogenated products $C_{60}H_{36}$ and $C_{60}H_{18}$, which reflects the strong electrophilicity of C_{60}. Cyclic voltammetry studies have shown that C_{60} can be highly reversibly converted to C_{60}^- and C_{60}^{2-}, which suggests that C_{60} may provide the material for a new generation of reversible batteries.

8.2.4.4 Reaction of fullerenes with free radicals

C_{60} has a strong ability to react with free radicals. It can react with 1–15 benzyl radicals or 1–34 methyl radicals to form adducts. This property of C_{60} gives it the title of "free radical sponge."

The reaction of C_{60} with a benzyl radical is considered to be carried out by a reaction formula represented by the following:

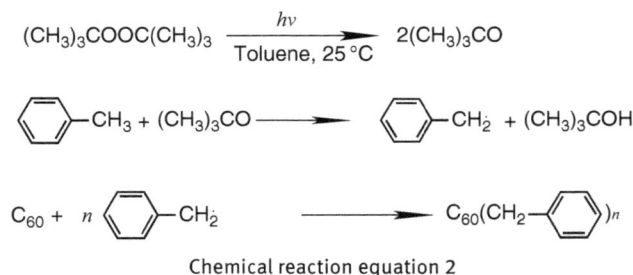

$$(CH_3)_3COOC(CH_3)_3 \xrightarrow[\text{Toluene, 25 °C}]{h\nu} 2(CH_3)_3CO$$

Chemical reaction equation 2

The electron spin resonance (ESR) spectrum of the adduct RC_{60} exhibits a special property that is different from the general compound. The ESR peak intensity of a typical compound decreases when the temperature rise, while the ESR peak intensity of RC_{60} tends to increase remarkably when the temperature rise. If the temperature decreases to a certain extent, the spectral peaks will disappear. According to the research, this is because the free radical adduct may conduct the following dimerization reaction:

$$RC_{60}C_{60}R \rightleftharpoons 2RC_{60}^{\cdot}$$

Chemical reaction equation 3

When temperature drops, the spectral peak is weakened; the temperature rises, the spectral peak enhances. This phenomenon can be repeated, and the strength of the bond between C_{60} in the dimer varies with R.

C_{60} can also react with NO_2, $RSiH_3$, and other substances under free radical conditions to form C_{60} (NO_2), H_nC_{60} $(SiRH_2)$, and so on.

8.2.4.5 Addition reaction of fullerenes

The addition reaction of C_{60} includes a nucleophilic addition reaction and an electrophilic addition reaction. It may be due to the influence of the introduction of the five-membered ring on the electron distribution of C_{60}. Although C_{60} has a spherical delocalized large π bond system, it is easy to carry out nucleophilic reaction rather than the electrophilic reaction which is easy for general aromatics. C_{60} is easy to form adducts with neutral reagents, such as amines (diethylamine, propylamine, morphine, etc.), and with phosphides and phosphates. Under the action of the format reagent, C_{60} can react with CH_3I to form various alkylation products.

Fullerenes can react with halogens and form a variety of useful halides. C_{60} and C_{70} mainly react with fluorine gas to form $C_{60}F_{36}$ and $C_{70}F_{40}$, and perfluorinated compound $C_{60}F_{60}$ is only one hundred thousandth of $C_{60}F_{36}$. The chlorine gas was passed through a heat-sealing tube containing C_{60} at 250 °C, and after 5 h, C_{60} was converted to $C_{60}Cl_{24}$. When liquid chlorine is reacted with C_{60}, $C_{60}Cl_{12}$ can be obtained. When the chlorination product is heated to 550 °C or treated with triphenylphosphine, the chlorine can be removed and the C_{60} can be recovered. C_{60} can react with pure bromine at room temperature, and adducts of different degrees of bromination can be obtained such as $C_{60}Br_2$, $C_{60}Br_4$, $C_{60}Br_6$, $C_{60}Br_8$, and $C_{60}Br_{24}$. Therefore, $C_{60}F_n$ can be used as a high-temperature lubricating material, coating, and waterproof material.

C_{60} can also undergo methoxylation and hydroxylation reactions to form ethers and alcohols which are partially soluble in water. The hydroxy compound of C_{60} is also called fullerol. The fullerene containing multiple hydroxyl groups is easily soluble in water, and the synthesis of water-soluble C_{60} compound is a real application of nonpolar molecule C_{60} in biochemistry. Therefore, the synthesis of fullerol has great theoretical and practical significance for the study of bioactive materials and biochemistry.

There are multiple approaches to synthesize fullerol. For example, a solution of C_{60} in benzene (or toluene) is added to an aqueous KOH solution while a few drops of tetrabutylammonium hydroxide are added and stirred at room temperature under an air atmosphere. After a few minutes of reaction, the benzene solution changed from the original dark purple to colorless, and the aqueous solution changed from colorless to brownish red. The upper benzene solution was removed, the aqueous solution was filtered, and methanol was added to crystallize several times and washed to give a brown solid product $C_{60}(OH)_x$.

The reaction of C_{60} with HNO_3/H_2SO_4 mixed acid can obtain a transparent yellow–brown liquid. After diluted with water, filtered, and adjusted to pH > 9 with NaOH, an amorphous brown solid fullerol can be obtained. This fullerol has moderate solubility in water but is soluble in an acidic aqueous solution with pH > 5. The characterization results show that this fullerol has an average of 14–15 hydroxyl groups per C_{60}, up to a maximum of 22.

8.2.4.6 Polymerization of fullerenes

The fullerene molecule can undergo polymerization under external conditions. In 1992, Yeretzian et al. first successfully used the method of laser irradiation of C_{60} to polymerize five C_{60} into one giant fullerene molecule. The C_{60} molecule is more easily polymerized by UV light irradiation of the C_{60} film, and 20 C_{60} polymerized polyfullerene has been prepared by this method.

The C_{60} polymerization was initiated by a radical polymerization initiator azobisisobutyronitrile, and a polyfullerene composed of 28 C_{60} was also obtained. Further, it was initiated by an anionic polymerization initiator and induced by a Ziegler–Natta catalyst and various structures of polyfullerene were obtained. The work on this aspect is also described in the following sections., omitted here.

8.3 Structure and properties of carbon nanotubes and graphene

8.3.1 Structure and properties of carbon nanotubes

8.3.1.1 Basic structure of carbon nanotubes

CNTs are tubular carbon molecules. Each carbon atom on the tube is sp^2 hybridized, and carbon–carbon σ bonds are combined with each other to form a hexagonal honeycomb structure as the skeleton of the CNTs. A pair of p electrons that do not participate in hybridization on each carbon atom form a conjugated π electron cloud across the entire CNT. According to the number of walls of the tube, it is divided into SWCNTs and MWCNTs. The radius of the tube is very thin, only on the nanometer scale, and tens of thousands of CNTs together have only one hair strand wide, and the name of the CNTs comes from it. In the axial direction, it can be as long as several tens to hundreds of micrometers.

CNTs are not always straight, and localized unevenness may occur due to the pentagon and heptagon mixed in the hexagonal structure. Where the pentagon appears, the CNTs bulge outward due to the tension. If the pentagon appears at the top of the CNTs, a seal of the CNTs is formed. Where the hemisphere appears, the CNTs are recessed inward.

The molecular structure of CNTs determines its unique properties. Due to the large aspect ratio (the radial dimension is on the order of nanometers and the axial dimension is on the order of micrometers), CNTs behave as typical one-dimensional quantum materials. According to the theoretical predictions, CNTs have extraordinary strength, thermal conductivity, magnetoresistance, and the properties will change with the structure, from insulator to semiconductor, from semiconductor to metal. The magnetic flux passing through the CNTs having metal conductivity is quantized, exhibiting an Akhanov–Bohm effect (A–B effect).

8.3.1.2 Mechanical properties of carbon nanotubes

The carbon atoms in the CNTs are sp^2 hybridized. Compared with sp^3 hybridization, the composition of s orbital in sp^2 hybridization is relatively large, which makes CNTs have higher modulus and higher strength.

CNTs have good mechanical properties and its tensile strength can reach 50–200 GPa which is 100 times that of steel. Its density is only 1/6 of that of steel. Its modulus of elasticity is up to 1 TPa, which is equivalent to the modulus of elasticity of diamond, which is about 5 times that of steel. For SWCNTs with the perfect structure, the tensile strength is about 800 GPa. Although the structure of CNTs is similar to that of polymer materials, its structure is much more stable than that of polymer materials.

CNTs have the same hardness as diamonds, but have good flexibility and can be stretched. Among the reinforcing fibers currently used in the industry, a key factor determining the strength is the aspect ratio, that is, the ratio of length to diameter. At present, material engineers hope to find a material whose aspect ratio is at least 20:1, and CNTs has an aspect ratio of generally more than 1000:1. They are ideal high-strength fiber materials and are therefore known as "super fibers."

CNTs are currently the materials with the highest specific strength. If we combine other engineering materials and carbon fibers, the composite material can exhibit good strength, elasticity, fatigue resistance, and isotropy, which greatly improves the performance of the materials.

Researchers at the Moscow University have placed CNTs under a water pressure of 1,011 MPa (equivalent to a pressure of 10,000 m under water), and the CNTs are crushed due to the enormous pressure. After the pressure is removed, the CNTs immediately return to shape like a spring and exhibit good toughness. This suggests that people can use CNTs to make thin and light springs, which can be used as shock absorbers in cars and trains and can greatly reduce the weight.

In addition, the melting point of CNTs is the highest among currently known materials, which is about 3,652 to 3,697 °C.

8.3.1.3 Electrical properties of carbon nanotubes

The p electrons of the carbon atoms on the CNTs can form a wide range of delocalized π bonds. Due to the significant conjugation effect, CNTs have some special electrical properties.

Since the structure of the CNTs is the same as that of the graphite, it has good electrical properties. The theory predicts that its electrical conductivity depends on its diameter and the helix angle of the tube wall. When the diameter of the CNTs is larger than 6 nm, the electrical conductivity is degraded; when the diameter is less than 6 nm, the CNTs can be regarded as a one-dimensional quantum wire with good electrical conductivity. It is considered that the CNTs with a diameter of 0.7 nm are superconducting, although their superconducting transition temperature is only 1.5×10^{-4} K; it indicates the application prospect of CNTs in the field of superconductivity.

The vector C_h is commonly used to represent the direction in which atoms are arranged on the CNTs (see Fig. 8.7). Where $C_h = na_1 + ma_2$, denoted as (n, m). a_1 and a_2 represent two basis vectors, respectively (see Fig. 8.7). (n, m) is closely related to the electrical conductivity of CNTs. For a given (n, m) CNT, if there is $2n + m = 3q$ (q is an integer), then this direction shows metallicity, which is a good conductor, otherwise it behaves as a semiconductor. For the direction of $n = m$, the CNTs exhibit good electrical conductivity and the electrical conductivity is up to 10,000 times that of copper.

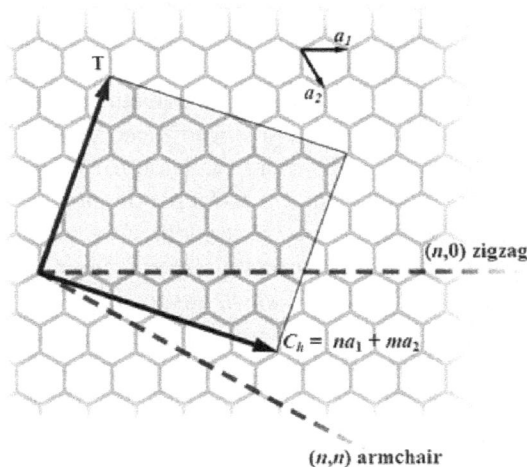

Fig. 8.7: Vector illustration of atoms arranged on carbon nanotubes.

8.3.1.4 Thermal conductivity of carbon nanotubes

CNTs have a very large aspect ratio, so that their heat exchange performance along the length direction is high, while the heat exchange performance in the vertical direction is relatively low. The theoretical thermal conductivity of CNTs is greater than 6,000 W/(m · K), which is twice that of diamond. The CNTs can synthesize a highly anisotropic heat conductive material by a suitable orientation. The study found that as long as the composite is doped with a small amount of CNTs (<1%), the thermal conductivity of the composite may be greatly improved.

8.3.2 Structure and properties of graphene

8.3.2.1 Basic structure of graphene

The structure of graphene is much simpler than fullerene and CNTs. In graphene, the carbon atom arrangement is exactly the same as the monoatomic layer of graphite and is a single-walled two-dimensional crystal in which carbon atoms are arranged in a honeycomb lattice in a sp^2 hybrid orbital. Graphene can also be thought of as an atomic grid formed by carbon atoms and their covalent bonds.

The name of graphene comes from graphite + ene (end of the olefin).

The structure of graphene is very stable, and the bond length of the carbon–carbon bond is 0.142 nm. The connection between the carbon atoms inside the graphene is very flexible. When an external force is applied to the graphene, the carbon atom plane is bent and deformed, so that the carbon atoms do not have to be rearranged to adapt to an external force, thereby maintaining the structural stability. This stable lattice structure gives graphene excellent thermal conductivity. In addition, when electrons in graphene move in orbit, they do not scatter due to lattice defects or introduction of foreign atoms. The interaction between atoms is very strong. At room temperature, even if the surrounding carbon atoms collide, the interference of the internal electrons of the graphene is very small.

8.3.2.2 Mechanical properties of graphene

Graphene is the most intensely known substance in humans. It is harder than diamonds and is 100 times stronger than the best steel in the world.

Physicists from Columbia University have conducted a comprehensive study of the mechanical properties of graphene. During the test, they selected some graphene particles with a diameter of 10–20 μm as the research object. The researchers first placed these graphene samples on a thin-crystal plate with a small hole drilled in the surface. The holes were between 1 and 1.5 μm in diameter. They then applied pressure to the graphene placed on the orifices using probes made of diamond to test their endurance. The study found that the graphene sample particles can withstand a maximum pressure of about 2.9 μN per 100 nm length before they begin to break. It is estimated that this result is equivalent to applying a pressure of 55 N to break the 1 m long graphene. If it is possible to produce graphene with a thickness equivalent to that of a general food plastic packaging bag (about 100 nm thick), it is necessary to apply a pressure of almost 20,000 N to tear it off. In other words, if the package is made of graphene, it will withstand about two tons of items.

8.3.2.3 Electrical properties of graphene

Graphene, unlike most common three-dimensional materials, is a semimetal or zero-gap semiconductor.

The biggest characteristic of graphene is that the speed of electron movement reaches 1/300 of the speed of light, far exceeding the speed of electron movement in a general conductor. This makes the properties of electrons in graphene very similar to relativistic neutrinos. Electron transmission measurement results show that graphene has a surprisingly high electron mobility at room temperature, exceeding 15,000 $cm^2/V \cdot s$ (the electron mobility of silicon material is 0.13 $cm^2/V \cdot s$ at room temperature of 300 K, and the electron mobility of germanium material is 0.38 $cm^2/V \cdot s$). And between 10 and 100 K, the mobility is almost independent of temperature. In addition, the resistivity of graphene is 10^{-6} $\Omega \cdot cm$, which is slightly smaller than the resistivity of silver which has the lowest resistivity at room temperature (1.59×10^{-6} $\Omega \cdot cm$).

Graphene has considerable opacity and can absorb about 2.3% of visible light. This is also the embodiment of the relativistic charge of the charge carriers in graphene. Due to the two-dimensional nature of graphene, scientists believe that charge fractionation (the apparent charge of a single quasiparticle of a low-dimensional material is less than a unit quantum) occurs in graphene. Therefore, graphene may be the most suitable material for making any subcomponents required for quantum computers.

8.4 Polymer modification of carbon nanomaterials

Due to the high symmetry of the molecular structure of fullerenes and the high dense packing of crystals, their solubility is poor and cannot be processed into any shape, and the practical application is greatly limited. Therefore, it is natural to think of chemical modification. Since the preparation of fullerenes from the Krätschmer et al. in 1990, a number of methods for chemically modifying fullerenes have been invented. The idea of polymerizing fullerenes has been widely recognized by scientists.

Comparing the structures of fullerenes, CNTs, and graphene, it can be found that the sp^3 hybrid components of carbon atoms are successively decreased, and the tension between carbon–carbon bonds is also sequentially decreased, so that the structural stability is sequentially increased. Compared with fullerenes, CNTs, and graphene are more difficult to chemically modify.

8.4.1 Polymerization of fullerenes

8.4.1.1 Fullerene macromolecular derivatives and their types

The polymerized derivatives of C_{60} can be basically divided into four categories: The first type is that C_{60} is suspended on the polymer chain (on-chain type), such as hanging C_{60} at one end or both ends of the main chain or hanging on the side chain of the polymer. The second type is that C_{60} is incorporated into the polymer backbone (in-chain type). The third type is to form a polymer network (polymer network type) with C_{60} as a node. The fourth type is C_{60} which is chemically bonded to the surface of a matrix material (such as an ion exchange resin and an inorganic material.) or forms a coating (matrix-bound type). Figure 8.8 illustrates some structural forms of the C_{60} macromolecular derivatives. Types 1,2,5,6,7 are on-chain type, types 3,4,8,11 are in-chain type, types 9,10 are polymer network type, and types 12,13 are matrix-bound type.

The combination of C_{60} and polymer will lead to the emergence of many new polymer materials. For example, researchers have reported that the introduction of C_{60} into common polymer compounds can produce good photoelectric effects. C_{60} has great application potential in the preparation of high-performance batteries, semiconductor materials, superconducting materials, and so on. The water-soluble C_{60} polymer derivatives that have been successfully developed have obtained surprising results in antiviral tests. Therefore, the study of C_{60} macromolecularization will also provide a basis for the development of life sciences and medical science.

However, as far as the current situation is concerned, the research on C_{60} macromolecular derivatives is still at a relatively preliminary stage. Many of the reaction conditions are still far from mature, and there is still insufficient means for the chemical structure and morphological structure characterization and performance studies of C_{60}-containing polymers.

8.4.1.2 Method of fullerene macromolecularization

In the past 10 years, various conventional polymerization methods have been used to prepare macrolide-containing polymerized derivatives. At present, there are two major types of methods for synthesizing fullerene-containing macromolecular derivatives: one starts from fullerenes themselves and polymerizable fullerene organic small molecule derivatives. It can be obtained by a conventional polymer reaction and can also be copolymerized with a conventional polymer monomer. This method is commonly used to prepare macromolecular derivatives containing fullerenes in the main chain. Another method is to introduce fullerene into a polymer chain by reacting a preprepared polymer precursor containing an active reactive group with fullerene. This method generally does not significantly alter the relative molecular mass of the precursor, so it is also known as the "polymer-analogous" reaction. Since various types of macromolecularization reactions can obtain numerous products, many types of fullerene-containing macromolecular derivatives have been synthesized. The existing methods for synthesizing fullerene-containing macromolecular derivatives can be classified into four types:

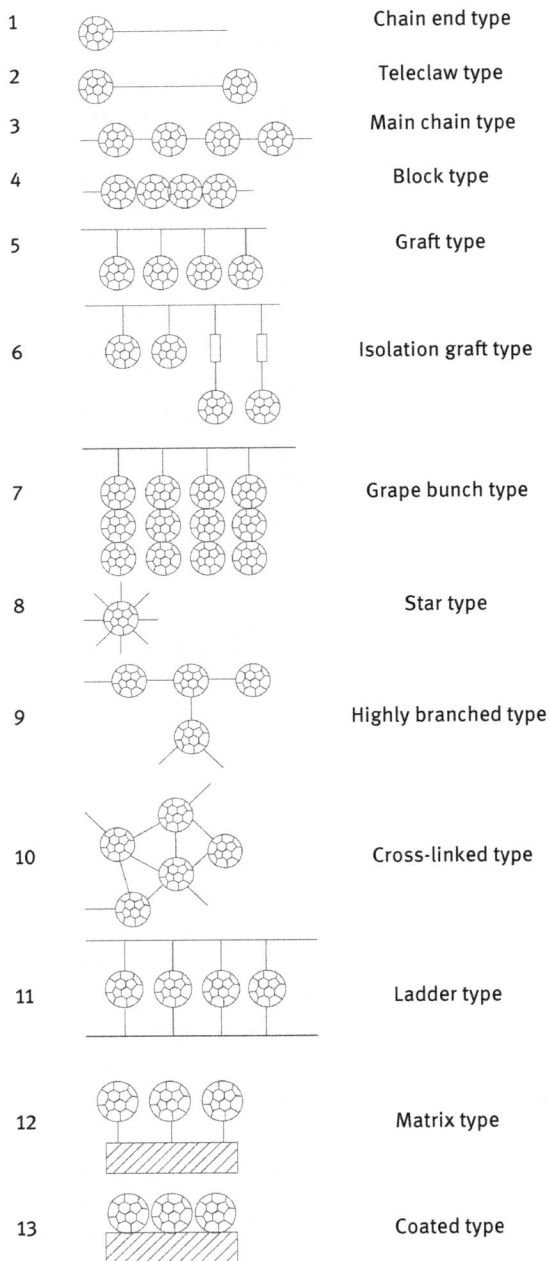

1	Chain end type
2	Teleclaw type
3	Main chain type
4	Block type
5	Graft type
6	Isolation graft type
7	Grape bunch type
8	Star type
9	Highly branched type
10	Cross-linked type
11	Ladder type
12	Matrix type
13	Coated type

Fig. 8.8: Schematic diagram of the structure C_{60}-containing macromolecular derivatives.

(1) Direct polymerization
This method utilizes multiple double bonds in fullerenes to directly polymerize fullerenes as monomers or copolymerizes with other monomers as comonomers to form in-chain-type or network cross-linking-type polymers.

In 1992, Loy et al. prepared a polymer whose main chain contained C_{60} by pyrolysis through free radicals reaction. Cao and Camp et al. introduced C_{60} into the polymer chain by free radical direct polymerization to prepare a soluble star polymer with C_{60} as the core. Ederle et al. used an anionic initiator to initiate the copolymerization of styrene with C_{60} to obtain a network-like C_{60}-containing macromolecule. Hong Han et al. used Ziegler–Natta catalyst to initiate copolymerization of vinyl ether monomer with C_{60} to obtain C_{60}-containing polyvinyl ether polymer. (The relative molecular mass is 3,000 to 4,000, the distribution is wide, and the polydispersity index is 5.2.)

Since C_{60} itself has many double bonds, they are quite active and the activity of each double bond is the same. Therefore, the macromolecularization with C_{60} as a monomer will easily produce multisubstituted derivatives, and sometimes even forms a three-dimensional network cross-linked structure, the product is very complicated, and separation is difficult.

(2) Modifying the surface of the fullerene sphere first, followed by polymerization
In this method, the fullerene sphere is first surface-modified. Polymerizable groups such as a double bond, a carboxyl group, a hydroxyl group, an amino group, an isocyanate group, or the like is introduced. Then, a C_{60}-containing "on-chain"- or "in-chain"-type polymer derivative is obtained by a conventional polymerization reaction such as a chain polymerization reaction or a stepwise polymerization reaction. This method has become one of the important methods for preparing fullerene-containing polymers. For example, Shi et al. attached a phenol group to C_{60}, followed by polycondensation with an acid chloride compound, or addition polymerization with an isocyanate to prepare an "in-chain"-type C_{60}-containing macromolecular derivative. Maswmi et al. bisphenolized C_{60} and then reacted it with an acid chloride compound to prepare an "on-chain"-type linear polymer.

(3) Reaction of polymer precursors containing active end groups with fullerenes
The double bond on the surface of fullerene lacks electrons, so it can become a quencher for active centers such as anions and radicals. Accordingly, a polymer precursor with a reactive group was first designed, and fullerene was used as a terminator to obtain an "on-chain"-type fullerene polymer derivatives with a fullerene sphere as a terminal group. For example, Wang et al. used TEMPO as an initiator for living radical polymerization to prepare polystyrene (PS) with active radical end groups, and then reacted with C_{60} to obtain a polymer whose each backbone contains only one C_{60} under suitable ratio conditions. Haruyuki also did a similar job.

(4) Introducing fullerenes into polymers by polymer side group reaction

This method is one of the most commonly used methods for preparing fullerene-containing macromolecular derivatives. This method allows the introduction of fullerenes while maintaining the original structure of the polymer chain, so that the advantages of both can be truly combined.

Gu Tao et al. used Friedel–Crafts reaction of C_{60} with polymethylphenylsiloxane with $AlCl_3$ as a catalyst to synthesize C_{60}-containing polymethylphenylsiloxane on the side chain benzene ring. Further, by directly copolymerizing dichloromethylphenylsilane with C_{60}, a C_{60}-containing polymethylphenylsiloxane with C_{60} bonded to the main chain was obtained. The photoconductivity of these two types of C_{60}-containing macromolecular derivatives is completely different. The latter one has good photoconductivity, while the former one has almost no photoconductivity.

Lu et al. prepared a C_{60}-containing polymer with a side chain by reacting the azide group with C_{60}. They first synthesized a polymethacrylate with a double bond on the side chain and added Br atoms to the side group with HBr. Then, they azide them with NaN_3 to obtain a polymer containing an azide group on the pendant group. Finally, the azidated polymethacrylate was subjected to a cycloaddition reaction with C_{60} to prepare a C_{60}-containing linear polymethacrylate with a relatively narrow molecular weight distribution. Others, like Prato and Wang Changchun also prepared a C_{60}-containing polymer on the side groups by azidation method.

8.4.1.3 Preparation of macrolide-containing derivatives

(1) Preparation of "on-chain"-type fullerene-containing macromolecular derivatives
With the gradual deepening of the research on the reaction of small organic molecules with fullerenes, people have a deeper understanding of the reaction regularity of fullerenes. On this basis, "on-chain"-type fullerene-containing macromolecular derivatives were successively synthesized. Since such reactions generally produce a long polymer chain containing a polymer capable of reacting with fullerene, the relative molecular mass of such materials can be made higher. Since such reactions generally produce a long polymer chain which is capable to react with fullerene, the relative molecular mass of such materials can be made higher. At the same time, due to the influence of macromolecular chains on the microenvironment of the reaction system in the solution, the products generally have good stereoselectivity and electronic environment. However, the activity of the polymer functional group also has large influence on the structure of the derivative. If the activity of the polymer functional group is too high, the product is very easily cross-linked since the fullerene molecule itself contains a plurality of reaction sites. The cross-linked product is difficult to process because it is insoluble, infusible, and has poor processability. Therefore, the synthesis of soluble C_{60}-containing macromolecular derivatives is a crucial step.

The preparation of "on-chain"-type C_{60}-containing macromolecular derivatives by C_{60}-containing monomers was first reported by Shi et al. They prepared a polymeric derivative containing C_{60} by condensation polymerization and addition polymerization, respectively. The reaction process is shown as following:

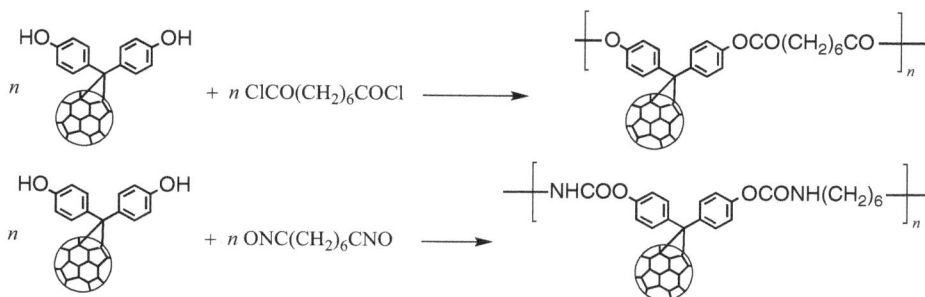

Chemical reaction equation 4

The C_{60} derivatives of poly(alkylene amine) and poly(alkylene imide) are two useful soluble C_{60} macromolecular derivatives. Many of their applications have been discovered, such as in metallurgy, wastewater treatment, radiochemistry, pharmaceuticals, analytical, and environmental chemistry.

The addition reaction of small molecule ammonia with C_{60} has been studied deeply. The reaction of a soluble polymer containing amino groups with the double bond of C_{60} to obtain a polymerized derivative containing C_{60} has been reported. This C_{60} polymerized derivative is prepared by reacting a polymer containing an amino group with C_{60} under room temperature. Since the polymer skeleton with amino group can be controlled by different reactions, it is also possible to synthesize other C_{60}-containing macromolecular derivative with different properties, for example, a water-soluble C_{60}-containing macromolecular derivative. This derivative can be used in biological and pharmaceutical research. The reaction formula is as follow:

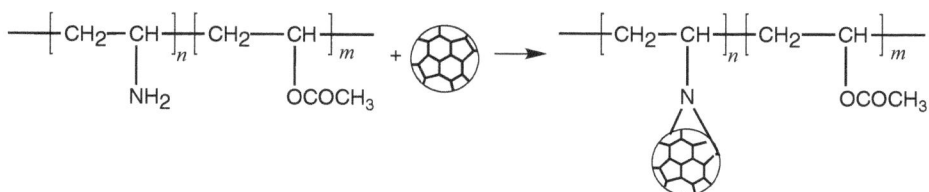

Chemical reaction equation 5

Fu shoukuan research group from Fudan University prepared two types of "on-chain"-type C_{60} macromolecular derivatives using PS as the precursor. One type is a derivative of PS with C_{60} attached to side groups. The method is first to chloromethylate PS, then azide the chloromethylated PS. The azidated PS was subjected to a cycloaddition reaction with C_{60} to prepare a PS derivative with C_{60} attached to the pendant group, as shown in following chemical reaction equation. These derivatives have similar solubility properties to PS and can be dissolved in many common solvents.

Chloromethylation

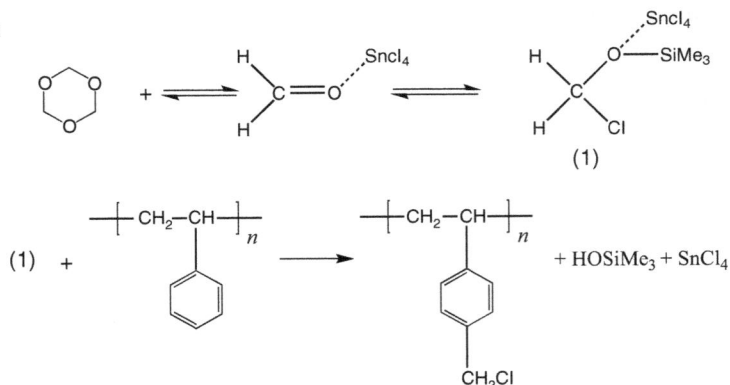

(1)

Azide

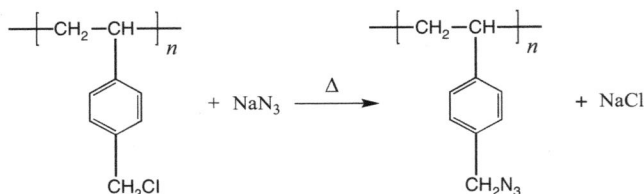

Introducing C$_{60}$

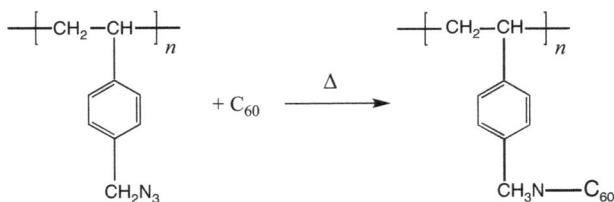

Chemical reaction equation 6

The other class is a monosubstituted PS and a C$_{60}$ derivative of polymethyl methacrylate. The preparation method is to first prepare a polymer long chain terminated with TEMPO by living radical polymerization. Then, using TEMPO to produce active radicals under heating conditions, it reacts with C$_{60}$ to obtain a polymer C$_{60}$ derivative capped with C$_{60}$. The reaction process is shown in Fig. 8.9.

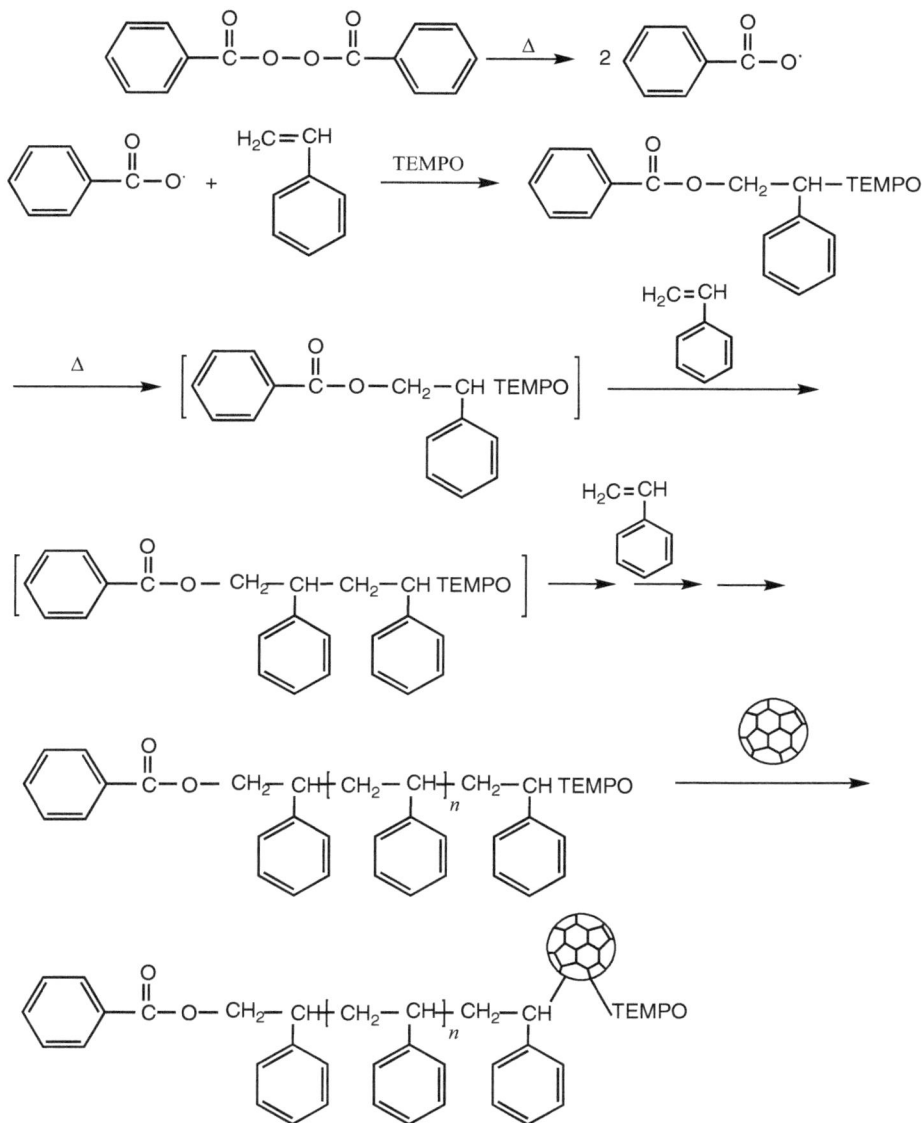

Chemical reaction equation 7

Wang Guojian et al. from Tongji University directly prepared linear PS and hyper-branched polymers containing benzyl chloride groups by ordinary free radical poly-merization and atom transfer radical polymerization with p-chloromethylstyrene as monomer. Thereafter, an azide reaction was carried out, and then the azide product was reacted with C_{60} to prepare various types of "on chain"-type C_{60}-containing macromolecular derivatives, as shown in Fig. 8.9.

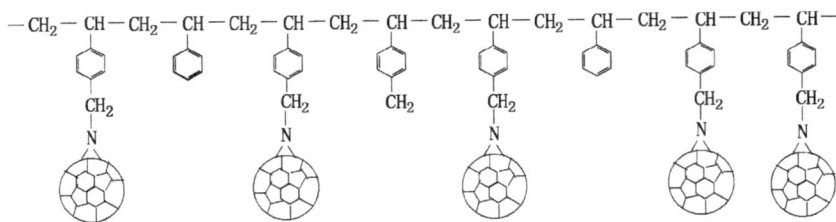

(a)

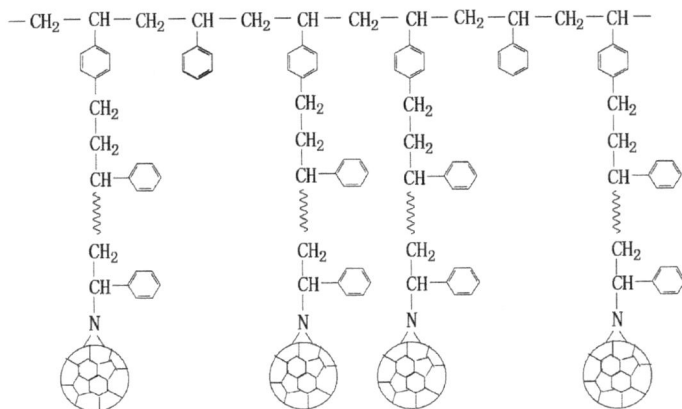

(b)

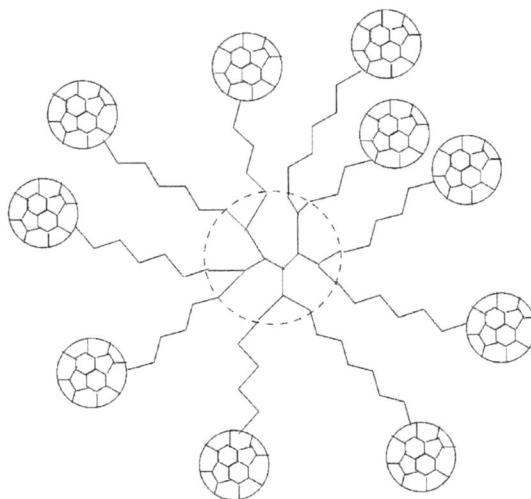

(c)

Fig. 8.9: Various "on-chain"-type C_{60} polystyrene derivatives prepared by *p*-chloromethylstyrene as a monomer.

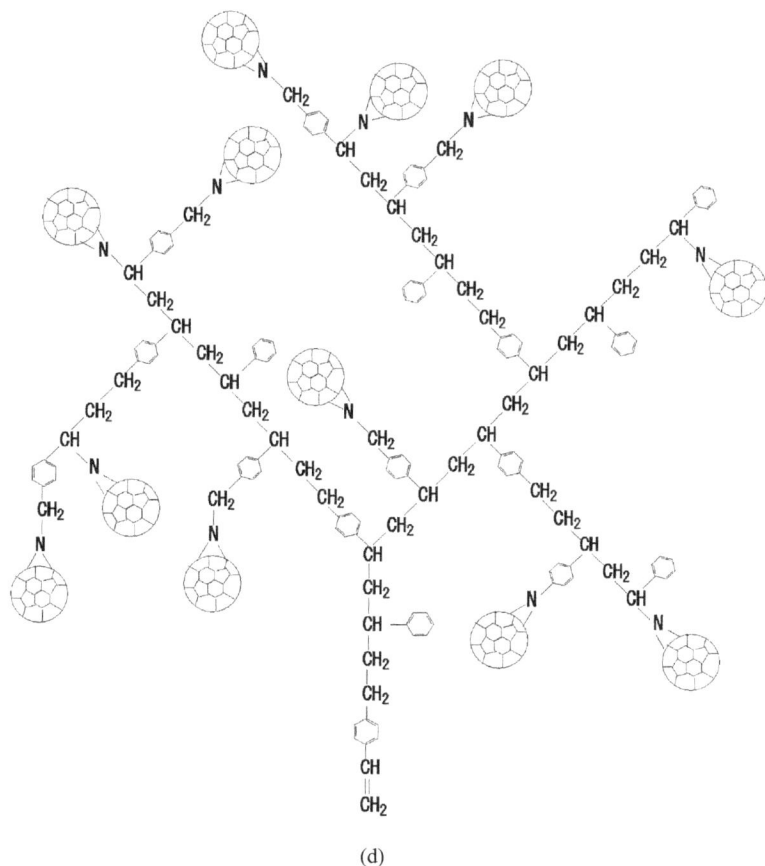

(d)

Fig. 8.9 (continued)

Hawker and Wcoley et al. obtained a single-added dendritic C_{60} macromolecular derivative by cycloaddition reaction of a dendritic azide with C_{60}. The structure is shown in Fig. 8.10.

(2) Preparation of "in-chain"-type fullerene-containing macromolecular derivatives
The work of polymerizing fullerene into a polymer chain to prepare an "in-chain"-type fullerene-containing macromolecular derivative has begun in the early stage of fullerene macromolecularization. However, compared with the study of "on-chain"-type fullerene-containing macromolecular derivatives, the progress seems to be much slower because this is a more complicated job. There have been many articles concerning the preparation of such compounds, but the work of actually preparing a polymer having a C_{60} in the main chain was initiated by Loy and Assink et al. Their famous work shows reacting C_{60} with p-xylene diradical to produce a copolymer of

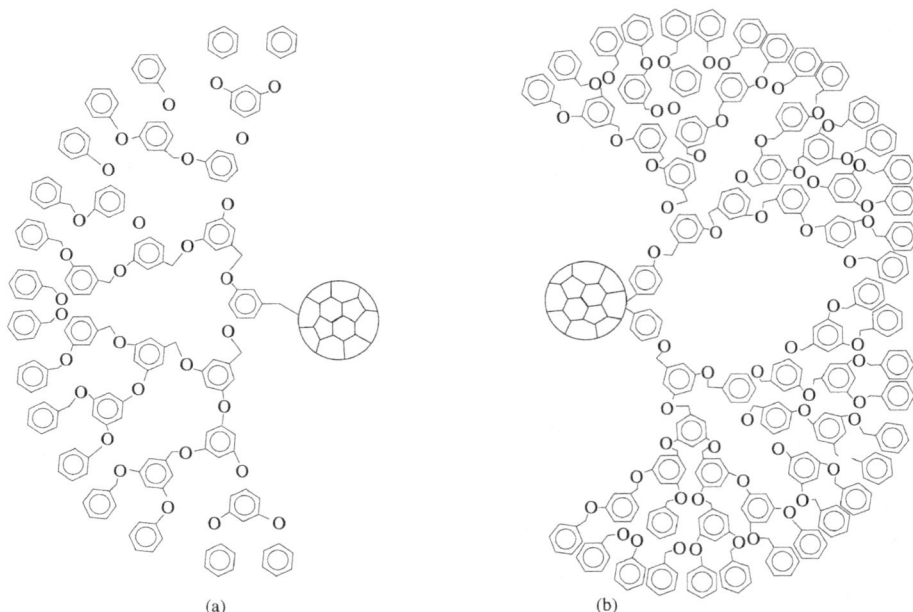

Fig. 8.10: Dendritic C_{60} macromolecular derivatives. (a) Side group suspension type; (b) side chain suspension type; (c) star type; and (d) tree type.

C_{60} and p-xylene. The p-xylene diradical is obtained by thermal cracking of 1,1,4,4-cyclodiethylenedibenzene, and the reaction scheme is shown as follows. Although the resulting copolymer is unstable and is very easy to react with O_2 to form a cross-linked product, this method provides us with a possible route for the preparation of C_{60}-containing macromolecular derivatives by free radical polymerization of C_{60}.

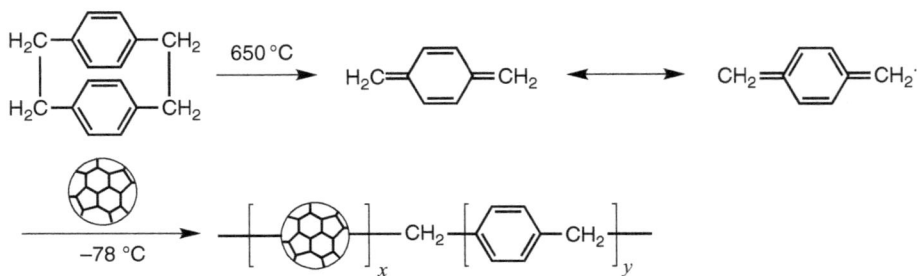

Chemical reaction equation 8

Inspired by these efforts, the work of simple free radical polymerization of C_{60} with PS to prepare stable soluble C_{60}-containing PS derivatives has been reported. The ratio of C_{60} to styrene monomer in the resulting copolymer can be as high as 50:50.

The product may be a mixture of a plurality of structural copolymers. The possible polymerization mechanism and polymer structure are shown:

$$PS \cdot \ + \ C_{60} \longrightarrow PS-C_{60} \cdot \xrightarrow{\ PS \cdot\ } PS-C_{60}-PS$$

$$2\ PS-C_{60} \cdot \longrightarrow PS-C_{60}-C_{60}-PS$$

$$R \cdot \ + \ PS-C_{60}-C_{60}-PS \longrightarrow PS-C_{60}-\underset{\underset{R}{|}}{C_{60}} \cdot -PS \xrightarrow{\ PS-C_{60} \cdot\ } PS-C_{60}-\underset{\underset{\underset{\underset{PS}{|}}{C_{60}}}{\overset{\overset{R}{|}}{C_{60}}}-PS$$

$$2\ PS-C_{60}-\underset{\underset{R}{|}}{C_{60}} \cdot -PS \longrightarrow PS-\underset{\underset{\underset{\underset{PS}{|}}{C_{60}}}{\overset{\overset{R}{|}}{C_{60}}}-\underset{\underset{\underset{\underset{PS}{|}}{C_{60}}}{\overset{\overset{R}{|}}{C_{60}}}-PS$$

Chemical reaction equation 9

In addition to free radical polymerization, the anionic polymerization of C_{60} and PS anion active species can also get copolymer of C_{60} and styrene. The molar ratio of PS chain to C_{60} is from 0.8:1 to 30:1. Each C_{60} molecule generally contains 4–10 PS chains, each of which has a degree of polymerization of 1 to 10. Therefore, the co-polymer is likely to be a star-shaped structure with a C_{60} core and a plurality of styrene chains extending outward. This copolymer has excellent solubility and can be dissolved in most common solvents.

In addition to styrene, copolymers of monomers such as methyl methacrylate, acrylic acid, vinyl acetate, and polyvinyl carbazole with C_{60} have also been prepared. It is found that although this kind of copolymer has more damage to the double bond structure of C_{60}, it still has very good photoconductivity. For example, the copolymer of polyvinylcarbazole and C_{60} can have a photoconductivity half-life as small as $t_{1/2} = 0.125$ s.

A water-soluble C_{60}-containing macromolecular derivative can be prepared by copolymerizing acrylic acid with C_{60}. Further, a water-soluble C_{60}-containing mac-romolecular derivative can also be obtained by a polymer modification reaction. For example, a water-soluble polymer C_{60} derivative can be obtained by using a polyethylene oxide as a main chain, an amino group, and a nucleophilic addition reaction with C_{60}. Moreover, various forms of C_{60}-containing macromolecular deriv-atives can be obtained by changing the polymer chain structure. Such derivatives are generally soluble in both water and toluene, but due to the high number of C_{60} reaction points, improper control can form cross-linked products.

It is an interesting work to directly initiate the polymerization of ethylene oxide using C_{60} as the initiation source. First, C_{60} is turned into a stable cation $(C_{60})^{n+}$, and then a stepwise addition of ethylene oxide is initiated. This reaction is named "ball-and-chain" polymerization. At present, only 6 ethylene oxide molecules can be continuously added to the C_{60} by this method, so it is not a true polymer chain. However, this reaction may be as valuable as C_{60} macromolecular derivatives by radical polymerization in the original C_{60}, which will provide valuable insights for further research.

(3) Preparation of "matdx-bound"-type fullerene-containing macromolecular derivatives
In 1993, K. Chen et al. first reported the research work of C_{60} on the formation of two-dimensional array on the surface of the substrate. They used an amino group to immobilize C_{60} on the surface of indium-calcium chloride to prepare a C_{60} macromolecular derivative with a supramolecular structure. Most of the current research work is to introduce a number of amino groups into the surface of the substrate, and then react C_{60} with it to form a C_{60} monolayer. A similar method for the preparation of C_{60} derivatives has also been reported by Chupa et al. They produced a self-assembled monolayer C_{60} membrane terminated by pyridyl on the surface of silica.

There is also another method for preparing a "matrix-bound"-type C_{60}-containing macromolecular derivative. That is the reaction based on diphenylmethyl terminated ethylene oligomers. An anionic active species is obtained by adding butyl lithium to the earlier oligomer. Then, it was reacted with C_{60}, and finally terminated with methanol to obtain a polymerized derivative having C_{60} attached to the surface of the resin.

8.4.2 Polymer modification of carbon nanotubes

8.4.2.1 Mechanism of surface modification of carbon nanotubes

CNTs can be considered as convolution of two-dimensional graphene sheets. The ideal structure is a seamless, hollow tube surrounded by a hexagonal carbon atom grid, covered at both ends by hemispherical large fullerene molecules. Studies have shown that fullerene chemistry is characterized by addition reaction. It is relatively easy for fullerenes to undergo such reactions. From the geometrical point of the sphere, the conversion from a carbon triangular bond into a tetrahedral bond will produce a large amount of energy, when fullerene is chemically reacted and modified; it is a reaction in which energy is relatively lowered. Systematic development of fullerene chemistry has shown that their addition reactivity greatly depends on the curvature of fullerenes. The greater the curvature of the carbon structure, the easier it is to react with other groups. Therefore, the preferred reaction site of CNTs is the "end cap" with the largest curvature, indicating that the two ends of the CNTs are the most susceptible sites for chemical reactions. Many studies have used the preferential reaction at this "end cap" to open both ends of the CNTs, allowing other substances to enter, while introducing reactive groups at both ends.

The defectless, structurally intact, closed CNT tube wall is composed of a hexagonal carbon ring grid. However, in the preparation of CNTs, sp^3 hybridized carbon atoms are also mixed, that is, there are a large number of geometric defects in the CNT wall. For example, the bond rotation defect and the paired five-membered ring/seven-membered ring, the carbon atoms forming the defect are different from the general carbonization of the tube wall, and the chemical reaction activity is also different, so the CNT surface still has large reactivity.

Most of the CNT sidewalls are composed of a six-membered ring of carbon. The carbon atoms in each six-membered ring forms carbon–carbon σ bond by overlapping the sp^2 hybrid orbitals with the sp^2 hybrid orbitals of carbon atoms on adjacent six-membered rings. The symmetry axes of the three sp^2 hybrid orbitals of each carbon atom are distributed in the same plane, and the angle between the two axes of symmetry is 120°, thus forming a regular hexagonal carbon skeleton. Semiquantitative information about the end of the CNTs (carbon–carbon double-bending bond) and the wall of the tube (carbon–carbon single-bending bond) can be obtained by calculation. The heat generated by the reaction of the tube wall is 125.5 to 209 kJ/mol less than the heat of formation of the end reaction. This suggests that the wall modification will require a more reactive reaction precursor than the end modification of the tube. Moreover, whether SWNTs or MWNTs, they have certain functional groups bonded to their surfaces, and the CNTs obtained by different preparation methods and posttreatment processes have different surface structure. In general, SWNTs are highly chemically inert and their surface is pure. The surface of MWNTs is much more active, combined with a large number of surface groups.

In addition, each carbon atom also has a p orbital perpendicular to this plane that forms a highly delocalized large π bond. These π electrons can be used to combine with other compounds containing π electrons by π–π noncovalent bonding, and modified CNTs can be obtained.

Therefore, from the above three aspects, the surface modification methods of CNTs are mainly divided into two major directions: covalent bond modification of the tube wall and noncovalent bond coating of foreign substances. The covalent bond modification of the tube wall includes the end of the CNT and defect modification and sidewall modification.

8.4.2.2 Covalent bond modification of carbon nanotubes

The method for covalent bond modification of CNTs is mainly to first introduce a carboxylic acid group on the surface of the CNT by a carboxylation treatment, and then perform acid chloride, alcoholation, or amination, thereby introducing a polymer molecule on the surface of the CNT. The acidification treatment of the CNTs can remove amorphous carbon, SiO_2, and catalyst particles such as Fe and Ni, and introduce active groups at the ends and sidewalls of the CNTs.

In 1994, Tsang et al. found that open SWNTs were obtained by chemically cutting SWNTs with strong acids. Subsequently, Lago et al. found that the top of this open SWNTs contains a certain number of reactive groups, such as hydroxyl groups, and carboxyl groups, and it is predicted that these reactive groups can be used to organically modify SWNTs. It has been found that defects on the sidewalls of CNTs are also important sites for chemical modification.

For example, short tube with a length of 100 to 300 nm can be obtained by treating SWNTs with a mixed acid of sulfuric acid and nitric acid. The terminal carboxylated SWNTs can be obtained by secondary oxidation treatment with concentrated sulfuric acid and hydrogen peroxide. Treating the terminal carboxylated SWNTs with thionyl chloride can convert the carboxyl group to a more reactive acid chloride group for further chemical modification. The thiol group can be introduced into the SWNTs by reaction with an acid chloride group and NH_2-$(CH_2)_{11}$-SH. The earlier reaction process is shown as follow:

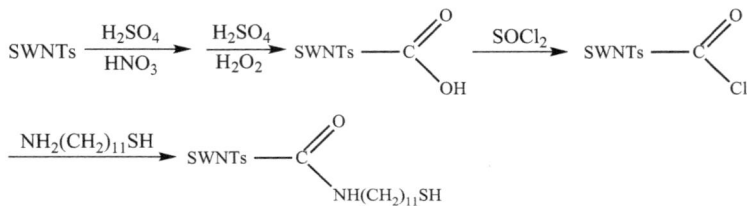

Chemical reaction equation 10

The hydroxyl group was introduced by reacting ethylene glycol with single-walled carbon nanotubes (SWNT) at 120 °C, and then reacted with 2-bromo-2-methylpropionic acid bromide to introduce an initiator group of atom transfer radical polymerization (ATRP) (SWNT-Br). The modified CNTs can be directly subjected to ATRP reaction of methyl methacrylate or styrene to obtain polymethyl methacrylate or PS grafted SWNTs. The thickness of the polymer layer on the surface of the SWNTs can be adjusted according to the ratio of the introduced polymer monomer to the SWNT-Br. The following shows the reaction formula of PS grafted MWNTs:

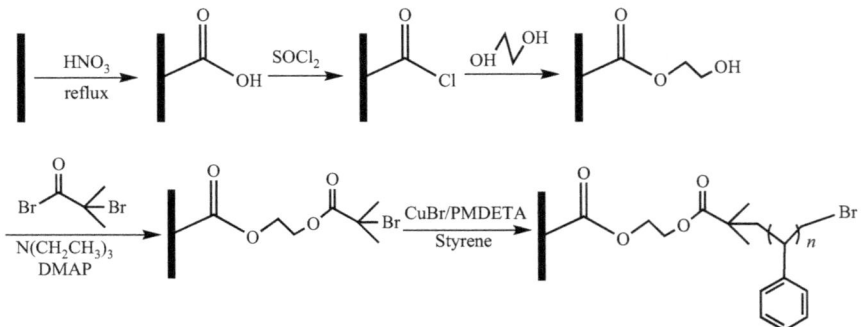

Chemical reaction equation 11

In addition to the earlier acidification treatment, the side wall of the CNT mainly composed of sp^2 hybridized carbon atoms can also undergo an addition reaction with some active substances such as halogen, carbene, free radical, azide group, and the like. The sidewall modified CNTs form a functionalized active surface that creates conditions for further reaction. The increase of the reactivity of the side wall of CNTs is related to its diameter. The sidewalls of small diameter CNTs have higher reactivity due to increased curvature and high rehybridization energy. These reactions also provide the possibility of polymer modification of CNTs.

For example, PS with a Br atom at the end (PS-Br) is first prepared by an APRT method, and then PS-Br is converted into PSt-N$_3$ by the reaction between sodium azide and halogen atoms. This azide-containing PS can react directly with the CNT sidewalls. The cycloaddition reaction is first carried out to form a five-membered ring intermediate, and the intermediate loses N$_2$ under heating (130 °C). Thereby, the addition of PS-Br to the CNTs is completed to obtain PS-modified CNTs (as follows).

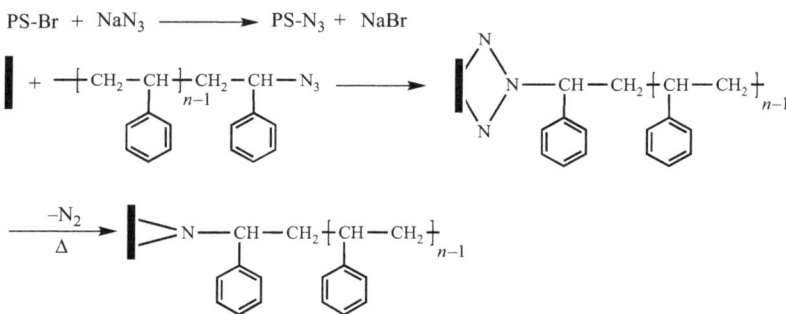

Chemical reaction equation 12

The polymer-modified CNTs prepared by the earlier method exhibit amphiphilicity because the CNTs and the polymer have different compatibility in different solvents. This allows it to form a neatly arranged self-assembly in a suitable solvent.

8.4.3 Chemical modification of graphene

8.4.3.1 The Basis of chemical modification of graphene
Natural graphite is neither hydrophilic nor oleophilic. If the sheet graphene obtained by stripping natural graphite is not chemically treated before reduction, it will quickly aggregate during the reduction process and reform the bulk graphite, so it is difficult to obtain a sheet of graphene. Even a small amount of lamellar graphene obtained has a property of being nonhydrophilic and nonlipophilic. Such properties also make it difficult to recombine with other materials, which limits the

wide application of graphene, and the composite material cannot fully exert the superior performance of graphene.

At present, graphene is prepared by various methods such as micro mechanical peeling, orientation epitaxy, chemical vapor deposition, and chemical stripping. The former methods have poor controllability, high preparation costs, and are difficult to mass produce. The chemical stripping method oxidizes graphite with a strong acid to obtain graphite oxide (GO), and then peel it into a single piece of graphene oxide by heat treatment or ultrasonic treatment, and finally obtain graphene by reduction. Compared with the former three methods, the chemical stripping method is an effective method for achieving low-cost, high-volume preparation of graphene.

Graphene with a complete crystal structure is not active except for its ability to adsorb some atoms and molecules (CO, NO, NO_2, O_2, N_2, CO_2, and NH_3). However, the graphene prepared by the chemical stripping method is obtained by reduction of graphite oxide, and the sheet structure contains a certain amount of carbonyl groups and carboxyl groups. The middle portion of the sheet contains highly reactive oxygen-containing functional groups such as hydroxyl groups and epoxy groups, and sometimes there are defects left by the removal of the epoxy group at a high temperature. These functional groups and defects can be utilized for the chemical treatment of graphene by grafting, coating, and so on to achieve the purpose of improving the surface properties of graphene.

Like the modification of CNTs, graphene can also be modified by covalent bonds and noncovalent bonds. In the noncovalent surface modification, the surface of the graphene is covered with a polymer to reduce the mutual attraction between the graphene sheets, thereby improving the dispersibility and stability of the graphene.

8.4.3.2 Noncovalent bond chemical modification of graphene

Because graphene is extremely easy to agglomerate due to its extremely large surface energy, it can be surface-modified with a surfactant or a polymer to improve its dispersibility in water or organic solvents. A lot of results have been achieved in this area of research. For example, the graphene is ultrasonically treated in the presence of the cationic surfactant cetyltrimethylammonium bromide to prepare a stable aqueous graphene dispersion. The graphene aqueous dispersion can be well mixed with the styrene-butadiene latex, and a graphene/styrene-butadiene rubber nanocompositeurther can be obtained after further aggregation.

The graphene oxide is reduced in the presence of an amphoteric water-soluble polymer sodium poly(p-styrenesulfonate) (PSS). The π–π interaction between PSS and graphene prevents the aggregation of graphene sheets. The resulting product has good water dispersibility. The poly[(m-phenylacetylene)-co-(2,5-dioctyloxy-p-phenylacety-lene)] (PmPV)-modified graphene prepared by using the π-π interaction between poly [(*m*-phenylacetylene)-co-(2,5-dioctyloxy-*p*-phenylacetylene)] and graphene in an organic solvent has a good dispersion in organ solvent.

8.4.3.3 Covalent bond chemical modification of graphene

The large number of functional groups generated on the surface of graphene after oxidation increases the activity of graphene and lays a foundation for covalent modification. Surface modification of graphene can be achieved by using reagents such as isocyanate, silane coupling agent, organic amine, and the like.

By utilizing the high reactivity of isocyanate with hydroxyl groups and carboxyl groups, the surface properties of graphene oxide can be changed to enable stable dispersion in organic solvents such as N,N-dimethylformamide (DMF). By using a bifunctional isocyanate as a bridging agent, polymers with amphiphilic functional group can be grafted on the surface of the graphite oxide, so that the obtained graphene oxide has good dispersion in both water and organic solvent. In addition, $SOCl_2$ can react with the −OH and −COOH functional groups on the surface of graphene oxide, and then the product reacts with long-chain alkylamines to form graphene with stable dispersion in nonpolar solvents such as xylene and tetrachloromethane. Graphene composites can also be prepared by selecting different surfactants to modify graphene oxide. For example, a SnO_2-graphene composite can be obtained by modifying GO with dibutyltin oxide and graphene with potential applications in the biological field can be obtained by biomolecule modification of graphene oxide.

Shen et al. prepared a styrene-polyacrylamide block copolymer (PS-PAM) by in situ active radical polymerization and used it to modify graphene. Since PS and polyacrylamide have good solubility in nonpolar solvents and polar solvents, the graphene can be dispersed in water and xylene, so PS-PAM modified graphite can be dispersed in a variety of polymer matrices.

Salavagione et al. prepared a polyvinyl alcohol (PVA)-modified graphene by esterification of a carboxyl group on graphene oxide with PVA followed by reduction. The product was well dispersed in dimethyl sulfoxide and water. The esterification reaction provides a simple and feasible method for graphene modification and is suitable for other hydroxyl group containing polymers.

Since graphene has a similar chemical structure to that of fullerenes and CNTs, many methods developed by chemical modification of fullerenes and CNTs are also suitable for chemical modification of graphene. However, since the structure of graphene tends to be more stable than fullerenes and CNTs, the difficulty of chemical modification is correspondingly increased, which requires us to further explore and practice.

References

[1] Baggott, J. Perfect symmetry: the accidental discovery of buckminsterfullerence. London: Oxford University Press, 1995.

[2] Kroto, HW, Heath, JR, O'Brien, SC, et al. C_{60}: Buclminsterfullerene. Nature. 1985;318:162–164.

[3] Klein, DJ, Schmaly, TG, Hite, TG, et al. Resonance in C_{60}, Buckminsterfullerene. J Am Chem Soc. 1986;108:1301–1302.

[4] Taylor, R, Walton, DRM. The chemistry of fullerenes. Nature. 1993;363:685–693.

[5] Diederich, F, Rubin, Y. Strategien zum Aufbau Molekularer und Polymerer Kohlenstoffallotrop. Angew Chem. 1992;104(9):1123–1146.

[6] Geckeler, KE, Stirn, J. Polyreaktionen-Mechanismen, Systematik, Relevanz. Naturewissenschaften. 1993;80:487–500.

[7] Diederich, F, Thilgen, C. Covalent fullerene chemistry. Science. 1996;271:317–323.

[8] Jiang, M, Fu, SK. Recent Topics of Polymer Science. Shanghai: Fudan Univ Press. 1998; 231–237.

[9] Loy, DA, Assink, RA. Synthesis of a C_{60}-p-xylylene Copolymer. J Am Chem Soc. 1992;114: 3977–3978.

[10] Cao, T, Wedder, SE. Free-radical copolymerization of fullerene with styrene. Macromolecules. 1995;28:3741–3743.

[11] Cao, T, Webber, SE. Free radical copolymerization of styrene and C_{60}. Macromolecules. 1996;29:3826–3841.

[12] Sun, YP, Lawson, GE, Bunker, CE, et al. Preparation and characterization of fullerene-styrene copolymer. Macromolecules. 1996;29:8441–8448.

[13] Kojima, Y, Matsnoka, T, Takahashi, H, et al. Optical limiting property of polystyrene-bounol C_{60}. Macromolecules. 1995;28:8868–8869.

[14] Shi, S, Khemani, KC, Li, QC, et al. A polyester and polyurethane of diphenyl C_{60}: retention of fulleroid properties in a polymer. J Am Chem Soc. 1992;114:10656–10657.

[15] Maswmi, T, Shoui, T, et al. Synthesis of polyesters containing the [60]fullerene moiety in the main chain. Polym J. 1997;29:1020–1022.

[16] Wang, C, He, J, Fu, S, et al. Synthesis and characterization of the narrow polydisperity fulleren-end-capped polystyrene. Polymer Bull. 1996;37:305–311.

[17] Wang, CC, Shu, C, Fu, SK. Synthesis of polystyrene with C_{60} group on side group. Chem Chin Univ. 1996;17(9):1477–1481.

[18] Wang, CC, Shu, C, Fu, SK. Characterization of polystyrene with C_{60} group on side group. Chem Chin Univ. 1996;17(10):1634–1637.

[19] Prato, M, Li, CQ, Wudl, F. Addition of azides to C_{60}: synthesis of azafulleroids. J Am Chem Soc. 1993;115:1148–1150.

[20] Lu, ZH, Coh, SH, Lee, SY. Synthesis of fullerene-containing poly(alkyl methacrylate)s with narrow polydispersities. Polym Bull. 1997;39:661–667.

[21] Liu, ZW. Modern Inorganic Synthesis. Beijing: Chemical Industry Press, 1999;142–156.

[22] Cox, DM, Behal, S, Disko, M, et al. Characterization of C_{60} and C_{70} clusters. J Am Chem Soc. 1991;113:2940–2944.

[23] Haddon, RC, Hebard, AF, Rosseinsky, MJ, et al. Conducting films of C_{60} and go by alkyl-metal doping. Nature. 1991;350:320–322.

[24] Hebard, AF, Rosseinsky, MJ, Haddon, RC, et al. Super-conductivity at 18K in potassium-doped C_{60}. Nature. 1991;350:600–601.

[25] Friedman, S H, DeComp, DL, Sijbesma, RP, et al. Inhibition of the HIV-I protease by fullerene derivatives: model building studies and experimental verification. J Am Chem Soc. 1993;115 (15):6506–6509.

[26] Wang, CC, Fu, SK. Preparation of C_{60} terminated polystyrene by living free radical polymerizatio. J Funct Polym. 2000;13(2):125–128.

[27] Kraus, A, Müllen, K. [60]Fullerene-containing poly(dimethylsiloxane)s: Easy access to soluble polymers with high fullerene content. Macromolecules. 1999;32:4214–4219.

[28] Gu, T, Chen, WX, Xu, ZD. Preparation and characterization of the fullerenated polymers. Polymer Bull. 1999;42:191–196.

[29] Chen, Y, Wang, JX, Lin, YH, et al. Synthesis and characterization of polystyrene-substituted [70]fullerene. Polymer. 2000;41:1233–1236.

[30] Yoshino, K, Lee, S, Fujii, A, et al. Near IR and UV enhanced photoresponse of C_{60}-doped semiconducting polymer photodiode. Adv Materials. 1999;11(16):1382–1385.

[31] Chikamatsu, M, Kikuchi, K, Nishikawa, H, et al. Photoconductivity of C60 derivatives with a long alkyl chain. Synthetic Metals. 1999;103:2403–2406.

[32] Camp, AG, Lary, A, Ford, WT. Free-radical polymerization of methyl methacrylate and styrene with C60. Macromolecules. 1995;28:7959–7961.

[33] Weis, C, Friedrich, C, Mulhaupt, R, et al. Fullerene-end-capped polystyrenes monosubstituted polymeric C60 derivatives. Macromolecules. 1995;28:403–405.

[34] Wang, ZY, Kuang, L, Meng, XS, et al. New route to incorporation of [60] fullerene into polymers via the benzocyclobutenone group. Macromolecules. 1998;31(16):5556–5558.

[35] Tang, BZ, Leung, SM, Peng, H, et al. Direct fullerenation of polycarbonate via simple polymer reactions. Macromolecules. 1997;30:2848–2852.

[36] Patil, AO, Brois, SJ. Fullerene grafted hydrocarbon polymers. Polymer. 1997;38(13): 3423–3424.

[37] Xin, F. Modification of carbon nanotubes and their composites. Beijing: Chemical Industry Press, 2012.

[38] Tasis, D, Tagmatarchis, N, Bianco, A, et al. Chemistry of carbon nanotubes. Chem Rev. 2006;106(3):1105–1136.

[39] Forrest, GA, Alexander, AJ. A model for the dependence of carbon nanotube length on acid oxidation time. J Phys Chem C. 2007;111(29):10792–10798.

[40] Saito, T, Matsushige, K, Tanaka, K. Chemical treatment and modification of multi-walled carbon nanotubes. Physica B Condensed Matter. 2002;323(1):280–283.

[41] Huang, W, Lin, Y, Taylor, S, et al. Sonication-assisted functionalization and solubilization of carbon nanotubes. Nano Letters. 2002;2(3):231–234.

[42] Saini, R K, Chiang, I W, Peng, H Q, et al. Covalent sidewall functionalization of single wall carbon nanotubes. J Am Chem Soc. 2003;125(12):3617–3621.

[43] Kongkanand, A, Kamat, PV. Interactions of single wall carbon nanotubes with methyl viologen radicals. quantitative estimation of stored electrons. J Phys Chem C. 2007;111(26): 9012–9015.

[44] Umek, P, Seo, JW, Hernadi, K, et al. Addition of carbon radicals generated from organic peroxides to single wall carbon nanotubes. Chem Mater. 2003;15(25):4751–4755.

[45] Bahr, JL, Tour, JM. Highly functionalized carbon nanotubes using in Situ generated diazonium compounds. Chem Mater. 2001;13(11):3823–3824.

[46] Ruther, MG, Frehill, F, O'Brien, JE, et al. Characterization of covalent functionalized carbon nanotubes. J Phys Chem B. 2004;108(28):9665–9668.

[47] Coleman, KS, Bailey, SR, Fogden, S, et al. Functionalization of single-walled carbon nanotubes via the Bingel reaction. J Am Chem Soc. 2003;125(29):8722–8723.

[48] Viswanathan, G, Chakrapani, N, Yang, H, et al. Single-Step in situ synthesis of polymer-grafted single-wall nanotube composites. J Am Chem Soc. 2003;125(31):9258–9259.

[49] Chen, S, Shen, W, Wu, G, et al. A new approach to the functionalization of single-walled carbon nanotubes with both alkyl and carboxyl groups. Chem Phys Lett. 2005;402 (4–6):312–317.

[50] Tong, X, Liu, C, Cheng, HM, et al. Surface modification of single walled carbon nanotubes with polyethylene via in situ Ziegler-Natta polymerization. J Appl Polym Sci. 2004;92(6): 3697–3700.

[51] Zhu, HW, Xu, ZP, Xie, D, et al. Graphene: structure, preparation and characterization. Beijing: Tsinghua University Press, 2011.

[52] Stankovich, S, Piner, RD, Nguyen, SBT, et al. Synthesis and exfoliation of isocyanate-treated graphene oxide nanoplatelets. Carbon. 2006;44(15):3342–3347.

[53] Zhang, DD, Zhu, SZ, Hao, BH. Inorganic-organic hybrid porous materials based on graphite oxide sheets. Carbon. 2009;47(13):2993–3000.

[54] Niyogi, S, Bekyarova, E, Itkis, ME, et al. Solution properties of graphite and graphene. J Am Chem Soc. 2006;128(24):7720–7721.

[55] Matsuo, Y, Miyabe, T, Fukutsuka, T, et al. Preparation and characterization of alkylamine-intercalated graphite oxides. Carbon. 2007;45(5):1005–1012.

[56] Si, YC, Samulski, ET. Synthesis of water-soluble grapheme. Nano Letters. 2008;8(6): 1679–1682.

[57] Shen, JF, Shi, M, Ma, HW, et al. Synthesis of hydrophilic and organophilic chemically modified graphene oxide sheets. J Colloid Interface Sci. 2010;235(2):366–370.

[58] Park, S, An, J, Piner, RD, et al. Aqueous suspension and characterization of chemically modified graphene sheets. Chem of Mater. 2008;20(21):6592–6594.

[59] Worsley, KA, Ramesh, P, Mandal, SK, et al. Soluble graphene derived from graphite fluoride. Chem Phys Lett. 2007;445(1–3):51–56.

[60] Hu, HT, Wang, XB, Wang, JC, et al. Preparation and properties of graphene nanosheets–polystyrene nanocomposites via in situ emulsion polymerization. Chem Phys Lett. 2010;484(4–6):247–253.

[61] Shen, JF, Hu, YZ, Chen, L, et al. Synthesis of amphiphilic graphene nanoplatelets. Small. 2009;5(1):82–85.

[62] Salavagione, HJ, Gomez, MA, Martinez, G. Polymeric modification of graphene through esterification of graphite oxide and poly(vinyl alcohol). Macromolecules. 2009;42(17): 6331–6334.

Exercises

1. **What is the relationship between diamond, carbene, and graphite?**
    Diamond with stereostructure is sp3 hybridized, while graphite with planar structure is sp2 hybridized, and the carbene with one-dimensional linear structure is sp hybridized.

2. **What is the relationship between C_{60} and carbon nanotube?**
    As the value of n increases, the fullerene is no longer a football-like circle, but gradually develops toward an ellipse. When n is large enough, the shape of Cn actually becomes a carbon tube with a tubular shape in the middle and a hemispherical shape at both ends.

3. **Under what condition do carbon nanotubes have good conductivity?**
    The vector C_h is commonly used to represent the direction in which atoms are arranged on the carbon nanotubes, where $C_h = na_1 + ma_2$, denoted as (n, m). a_1 and a_2 represent two basis vectors, respectively. (n, m) is closely related to the electrical conductivity of carbon

nanotubes. For a given (n, m) carbon nanotube, if there is $2n + m = 3q$ (q is an integer), then this direction shows metallicity, which is a good conductor; otherwise, it behaves as a semiconductor. For the direction of $n = m$, the carbon nanotubes exhibit good electrical conductivity and the electrical conductivity is up to 10,000 times that of copper.

4. **Which part of carbon nanotube is easier to be modified? The end cap or the wall? Why?**
    The end caps
    The greater the curvature of the carbon structure, the easier it is to react with other groups. Therefore, the preferred reaction site of carbon nanotubes is the "end cap" with the largest curvature, indicating that the two ends of the carbon nanotubes are the most susceptible sites for chemical reactions.

5. **As the research progresses, more and more novel carbon nanomaterials are prepared. List several new carbon nanomaterials.**
    Graphyne, graphdiyne, carbon nanosol, and so on.

Index